EUROPEAN SOURCES OF SCIENTIFIC AND TECHNICAL INFORMATION

REFERENCE BOOKS AVAILABLE FROM LONGMAN GROUP UK LIMITED

Longman Reference on Research Series

Agricultural Research Centres
Directory of Technical and Scientific Directories
Earth and Astronomical Sciences Research Centres
Electronics Research Centres
Engineering Research Centres
European Research Centres
European Sources of Scientific and Technical Information
Industrial Research in the United Kingdom
International Who's Who in Energy and Nuclear Sciences
International Medical Who's Who
Materials Research Centres
Medical Research Centres
Who's Who in Science in Europe
Who's Who in World Agriculture
World Energy Directory
World Nuclear Directory

Longman Guide to World Science and Technology

Science and Technology in the Middle East
    by Ziauddin Sardar
Science and Technology in Latin America
    by Latin America Newsletters Limited
Science and Technology in China
    by Tong B. Tang
Science and Technology in Japan
    by Alun M. Anderson
Science and Technology in the USA
    by Albert H. Teich and Jill H. Pace

In preparation
Science and Technology in the United Kingdom

Science and Technology in South-East Asia
    by Ziauddin Sardar
Science and Technology in Australasia, Antarctica and the Pacific
    by Jarlath Ronayne
Science and Technology in Africa
    by John W. Forje
Science and Technology in USSR
    by Mike J. Berry
Science and Technology in Western Germany
    by Christoph Schneider
Science and Technology in France and Belgium

EUROPEAN SOURCES OF SCIENTIFIC AND TECHNICAL INFORMATION

SEVENTH EDITION

Editor: Anthony P. Harvey

EUROPEAN SOURCES OF SCIENTIFIC AND TECHNICAL INFORMATION

Longman Group UK Limited, Professional Reference and Information Publishing
Division, Longman House, Burnt Mill, Harlow, Essex CM20 2JE, UK
Telephone: Harlow (0279) 442601

Distributed exclusively in the USA and Canada by Gale Research Company, Book
House, Detroit, MI 48226, USA

First edition 1957
Fourth edition 1976
Fifth edition 1981
Sixth edition 1984
Seventh edition 1986

British Library Cataloguing in Publication Data
European sources of scientific and technical information.—7th ed.—(Reference on
research)
   1. Science — Information services — Europe — Directories
   2. Technology — Information services — Europe — Directories
   I. Harvey, Anthony P. II. Series
   507'.04   Q224.3.E85

ISBN 0-582-90153-7

Typeset by Associated Publishing Services UK

Printed and bound in Great Britain by Biddles Ltd, Guildford and Kings Lynn

FOREWORD

The first edition of this work was published in 1957 by the OECD in Paris under the title *Guide to European Sources of Technical Information*. Successive editions have built on the original plan. Coverage was initially extended to include the countries of Eastern Europe and in the fifth edition to include science as well as technology. The seventh edition continues this evolutionary process by including many more information services and consultancies which charge for their services. This addition reflects the increasing importance of value added information.

The main objective of the directory continues to be to provide a contact point within each European country for each subject area. It is not intended as an exhaustive listing of European information centres but as a means of identifying key national centres which might be reasonably expected to answer questions from all types of enquirer, either from their own resources or by referral. The entries in this edition are based on those in the previous ones but once again a wide search has been undertaken to locate relevant organizations which have not been included before.

Our thanks are due to the many organizations throughout Europe which have completed our questionnaire. Our particular thanks are due to those organizations and individuals who suggested suitable institutes in their countries for inclusion. Every effort has been made to obtain up-to-date information and it is regretted that a small number of seemingly relevant organizations have not replied to our repeated requests for information on their activities.

Meanwhile, it is to be hoped that this seventh edition will, like its predecessors, continue to provide a focal point for enquiries on all aspects of science and technology within Europe and worldwide.

Anthony P. Harvey

Broad Oak, Heathfield, UK
May 1986

PUBLISHER'S INTRODUCTION

The seventh edition of **European Sources of Scientific and Technical Information** supersedes the sixth edition which was published in 1984. This edition has been completely revised and updated using questionnaires and telephone enquiries.

In this edition establishments and libraries are placed into one or, if pertinent, two topic chapters. Each chapter provides one or more enquiry points for each of the European countries. The European countries included are all those in the EEC, the EFTA, and the Comecon, but exclude USSR. In this edition there is no 'International' category within each chapter. Instead, every information centre is entered under the country which houses its major office. If there is more than one information centre they are listed alphabetically.

The original language is used except where it does not use a roman alphabet; then English is used. If the reader cannot locate a particular organization in the relevant chapter, he is invited to use the Titles of establishments index.

Each entry is introduced with the full title of the organization accompanied by its acronym if used, and an English translation where applicable. It should be noted that the translated title is not a transliteration of the original language, but an English version intended to give an indication of the work of that particular organization. Introduced by the word or phrase typeset in bold below, the following details, where available, are given in an entry: title; acronym, if regularly used; the English translation of the title; full postal **address**; **telephone** number; **telex**; **facsimile**; parent body or **affiliation**, indicating the administrative links; year **founded**; name of the person to whom **enquiries** should be addressed; **subject coverage**; **library facilities** and whether charges are made, including loan and reprographic services; **library holdings**, indicating the number of volumes, periodicals and reports held by the library; **information services** with details of bibliographic services, translation services, literature searches, and access to on-line information retrieval systems, including whether or not a charge is made; details of **consultancy** and advice services and whether these are charged or not; **courses** provided for non-staff users; and **publications**, giving details of titles and frequency of information reports, journals, etc.

Where a fee for a facility or service is charged, the abbreviation 'c' (charge) has been included; if no charge is made, the abbreviation 'nc' has been used. An asterisk (*) appearing after the title indicates that a reply was not received in time for inclusion in the book. The editor believes, however, that these organizations are important sources of information and they have been entered in the book with details available from other sources, including the Longman directories **Industrial Research in the United Kingdom**, and **European Research Centres: a directory of organizations in science, technology, agriculture, and medicine**.

The Titles of establishments index will be of particular use to the reader in identifying narrower fields of interest than are indicated in the subject area heading, as English language titles have been revolved so that keywords are brought to the front and listed in alphabetical sequence. An establishment with a title in a language other than English is entered under its original language title, its English translation, and under the main keyword in the English translation if applicable. Translated titles are printed in italics. The acronyms of major organizations are also included in this index.

The Subject index is compiled primarily from the 'Subject coverage' section of each entry and the terms are based on the controlled vocabulary of the British Standards Institution ROOT Thesaurus (1986).

Both the Titles of establishments index and the Subject index direct the user by chapter number and entry number to the full information on that establishment.

Much of the information in this directory has been elicited by questionnaire and we offer our sincere thanks to all who made this publication possible by completing the form sent to them. It is hoped to produce updated versions of this directory at regular intervals in the future, and the publishers would appreciate hearing from any users who may have suggestions to make for the improvement of future editions, or who are able to point out any errors of omission or commission.

Colin P. Taylor   July 1986
Publisher

CONTENTS

1 SCIENTIFIC AND TECHNICAL INFORMATION CENTRES

ALBANIA

Centre for Scientific and Technical Information * 1.1

Address: Tiranë,
Enquiries to: Director general, G. Xhuvani

AUSTRIA

Österreichische Akademie der Wissenschaften 1.2

[Austrian Academy of Sciences]
Address: Dr Ignaz Seipel-Platz 2, A-1010 Wien, Austria
Telephone: (0222) 52 96 81
Enquiries to: Amtsdirektor, Beate Amstädter
Founded: 1847
Library facilities: Open to all users; loan services (c); reprographic services (c).
Loans are made to institutions only.

BELGIUM

Centre National de Documentation Scientifique et Technique 1.3

– CNDST
[National Centre for Scientific and Technical Documentation]
Address: Bibliothèque Royale Albert Ier, Boulevard de l'Empereur 4, Mont des Arts, B-1000 Bruxelles, Belgium
Telephone: (02) 519 56 60
Telex: 221157
Parent: Royal Library of Belgium
Enquiries to: Director, Dr A. Cockx
Founded: 1964
Subject coverage: Science, technology, biomedical sciences and agronomy. Human sciences are excluded.
Library facilities: Open to all users (c); loan services (limited to interlibrary loans to institutions - postage charge); reprographic services (c).
Library holdings: Approximately 4 million books; 24 000 periodicals; deposit library for national and international publications: reports, congress reports, technical reports etc.
Information services: Bibliographic services (computerized services are charged); translation services (c); literature searches (c); access to on-line information retrieval systems (c) (worldwide).
Also: document delivery, referral and technological forecasting.
Consultancy: Telephone answers to technical queries (c); writing of technical reports (c); no compilation of statistical trade data; market surveys in technical areas (c).
Courses: Approximately 28-30 training seminars for 400 terminal users in Belgium. Also provides individual consultancy on networks and data bases.

BULGARIA

Bulgarian Academy of Sciences Central Library 1.4

– CB BAN
Address: 7 Noemvri 1, 1040 Sofia, Bulgaria
Telephone: 8 41 41
Telex: 22424 ban sf bg
Enquiries to: Director, Professor Elena Savova
Founded: 1869
Subject coverage: Natural, mathematical and social sciences.

Library facilities: Open to all users (nc); loan services (nc); reprographic services (nc).
Library holdings: 1 577 703 bound volumes; 5391 current periodicals.
Information services: Bibliographic services (nc); no translation services; literature searches (nc); access to on-line information retrieval systems (nc).
Consultancy: Telephone answers to technical queries (nc); no writing of technical reports; compilation of statistical trade data (nc); no market surveys in technical areas.
Publications: *Bio-Bibliographies of Bulgarian Scientists*; *Problems of Special Libraries* (both are serial publications).

Central Institute for Scientific and Technical Information 1.5

Address: Boul, G.A. Našar 52A, 1040 Sofia, Bulgaria
Telephone: 71 70 24
Telex: 22404
Enquiries to: Director general, P. Kiracov
Library facilities: Open to all users; reprographic services.
Information services: Translation services; literature searches.

CZECHOSLOVAKIA

Ústředí Vědeckých Informací 1.6 Československé Akademie Věd, Základní Knihovna
– ÚVIČSAV - ZK
[Scientific Information Centre of the Czechoslovak Academy of Sciences, Main Library]
Address: Národní Trída 3, 115 22 Praha 1, Czechoslovakia
Telephone: 24 34 41
Telex: 12 10 40 Acad c
Enquiries to: Director, Jiri Zahradil
Founded: 1952
Subject coverage: Fundamental research in natural and social sciences. The centre coordinates the scientific information system of 75 ČSAV institutes.
Library facilities: Not open to all users; loan services (nc); reprographic services (nc).
Library holdings: 900 000 bound volumes; 3900 current periodicals.
Information services: Bibliographic services (nc); no translation services; literature searches (nc); access to on-line information retrieval systems (nc).
Consultancy: Telephone answers to technical queries (nc); no writing of technical reports; no compilation of statistical trade data; no market surveys in technical areas.

Courses: Annual training course (50 hours); seminars (10 per year). Themes of course and seminars: new information technology.

Ústředí Vědeckých Technických a 1.7 Ekonomických Informací
– ÚVTEI
[Central Office of Scientific, Technical and Economic Information]
Address: Klementinum, nám dr Vacka 5, 113 07 Praha 1, Czechoslovakia
Telephone: (2) 265721
Telex: 122214 UVTEI PRAHA
Enquiries to: Director, Dr E. Sošková
State Technical Library
Founded: 1966
Subject coverage: Scientific, technical and economic information.
Library facilities: Open to all users (nc); loan services (nc); reprographic services (c).
Library holdings: 938 513 bound volumes; 3697 current periodicals.
Information services: Bibliographic services (nc); no translation services; literature searches (c); access to on-line information retrieval systems (nc).
Consultancy: Telephone answers to technical queries (nc); writing of technical reports (c); compilation of statistical trade data (c); market surveys in technical areas (c).
ÚVTEI provides these consultancy services at the following address: Konviktská 5, 113 57 Praha 1.
Courses: Courses for information and library staff twice a year; specialized courses for information and library staff as needed; courses for users of data bases as needed.

DENMARK

Danmarks Tekniske Bibliotek 1.8 med Dansk Central for Dokumentation *
– DTB
[National Technological Library of Denmark and Danish Centre for Documentation]
Address: 1 Anker Engelunds vej, DK-2800 Lyngby, Denmark
Telephone: (02) 883088
Telex: 37148 dtbc dk
Affiliation: Technological University of Denmark
Founded: 1942
Subject coverage: Science and technology.
Library facilities: Open to all users; loan services; reprographic services.

Information services: Bibliographic services; translation services (the library maintains a file of translators within science and technology); literature searches (searches are done not only in own on-line catalogue, but also in IRS, SDC, Lockheed, etc after previous arrangement); access to on-line information retrieval systems.

The catalogue of the library is on-line accessible via Datapak (Scannet) and by direct call on (02) 888211 (300 baud) and (02) 888284 (1250 baud), data base name ALIS, data base use free of charge.
Consultancy: Yes.

Jysk Teknologisk   1.9
[Jutland Technological Institute]
Address: Teknologiparken, DK-8000 Århus C, Denmark
Telephone: (06) 142400
Telex: 68722 jytek dk
Enquiries to: Library
Founded: 1943
Subject coverage: Energy sciences; civil engineering; electrical engineering; mechanical engineering; pollution studies; food science; fisheries; management science; acoustics; materials testing and plastics; timber technology.
Library facilities: Open to all users; loan services (c); reprographic services (c).
Library holdings: 10 000 bound volumes; 500 current periodicals.
Information services: Bibliographic services (c); no translation services; literature searches (c); access to on-line information retrieval systems (c).
Consultancy: Telephone answers to technical queries (c); writing of technical reports (c) compilation of statistical trade data (only for customers); market surveys in technical areas (only for customers).

Teknologisk Institut   1.10
[Technological Institute]
Address: Box 141, Gregersensvej, DK-2630 Tåstrup, Denmark
Telephone: (02) 99 66 11
Telex: 334 16 ti dk
Enquiries to: Head, Torben Colding
Information Department
Founded: 1906
Subject coverage: Production engineering; industrial metallurgy; automotive engineering; industrial automation; wood technology; building technology; heating, ventilation, sanitation, energy; coatings technology; chemical technology; leather technology; plastics technology; industrial psychology; educational technology.
Library facilities: Open to all users (c); loan services (c); reprographic services (c).
Library holdings: 10 000 bound volumes; 450 current periodicals.

Information services: Bibliographic services (c); no translation services; literature searches (c); access to on-line information retrieval systems (c).
Consultancy: Telephone answers to technical queries (nc); no writing of technical reports; no compilation of statistical trade data; no market surveys in technical areas.
Courses: Courses are available.
Publications: Research reports.

FINLAND

Helsingin Yliopiston Kirjasto,   1.11
Luonnontieteidenkirjasto
[Helsinki University Library, Science Library]
Address: Tukholmankatu 2, SF-00250 Helsinki 25, Finland
Telephone: (90) 410566
Telex: 122785 tsk sf
Enquiries to: Head of information services, Marita Rosengren
Founded: 1979
Subject coverage: Natural sciences, especially biological sciences.
Library facilities: Open to all users (nc); loan services (nc); reprographic services (c).
The library participates in the interlibrary lending scheme.
Library holdings: The library has 6000 current serials, acquired on an exchange basis. In addition to its own collection, the library keeps a card index of titles held by the science institutes of the university.
Information services: Bibliographic services (c); no translation services; literature searches (c, but no charge for searches taking less than an hour); access to on-line information retrieval systems (c - Dialog, ESA/IRS, STN-Intenational, Pergamon InfoLine, Libris, Alis Recodex, DIMDI, KDOK/MINTTU, etc).
Consultancy: Telephone answers to technical queries; no writing of technical reports; no compilation of statistical trade data; no market surveys in technical areas.
Courses: Courses on information retrieval and library use (for various institutions of the Faculty of Natural and Mathematical Sciences of the university, once a year for each institution).
Publications: Current microfiche publications (list on request) include: *Fennica*, Finnish national bibliography 1978-, microfiche edition; *Fennica ISBN; Fennica ISDS*, Finnish serials with ISSN; *Finuc-S, Union Catalogue of Foreign Serials in the Research Libraries of Finland; Union Catalogue of Foreign Literature in the Research Libraries of Finland, (A: Books)*.

Oulun Yliopiston Kirjasto — OYK 1.12
[Oulu University Library]
Address: PL 186, Kasarmintie 7, SF-90101 Oulu 10, Finland
Telephone: (981) 223455
Telex: 32256 Oyk sf
Enquiries to: Chief librarian, Vesa Kautto
Founded: 1959
Subject coverage: Science, technology, medicine, education and humanities.
Library facilities: Open to all users (nc); loan services; reprographic services (c).
The library participates in the interlibrary loan scheme.
Library holdings: 550 000 bound volumes; 14 367 current periodicals.
Information services: Bibliographic services (nc); no translation services; literature searches (c); access to on-line information retrieval systems (c).
The library has access to the following on-line information retrieval systems: Dialog; IRS; FIZ; Pascal SDC; and national data banks.
Consultancy: No consultancy.
Publications: List on request.

Tampereen Teknillisen Korkeakoulun Kirjasto 1.13
[Tampere University of Technology Library]
Address: Box 537, SF-33101 Tampere, Finland
Telephone: (931) 162111
Telex: 22313 ttktr sf
Enquiries to: Information specialist, Arja Valta
Founded: 1958
Subject coverage: Engineering and technology; physics and chemistry; mathematics and computer science; architecture and regional planning.
Library facilities: Open to all users; loan services (nc); reprographic services (c).
Local loans are free of charge, interlibrary loans within Finland at cost, and abroad, charged. The library has on-line catalogues, open to all users on site.
Library holdings: 110 000 books; 1500 current periodicals.
Information services: Bibliographic services (c); no translation services; literature searches (c); access to on-line information retrieval systems (c).
Consultancy: Telephone answers to technical queries (c); writing of technical reports (nc); no compilation of statistical trade data; no market surveys in technical areas.
Courses: Ten courses are available for non-staff users.

Teknillinen Korkeakoulu Kirjasto 1.14
[Helsinki University of Technology Library]
Address: Otaniementie 9, SF-02150 Espoo 15, Finland
Telephone: (0) 4512812
Telex: 12 1591
Facsimile: (0) 4512832
Enquiries to: Director, Professor Elin Törnudd
Founded: 1849
Subject coverage: Science and technology.
Library facilities: Open to all users (nc); loan services (nc); reprographic services (c).
Telefax transmission. The library is the national resource library for technology and allied sciences.
Library holdings: 600 000 volumes; 5000 current periodicals; 600 000 reports.
Information services: Bibliographic services (c); translation services (c); literature searches (c); access to on-line information retrieval systems (c).
The library has access to the following on-line information retrieval systems: IRS, SDC, Lockheed, SDI. The library's catalogues are on-line data bases and accessible by Scannet, Euronet, Finpac and by direct call. The system Tenttu covers the book catalogue, the serials catalogue, and a data base of Finnish technical journal articles. The catalogues are also available on COM.
The library supervises Finnish input into the following data bases: *Nordic Energy Index*; *International Nuclear Information System R*; *Energy Database*; *Nordic Energy Research in Progress*.
Consultancy: Telephone answers to technical queries (nc for the first half hour); no writing of technical reports; no compilation of statistical trade data; no market surveys in technical areas.
Courses: Annual information retrieval courses covering the following topics: physics and mathematics; mechanical engineering; chemistry; urban planning; electrical engineering; forest products; mining and metallurgy; industrial management. There is also user training in the use of the library's OPAC once a month.
Publications: Dissertations of the university; research papers; OTA-kirjasto A, B, C; list of publications issued on request.

Turun Yliopiston Kirjasto 1.15
[Turku University Library]
Address: Yliopistonmäki, SF-20500 Turku, Finland
Telephone: (921) 645176
Telex: 62123 tyk sf
Enquiries to: Librarian, Marjaana Peltonen
Founded: 1920
Subject coverage: Humanities, social sciences, natural sciences and mathematics, physics, chemistry, medicine, dentistry.
Library facilities: Open to all users (nc); loan services (c); reprographic services (c).

Loans are subject to postage charges; the library receives a free copy (dépot legal) of all publications published in Finland.
Library holdings: 1 400 000 bound volumes; 4500 current periodicals.
Information services: No bibliographic services; no translation services; literature searches (nc); access to on-line information retrieval systems.
Consultancy: No consultancy.

Valtion Teknillinen Tutkimuskeskus * 1.16
– VTT
[Technical Research Centre of Finland]
Address: Vuorimiehentie 5, SF-02150 Espoo 15, Finland
Telephone: 358 0 4561
Telex: 122972
Affiliation: Ministry of Trade and Industry
Enquiries to: Director, Sauli Laitinen
Information Service
Founded: 1942
Subject coverage: Building and community technology; energy technology; information technology; process technology; manufacturing technology.
Library facilities: Open to all users; loan services; reprographic services.
Loan services are free of charge if collected from the library: postal loans are charged. Copies also taken from microforms.
The Information Service has long experience in quickly identifying the fastest supplier of a document and procuring the desired document from abroad.
Information services: Bibliographic services (c); no translation services; literature searches (c); access to on-line information retrieval systems (c).
The service conducts literature searches, made on-line from more than 400 data bases, with access to approximately 50 on-line systems. SDI is also offered from all data bases in use.
Consultancy: Yes (c).
Publications: List available on request.

FRANCE

Association Nationale de la Recherche Technique 1.17
[National Association for Technical Research]
Address: 101 avenue Raymond Poincaré, 75116 Paris, France
Enquiries to: President, G. Worms
Publications: *Progrès Technique*, quarterly.

Centre de Documentation Scientifique et Technique du CNRS 1.18
– CDST
[CNRS Scientific and Technical Documentation Centre]
Address: 26 rue Boyer, 75971 Paris Cedex 20, France
Telephone: (1) 43 58 35 59
Telex: CNRSDOC 220 880F
Enquiries to: Director, Daniel Confland
Founded: 1940
Subject coverage: The centre publishes information in science and technology, extracted from the Pascal scientific and technical bibliographic data base; Pascal covers exact sciences, life sciences, earth sciences, and technology, and receives information from scientific sources worldwide.
Library facilities: Open to all users; no loan services; reprographic services.
Reprographic services include photocopy, microfiche and microfilm.
Information services: Bibliographic services; translation services; literature searches; access to on-line information retrieval systems.
Pascal products and services include: Pascal on-line; Pascal on magnetic tapes; retrospective bibliographies; documentary profiles; bibliographic journal; reprography research department; research and development of new products.
Publications: *Map Indexes of Science and Technology*; publications list on request.

Committee on Data for Science and Technology 1.19
– CODATA
Address: 51 boulevard de Montmorency, 75016 Paris, France
Telephone: 45 25 04 96
Telex: 630553
Parent: International Council of Scientific Unions
Enquiries to: Executive secretary, Phyllis Glaeser
Founded: 1966
Subject coverage: Numerical, scientific and technical data: interdisciplinary extraction, compilation and evaluation of data for specialized fields; data compression; abstracting and indexing; bibliographic standards; storage and retrieval.
Library facilities: Open to all users; no loan services.
Information services: No bibliographic services; no translation services; no literature searches; no access to on-line information retrieval systems.
CODATA establishes directories of data sources (data banks and publications) in various disciplines of science. It sponsors a data bank on hybridomas and monoclonal antibodies.
Publications: *CODATA Newsletter*, irregular periodical in English; *CODATA Bulletin*, of which various issues have been compiled and published as

the *Directory of Data Sources for Science and Technology*; *CODATA Special Reports;* publications list available on request.

International Council for Scientific and Technical Information
1.20

– ICSTI
Address: 51 boulevard de Montmorency, 75016 Paris, France
Telephone: 45 25 65 92
Telex: 630553
Affiliation: International Council of Scientific Unions
Enquiries to: Executive secretary, Marthe Orfus
Founded: 1953
Subject coverage: To increase accessibility to, and awareness of, scientific and technical information. Information services for the topics covered in this directory are provided by ICSTI members.
Library facilities: No library facilities.
Information services: No information services.
Consultancy: No telephone answers to technical queries; writing of technical reports (under contract and/or for publication); no compilation of statistical trade data; no market surveys in technical areas.
Organization of conferences and other meetings; research in the field of science and technology information; publication of reports.

GERMAN DEMOCRATIC REPUBLIC

Zentralinstitut für Information und Dokumentation der DDR
1.21

– ZIID
[Central Institute for Information and Documentation of the German Democratic Republic]
Address: Köpenicker Strasse 80/82, 1020 Berlin, German Democratic Republic
Telephone: 2391 280
Telex: 114690
Enquiries to: Director, H. Och
Founded: 1963
Subject coverage: Information and documentation; information science; science and technology; management and planning; standardization.
Library facilities: Not open to all users; loan services; no reprographic services.
Information services: Bibliographic services; translation services; literature searches.
Consultancy: Yes.
Publications: *Informatik* (Information Science), bimonthly; *Information/Dokumentation - annotierte Titelliste* (Information/Documentation - annotated

List of Titles), quarterly; *Leitung und Planung von Wissenschaft und Technik* (Management and Planning of Science and Technology), quarterly; *Congress Calendar GDR* (notifications of GDR conventions with international participation), annual; *ZIID-Schriftenreihe*, irregular publications series; *Informationsdienst Übersetzungen*, irregular microfiche translations in 25 subject areas; *Service of Congress Dates - Convention Abroad (Socialist Countries)*, quarterly.

GERMAN FEDERAL REPUBLIC

Deutsche Forschungsgemeinschaft
1.22

– DFG
[Technical Science Information Agency]
Address: Postfach 20 50 04, Kennedyallee 40, D-5300 Bonn 2, German Federal Republic
Telephone: (0228) 8851
Telex: 17 228312 dfg
Enquiries to: Reinhard Rutz
Founded: 1920
Subject coverage: The DFG promotes and funds research projects in all branches of science and humanities; however it has no research facilities or personnel of its own.
In the field of information science, DFG gives support to the central specialist libraries for technology (Hannover), for economic sciences (Kiel) medicine (Köln), and agriculture (Bonn); also to special collections at 17 state and university libraries. DFG produces summaries of general and specialist catalogues, and is developing a central catalogue of German monographs and journal articles.
DFG promotes pilot schemes for the use of electronic data processing in library science, the development of unified systems and increased efficiency of interlibrary loans, and research projects in library science.
Library facilities: Limited library for internal use only.
Information services: No information services.
DFG is obliged by statute to advise parliaments and public authorities on scientific methods; in effect this takes the form of published recommendations and advice documents, principally in the fields of preventive medicine and environmental research.

Fachinformationszentrum Technik eV *
1.23

[Special Information Centre for Technology]
Address: Ostbahnhofstrasse 13, D-6000 Frankfurt am Main 60, German Federal Republic
Telephone: (0611) 4308 213
Telex: 4189 459

Enquiries to: Managing director, P. Genth
Founded: 1972
Subject coverage: Structural design; construction materials; materials properties and testing; water treatment; pollution; sanitary engineering; wastes; ocean technology; mining engineering; fuel technology; metallurgical engineering; mechanical engineering - general plant and power; fluid flow; hydraulics; pneumatics; heat and thermodynamics; marine engineering; materials handling and coating and finishes; plastics plant equipment and processes; industrial engineering; production planning and control, safety engineering.
Library facilities: Not open to all users; no loan services; reprographic services (on microfilm).
Information services: Bibliographic services; translation services; literature searches; access to on-line information retrieval systems.
The centre also provides profile and abstracts services, and access to the following on-line information retrieval systems: DIRS 3; Host in Euronet/Diane; Teletype; SDI and Magnetband services.
Consultancy: Yes.

Gesellschaft für Information und Dokumentation mbH   1.24
– GID-12
[Society for Information, Documentation and Practice]
Address: Lyoner Strasse 44-48, Postfach 71 03 70, D-6000 Frankfurt am Main 71, German Federal Republic
Telephone: (0611) 66 87
Telex: 4 14 351 gidfm d
Enquiries to: Information officer
Founded: 1977
Subject coverage: Information science and practice; methods in information and documentation (IuD); mechanized information processing; information systems and networks; user research; classification and thesaurus research; economics and education of IuD; information policy; IuD aspects of the following: standardization, terminology, linguistics, law, computer science, reprography, communications technology, librarianship, archives, publishing, and bookselling.
Library facilities: Open to all users; loan services; reprographic services.
Journals are not available for loan.
Library holdings: About 18 000 reports (annual increase of 1200 items); about 1000 microfiches; about 400 journals.
Information services: Bibliographic services (c); literature searches (c); access to on-line information retrieval systems (c) (- Infodata).
Consultancy: Yes.
Publications: Biennial list of German information centres; annual list of research projects in information

science; inventory lists of the library's thesauri and journals.

Informationsvermittlung Technik *   1.25
– IVT
[Technical Science Information Agency]
Address: Technische Universität Berlin, Universitätsbibliothek, Strasse des 17 Juni 136 MA5-8, D-1000 Berlin 12 (Charlottenburg), German Federal Republic
Telephone: (030) 314-4429
Telex: 01-83-872
Affiliation: Technische Universität Berlin
Enquiries to: Kurt Penke
Founded: 1963
Subject coverage: Information in all technical branches, especially electrical and mechanical engineering and physics, using European and US data bases.
Library facilities: Open to all users; loan services (nc); reprographic services (nc).
Information services: Bibliographic services (nc); no translation services; literature searches (nc); access to on-line information retrieval systems (nc).
Inspec; ZDE; DOMA; NTIS; Compendex 66; PHYS; ENERGY; DECHEMA; METADEX 79; MEATADEX 66; SDIM 1; SDIM 2; COAL; ENERGYLINE; MEDITEC; DKF; GEOLINE.
Consultancy: Yes (nc).

STN International: Scientific and Technical Information Network   1.26
Address: Postfach 2465, D-7500 Karlsruhe 1, German Federal Republic
Telephone: (07247) 824566
Telex: 17724710
Facsimile: (07247) 824639
Affiliation: Fachinformationszentrum Energie, Physik, Mathematik GmbH; American Chemical Society's Chemical Abstracts Service
Enquiries to: Marketing director, Dr B. Jenschke
Founded: 1983
Subject coverage: International on-line service giving direct access to scientific and technical data bases.
Library facilities: No library facilities.
Information services: No bibliographic services; no translation services; no literature searches; access to on-line information retrieval systems.
Data bases available include: CAS ONLINE Files (CA, REGISTRY, LCA, REGISTRY), MATH, NTIS, PHYS, VTB, DETEQ, DEQUIP, Biosis Previews, CLAIMS, nineteen ASC primary journals, Inspec, Compendex, ENERGY, Biomass, PATDPA.
Consultancy: Telephone answers to technical queries; no writing of technical reports; no compilation of statistical trade data; no market surveys in technical areas.

Courses: Fifty seminars per annum for beginners and experienced users: Messenger retrieval language and data base contents.
Publications: *STN News*, quarterly newsletter; leaflets; list on request.

Universitätsbibliothek Hannover und Technische Informations-bibliothek 1.27
– UB/TIB
[Hanover University Library and Technical Information Library]
Address: Welfengarten 1B, D-3000 Hannover 1, German Federal Republic
Telephone: (511) 762-2268
Telex: 922 168 tibhn d
Facsimile: (511) 715936
Enquiries to: Deputy librarian, Jobst Tehnzen
Founded: Universitätsbibliothek - 1831; Technische Informationsbibliothek - 1959
Subject coverage: Engineering/technology including the basic sciences, chemistry, mathematics and physics; all as comprehensive as possible and worldwide. The library is also developing its own information activities to a certain degree, in close coordination with the technical information centres. Particularly interesting specialist literature is evaluated by language and subject specialists of the staff and is announced in the literature services of the technical information centres. In addition, translations are made into Western languages of technical literature in Eastern languages or are commissioned to specialist translators. Unpublished German research reports are listed in a bibliography and included in the European data base SIGLE (System for Information on Grey Literature in Europe). The special activities of the UB include the 'Online Tec' information service, which offers on-line literature searches in domestic and foreign data bases. Searches are carried out by the library's subject specialists.
Library facilities: Open to all users (nc); loan services (c); reprographic services (c).
Library holdings: 2 800 000 bound volumes; 21 500 current periodicals; 870 000 reports.
Information services: No bibliographic services; translation services (c); literature searches (assistance for local users only - c); access to on-line retrieval systems (c - reduced charge for students). Document supplier for subjects covered to DIALOG; ESA; STN Internat; DIMDI; ZDB; DOCLINE; FTZ Technik (Data Star).
Consultancy: No consultancy.
Courses: UB/TIB Seminar für Bibliothekspraxis - twice annually since 1971.

HUNGARY

Budapesti Műszaki Egyetem Központi Könyvtára 1.28
– BMEKK
[Technical University of Budapest Central Library]
Address: Budafoki utca 4-6, H-1111 Budapest, Hungary
Telephone: 664 011/12 10
Telex: 22 59 31
Enquiries to: Director, Imre Lebovits
Founded: 1848
Library facilities: Open to all users (nc); loan services (nc); reprographic services (c).
The library maintains the union catalogue of books and periodicals of the university library system; interlibrary loan service; international exchange of publications; information services on special fields; centralized services for member libraries of the university.
Library holdings: 500 000 bound volumes; 1875 current periodicals; 130 000 reports in special collections.
Information services: Bibliographic services (c); no translation services; literature searches (for students only) no access to on-line information retrieval systems.
The Information and Methodological Department of the library provides a central information service in the fields of chemical, mechanical, electrical, civil, transport engineering, architecture and political sciences, in the form of consultancy and bibliographic services; provides literature searches for the staff of the university, and literature search instruction for students.
Consultancy: Telephone answers to technical queries (nc); no writing of technical reports; no compilation of statistical trade data; no market surveys in technical areas.
Publications: *A Budapesti Műszaki Egyetem Központi Könyvtárának Évkönyvei* (Yearbooks of the Central Library of the Technical University of Budapest), abstracts in English, every 5 years; *Műszaki Egyetemi Könyvtáros* (Technical University Librarian), semiannual journal on librarianship; *Felsőoktatási Szakirodalmi Tájékoztató A sorozat: Műszakié Természettudományok* (Special Literature Review in Higher Education. Series A: Science and Technology) semiannual abstracts and bibliography; *Módszertani Kiadványok* (Methodological Publications); *Tudományos Műszaki Bibliográfiák* (Scientific Bibliographies in Technology); *A Magyar Műszaki Egyetemeken Elfogadott Doktori Disszertációk Jegyzéke* (Guide to theses accepted by Hungarian technical universities for doctors' degrees), biannual, containing titles in English, Russian and German,

and abstracts in Hungarian; *Európai Műszaki Egyetemekés Főiskolák Szakosítasi Rendje* (Specialization Systems at European Technical Universities and Colleges); *A Szakirodalomkutatás Segédkönyvei* (Literature Search Manuals).

Magyar Tudományos Akadémia Könyvtára   1.29
– MTAK
[Hungarian Academy of Sciences Library]
Address: Akadémia utca 2, PO Box 7, H-1361 Budapest, Hungary
Telephone: 382 344
Telex: 224 132 aktar h
Enquiries to: General director, Dr György Rózsa
Founded: 1826
Subject coverage: Social sciences and humanities, science (basic research).
Library facilities: Not open to all users; loan services; reprographic services.
The library is open to a restricted circle of scientists and researchers.
Library holdings: 934 458 books; 261 810 periodicals; 527 115 manuscripts; 20 716 microfilms; 5500 current periodicals.
Information services: Bibliographic services; no translation services; literature searches; access to on-line information retrieval systems.
The library provides computer-based off-line SDI and IR services in the field of science.
Consultancy: No consultancy.
Publications: *Publicationes Bibliothecae Academiae Scientiarum Hungaricae; Catalogi Collectionis Manuscriptorum Bibliothecae Academiae Scientiarum Hungaricae; Informatics and Scientometrics Oriental Studies*; bulletins; publications list on request.

Országos Műszaki Információs Közpóntés Könyvtár   1.30
– OMIKK
[National Technical Information Centre and Library]
Address: Múzeum u. 17, PO Box 12, H-1428 Budapest, Hungary
Telephone: 336 300
Telex: 224944 omikk-h
Enquiries to: Head, Int. Department, Péter Szántó
Founded: 1883
Subject coverage: Natural sciences (mathematics, physics, chemistry); all kinds of technical literature of medium and high level; management; information science.
Library facilities: Open to all users (c); loan services (c); reprographic services (c).
The library provides interlibrary loan services for national and international users; photoreproductions of scientific materials on exchange basis, or upon order.

Library holdings: 1 400 000 bound volumes; 6400 current periodicals; 120 000 reports.
Information services: Bibliographic services (c); translation services (c); literature searches (c); access to on-line information retrieval systems (c).
Literature searches include import of unavailable primary documents. Verbal translation and advisory service in English, German and Russian.
OMIKK-Technoinform, the foreign trade department of the institution offers organization of lectures, conferences and meetings promoting technical-economic development in Hungary; and undertakes translations of scientific, technical and economic literature from foreign languages (English, French, German, Russian, and Spanish) into Hungarian and vice versa. Studies and forecasts.
Consultancy: Telephone answers to technical queries (nc); writing of technical reports (c); no compilation of statistical trade data; market surveys in technical areas (c).
The library provides consultancy on construction of information networks, and design of information systems for selected fields.
Courses: Courses for technical librarians and for information users.
Publications: *Hungarian R&D Abstracts*, selected Hungarian scientific and technical literature, in English, quarterly; *Szakirodalmi Tájékoztatók* (Technical Abstracts), monthly abstracts journal in 21 series; *Tudományosés Műszaki Tájékoztatás* (Scientific and Technical Information), monthly; *Műszaki Információ* (Technical Information), digests of foreign technical literature in 15 series; *Műszaki-Gazdasági Információ* (Technical-Economic Information), digest of foreign technical and economic articles and reports, in 5 series; *Műszaki-Gazdasági Tájékoztató* (Technical-Economic Information), reviews on the interrelations of current technical trends and economic development, compiled from foreign literature, issued monthly; scientific films and videocassettes (English). *Audio-Vizúalis Közlemények* (Review of Audio-Visual Techniques), bimonthly; *Uj kutatásiés műszaki fejlesztési jelentések az OMK-ban* (New Research and Development Reports in the Holdings of the National Technical Library), monthly additions list and yearly index; *Uj szakkönyvek az OMK-ban* (New technical books in the Holdings of the NTL), every two months, additions list in 12 series.

ICELAND

Idntaeknistofnun Islands 1.31
– ITI
[Technological Institute of Iceland]
Address: Keldnaholt, IS-112 Reykjavik, Iceland
Telephone: (91) 687000
Telex: 3020 Istech is
Enquiries to: Librarian, E. Arnvïdardóttir
Subject coverage: Industrial research and development; technical assistance; standardization.
Library facilities: Open to all users (nc); loan services (nc); reprographic services (nc).
Library holdings: 4700 bound volumes; 230 current periodicals; 1550 reports.
Information services: Bibliographic services (c); no translation services; literature searches (c); access to on-line information retrieval systems (c).
Consultancy: Telephone answers to technical queries (nc); no writing of technical reports; no compilation of statistical trade data; market surveys in technical areas (c).
Courses: Various courses are run through the year for non-staff users.

Rannsóknarád ríkisins* 1.32
[National Research Council]
Address: Laugaveg 13, Reykjavik, Iceland
Enquiries to: Scientific and Technical Information Service
Founded: 1978
Subject coverage: Science and technology.
Library facilities: Open to all users; no loan services; reprographic services.
Information services: Literature searches; access to on-line information retrieval systems.
The service has access to the following on-line information systems: ESA-IRS, Euronet, Scannet, Dialog.

IRELAND

Institute for Industrial Research and Standards * 1.33
– IIRS
Address: Ballymun Road, Dublin 9, Ireland
Telephone: (01) 370101
Telex: 25449
Enquiries to: Head, information services department, J. McClusky
Founded: 1947
Subject coverage: Science and technology (excluding agriculture, medicine and nuclear).

Library facilities: Open to all users; loan services (nc); reprographic services (c).
The library has a reference collection of national, ISO and other international standards.
Information services: Bibliographic services (c); no translation services; literature searches (c); access to on-line information retrieval systems.
The library has access to the following: Blaise, ESA/IRS, InfoLine, Dialog, SOC. It also maintains Biomass, an international information data base for the IEA.
Consultancy: Telephone answers to technical queries.
Publications: *Technology Ireland*, monthly; *Irish Construction Directory; Irish Engineering Directory; Survey of Scientific and Technical Information in Ireland*; full list on request.

ITALY

Centro Informazioni Studi Esperienze SpA * 1.34
– CISE
[Information, Studies and Experiments Centre]
Address: PO Box 12081, 39 Via Reggio Emilia, 20100 Milano, Italy
Telephone: (02) 2167 1
Telex: 311643 CISE I
Affiliation: ENEL (National Electricity Agency)
Enquiries to: Head, documentation office, Dr P.A. Comero
Centre of Bibliographic Information
Founded: 1946
Subject coverage: Energy conservation; solar energy; nuclear energy; thermohydraulics; industrial diagnostics; materials; environmental surveillance; lasers; electronics components and systems; data acquisition and processing; analytical chemistry.
Library facilities: Not open to all users; no loan services; reprographic services (c).
Access to non-staff users is granted case by case. The centre keeps KWOC indexes of books, periodicals and technical reports.
Information services: Bibliographic services (c); no translation services; literature searches (c); access to on-line information retrieval systems (c).
The centre has access to the following: ESA Quest; Dialog; SDC Orbit; INKA; Data-Star; Télésystèmes-Questel; Pergamon InfoLine; Echo.
Consultancy: Yes (c).

Istituto di Studi sulla Ricerca e Documentazione Scientifica

1.35

[Institute for Studies on Scientific Research and Documentation]

Address: Via Cesare de Lollis 12, 00185 Roma, Italy
Telephone: (06) 495 23 51
Affiliation: Consiglio Nazionale delle Ricerche
Enquiries to: Librarian, Giliola Negrini
Founded: 1968
Subject coverage: Economics and statistics in research; research organization and legislation; documentation (Italian edition of the UDC); computerized union catalogues; information retrieval.
Library facilities: Open to all users; loan services; reprographic services.
Information services: Bibliographic services; no translation services; no literature searches; access to on-line information retrieval systems.
The library produces bibliographies in specialized subject areas, and union catalogues. Access to on-line information retrieval systems is not provided for as a service but only for demonstration and methodological studies.
Consultancy: Yes.
Publications: *Quaderni,* quarterly journal; *Note di Bibliografia e di Documentazione Scientifica;* publications list on request.

LIECHTENSTEIN

Liechtensteinische Landesbibliothek

1.36

[National Library of Liechtenstein]

Address: Gerberweg 5, PO Box 385, 9490 Vaduz, Liechtenstein
Telephone: (075) 66 346
Enquiries to: Librarian, Dr Alois Ospelt
Founded: 1961
Subject coverage: The library is the national scientific library.
Library facilities: Open to all users (nc); loan services (nc); reprographic services.
Library holdings: 73 400 bound volumes; 162 current periodicals.
Information services: Bibliographic services; no translation services; literature searches; no access to on-line information retrieval systems.
Consultancy: No consultancy.
Publications: *Liechtenstein Bibliography;* publications list on request.

LUXEMBOURG

Commission of the European Communities, Information Market and Innovation *

1.37

– CEC

Address: Jean Monnet Building, rue Alcide de Gasperi, Kirchberg, Luxembourg
Telephone: 43011
Telex: 2752 Eurodoc
Enquiries to: Director general, Raymond Appleyard
Founded: 1958
Subject coverage: European networks (Euronet Diane) and development of information science industry; data communication in documentary applications; applications of new technology; transfer between community languages.
Library facilities: Not available to non-staff users.
Information services: Referral service for Euronet Diane (c). Euronet enquiry service is operated under contract by BPO in London. The Echo service offers several CEC.
Publications: Information pamphlets on Euronet Diane.

MALTA

Malta Development Corporation

1.38

Address: PO Box 571, House of Catalunya, Valletta, Malta
Enquiries to: Head, Research Services Department
Founded: 1967
Subject coverage: Promotion of industrial development in Malta.
Library facilities: Open to all users; no loan services; no reprographic services.
Publications: List on request.

University of Malta Library

1.39

Adress: Msida, Malta
Enquiries to: Librarian, Dr P. Xuereb
Founded: 1769

NETHERLANDS

Centrum voor Informatie en Dokumentatie TNO *

1.40

– CID-TNO

[Centre for Information and Documentation TNO]

Address: Postbus 36, Schoemakerstraat 97, 2600 AA Delft, Netherlands
Telephone: (015) 56 93 30

Telex: 38071 zptno nl
Parent: Netherlands Organization for Applied Scientific Research (TNO)
Enquiries to: Deputy head, Charles L. Citroen
Founded: 1977
Subject coverage: Science; technology; marketing.
Library facilities: Not open to all users.
Information services: No bibliographic services; no translation services; literature searches (c); access to on-line information retrieval systems.
All major on-line systems and services in Europe and the USA are accessed. The centre offers patent search services.
Consultancy: Yes.

Hoofdgroep Maatschappelijke Technologie TNO 1.41
– HMT-TNO
[TNO Technology for Society Division]
Address: Postbus 342, 7300 AH Apeldoorn, Netherlands
Telephone: (055) 773344
Telex: 36395 tnoap
Facsimile: (055) 419837
Parent: Netherlands Organization for Applied Scientific Research (TNO)
Enquiries to: Managing director, C.J. Duyverman
Founded: 1976
Subject coverage: Chemical analysis; biological environmental research; air pollution; environmental technology; wind nuisance; living and working atmosphere; biotechnology; organic chemistry and synthesis; chemical engineering; energy saving and combustion; safety in industry and coal technology.
Library facilities: Open to all users; loan services; reprographic services.
The library has microfiche and microfilm reader/printer facilities.
Information services: Bibliographic services; no translation services; literature searches; access to on-line information retrieval systems.
The library has access to the following information retrieval systems: Euronet, SDC, Dialog, Datastairs, Inis.

International Translations Centre 1.42
– ITC
Address: 101 Doelenstraat, 2611 NS Delft, Netherlands
Telephone: (015) 142242
Telex: 38104
Enquiries to: Director, D. v. Bergeijk
Founded: 1961
Subject coverage: The centre maintains a file of existing translations (from all source languages into Western European languages) in all fields of science and technology, and has access to over 900 000 translations.
Library facilities: Open to all users; no loan services; reprographic services.
The centre acts as a referral agency: it does not undertake or commission translations.
The centre maintains the *World Transindex*, established 1978, centralizing bibliographic data collected in conjunction with the Commission of the European Communities (1978-80) and with the documentation centre of the Centre National de Recherche Scientifique (CNRS), Paris, France. The centre and CNRS co-produce the WTI data base, using the Pascal system, available on-line through ESA/IRS, Frascati, Italy. It contains around 190 000 references.
Publications: *World Transindex*, 10 issues a year, announcing 25 000 translations and giving full bibliographic reference, author and source index; annual and eight-year cumulations of the *World Transindex* and the five-year cumulations of the former *World Index of Scientific Translations* published by ITC before 1978; *Journals in Translation*, an irregular publication containing over 1000 titles, published jointly with the British Library Lending Division.

Koninklijke Nederlandse Akademie van Wetenschappen * 1.43
– KNAW
[Royal Netherlands Academy of Arts and Sciences]
Address: Kloveniersburgwal 29, Postbus 19121, 1011 JV Amsterdam, Netherlands
Telephone: (020) 22 29 02
Telex: 18766 cobd nl
Enquiries to: Reference librarian, R.W.J. Pieters
Founded: 1808
Subject coverage: Science and medicine.
Library facilities: Open to all users; loan services (nc); reprographic services (c).
On-line document delivery via ESA-IRS and Lockheed Dialog.
Information services: Bibliographic services (c); no translation services; literature searches (c); access to on-line information retrieval systems (c).
On-line access to ESA-IRS; Lockheed Dialog; DIMDI; INKA; Datastar.
Consultancy: No consultancy.

Nederlands Orgaan voor de Bevordering van de Informatieverzorging 1.44
[Netherlands Organization for Information Policy]
Address: Burg. Van Karnebeeklaan 19, s'Gravenhague, Netherlands
Enquiries to: Secretary, C. Booster
Founded: 1971

Subject coverage: Coordinates and promotes scientific and technical information handling.

NV Kema, Library 1.45

Address: Postbus 9035, 6800 ET Arnhem, Netherlands
Telephone: (085) 45 70 57
Telex: 450 16
Enquiries to: A.J. te Grotennnuis
Founded: 1920
Subject coverage: Electrical engineering; nuclear science; environmental pollution; materials testing.
Library facilities: Not open to all users; loan services; reprographic services.
Information services: Literature searches; access to on-line information retrieval systems.
The library has on-line access to the following: SDC; ESA; Inis.

Technische Hogeschool Delft, 1.46 Bibliotheek

– BTHD
[Delft University of Technology Library]
Address: Postbus 98, Doelenstraat 101, 2600 MG Delft, Netherlands
Telephone: (015) 78 56 67
Telex: 38070
Enquiries to: Librarian, M.A.A. Eszer-van Dijck
Founded: 1905
Subject coverage: Technology and applied sciences; there are also specialist departmental libraries for: geodesy, civil engineering, building science, building materials science, electrotechnology, mining, marine engineering, metallurgy.
Library facilities: Open to all users (nc); loan services (nc); reprographic services (c).
Library holdings: 450 000 monographs; 300 000 serials; 9600 current periodicals; 700 000 reports.
Information services: Bibliographic services; no translation services; no literature searches; access to on-line information retrieval systems.
The library has access to most on-line bibliographic data bases on technology and applied sciences, available at host's price.
Consultancy: No consultancy.
Publications: *Aanwinsten Bibliotheek TH Delft* (BTHD Accessions), monthly; *Lijst van lopende tijdschriftabonnementen Bibliotheek TH Delft*, list of periodicals.

Voorlichtingsdienst Technische 1.47 Hogeschool Delft

[Information Centre, Delft University of Technology]
Address: Aula, Mekelweg 1, 2628 CC Delft, Netherlands

NORWAY

Norsk Senter for Informatikk 1.48 AS *

[Norwegian Centre for Informatics]
Address: Forskningsveien 1, Oslo 3, Norway
Telephone: (02) 452010
Telex: 72042
Enquiries to: Managing director, Hans Krog
Subject coverage: Central organization for technical information services and information technology systems.

Riksbibliotektjenesten * 1.49

[National Office for Research and Special Libraries]
Address: PO Box 2439, Dramensveien 42, N-0202 Oslo 2, Norway
Telephone: (02) 430880
Telex: 76078
Enquiries to: Director, G. Munthe

Statens Teknologiske Institutt * 1.50

– TEKNOL STI
[National Institute of Technology, Norway]
Address: Postboks 8116 Dep, Akersveien 24C, Oslo 1, Norway
Telephone: (02) 204550
Enquiries to: Librarian, Grethe Kjelldahl
Founded: 1917
Subject coverage: Cereal and meat industry; management; welding and metallurgy; civil engineering; plastics; electrical and electronic engineering; machine tool engineering; surface coatings and wood technology; automotive technology; printing.
Library facilities: Open to all users; loan services; reprographic services.
Information services: Literature searches (c); access to on-line information retrieval systems (c).
Consultancy: Yes (c).

Universitet i Trondheim, Norges 1.51 Tekniske Högskole *

– UNIT-NTH
[University of Trondheim, Norwegian Institute of Technology]
Address: Högskoleringen 1, N-7034 Trondheim NTH, Norway
Telephone: (07) 595110
Telex: 55 186 nthhb n
Enquiries to: Chief librarian, R. Gjersvik
Founded: 1910
Subject coverage: The library is the central technical library of Norway. Exact sciences, and technical aspects of architecture, the arts, economics and psychology, with emphasis on applied sciences and

technology.

Library facilities: Open to all users; loan services; reprographic services (c).

Information services: Bibliographic services (c); literature searches; access to on-line information retrieval systems (c).

The library has access to the following on-line information systems: Scannet; Lis; SDC; IRS; NLM; Technotec.

A specialized documentation department offers manual and on-line documentation services, which are charged.

Publications: *Meldinger og Boklister,* monthly accessions list; *Litteraturlister,* irregular subject bibliographies; *Facsimilia Scientia et Technica Norvegica.*

Universitetsbiblioteket i Bergen * 1.52
– UB Bergen

[Bergen University Library]

Address: Möhlenprisbakken 1, N-5000 Bergen, Norway

Telephone: (05) 213050

Telex: 42690 ubb n

Enquiries to: Director

Founded: 1948

Subject coverage: Biological sciences; earth sciences; energy sciences; medical sciences.

Library facilities: Open to all users; loan services (nc); reprographic services (c).

Information services: Bibliographic services (nc); no translation services; literature searches (nc); access to on-line information retrieval systems (c).

The library contributes to the Norwegian University library data bank Bibsys.

Consultancy: Yes (nc).

POLAND

Biblioteka Główna Politechniki Warszawskiej * 1.53

[Technical University of Warsaw, Library]

Address: 1 Plac Jedności Robotniczej, 00-661 Warszawa, Poland

Telephone: 21 13 70; 21 00 70

Telex: (Technical university) 81 3307; (library) 81 6467

Enquiries to: General director, Edward Domański

Founded: 1915

Subject coverage: Architecture; automation; chemistry; civil engineering; computers; electrical engineering; energetics; mechanical engineering; sanitary engineering; transport; geodesy; machine construction.

Library facilities: Loan services; reprographic services.

Information services: Bibliographic services.

Publications: *Wykaz nabytków zagranicznych* (List of Foreign Acquisitions), semiannual; *Wykaz Tytułów Czasopism i Innych Wydawnictw Cząglych,* periodicals list; *Bibliografia Adnotowana Prac Doktorskich i Habilitacyjnych* (Annotated Bibliography of Doctors' and Professors' Theses), annual; *Bibliografia Publikacji Pracowników Politechniki Warszawskiej,* bibliography of publications of the university, irregular.

Instytut Informacji Naukowej, 1.54 Technicznej i Ekonomicznej *
– IINTE·

[Scientific, Technical and Economic Information Institute]

Address: 3/5 Ulica Żurawia, 00-926 Warszawa, Poland

Telephone: 25 28 09

Telex: 813716

Affiliation: Centrum Informacji Naukowej, Technicznej i Ekonomicznej

Enquiries to: Director, Professor Jacek Bańkowski

Founded: 1949, reorganized 1972

Subject coverage: Research and development activities in the field of scientific, technical and economic information; provides professional advice and other forms of methodological and organizational assistance; involved in information software tools development, especially in CDS Isis package, in cooperation with Unesco.

Library facilities: Open to all users.

Library holdings: 75 000 volumes.

Information services: Access to on-line information retrieval systems.

The clearing house for thesauri and classification schemes at the institute collects and disseminates information on thesauri, classification systems and schedules, descriptors, key-words, and subject-headings lists edited in all languages and concerning various domains of human activity. More important tasks of the clearing house service include meeting users' needs through answering their questions, as well as through the preparation of publications.

Consultancy: Yes.

Publications: *Przegląd Dokumentacyjny Informacji Naukowej,* abstracting information service, monthly; *Bibliographic Bulletin of the Clearing house at IINTE,* annual; *Prace Instytutu INTE* (Reports of the Institute), irregular; *UKD - Zmiany i Uzupełnienia* (Corrections and Extensions to the UDC), quarterly; *Tablice UKD* (Schedules of UDC), irregular.

Instytut Wzornictwa Przemysłowego *   1.55

– IWP
[Institute of Industrial Design]
Address: Ulica Ǎwiętojerska 5/7, 00-236 Warszawa, Poland
Telephone: 31 22 21
Affiliation: Ministry of Science and Education
Enquiries to: Director, R. Terlikowski
Founded: 1950
Library facilities: Open to all users; loan services; reprographic services (c).
Information services: Bibliographic services; translation services; literature searches; no access to on-line information retrieval systems.
Consultancy: Yes.
Publications: *IWP News; Documental Review of Industrial Design; Works and Materials;* annual report.

Ośrodek Informacji Naukowej Polska Akademia Nauk *   1.56

– OIN PAN
[Scientific Information Centre of the Polish Academy of Sciences]
Address: Ulica Nowy Swiat 72, 00-330 Warszawa, Poland
Telephone: 26 84 10
Telex: 815414
Enquiries to: Director, Dr A. Gromek
Founded: 1953
Subject coverage: Social science and natural science; fundamental research.
Library facilities: Open to all users; loan services; reprographic services (c).
National and international interlibrary loan service and reprographic service of materials held by the library and other Polish sources; exchange of scientific materials with foreign research centres.
Information services: Bibliographic services; literature searches.
The library imports unavailable primary documents.
Consultancy: Consultancy services are offered on the design and construction of information services.
Publications: *Zagadnienia Informacji Naukowej* (Problems of Scientific Information), semiannual; *Przeglad Informacji o Naukoznawstwie* (Review of Information Science), quarterly annotated bibliography; *Wiadomości o Nauce* (News on Science), monthly; *Informator o Wynikach Badań Naukowych Zakonczonych* (Directory of Results of Completed Scientific Projects), annual.

Ośrodek Postępu Technicznego *   1.57
[Centre for Technical Progress]
Address: 1b Ulica Buczka, 40-955 Katowice, Poland
Telephone: 59 60 61
Telex: 0312458 OPT PL
Enquiries to: Managing director, Józef Żyła

Founded: 1963
Subject coverage: The centre keeps interested parties informed on products from the following industries: environment pollution control and health protection; transport; power industry; electrical and electronic industry; precision engineering industry; building industry; mining, steel, and chemicals industries; agriculture.
Library holdings: The library covers 35 000 catalogues from 10 000 firms all over the world, systematized in 17 main industries.
Information services: Patent descriptions from Poland, USSR, German FR, Belgium, Hungary, Netherlands, Czechoslovakia, and France.
Publications: *Informator Patentowy* (Patent Description); *Biuletyn Projektów Wynalazczych* (Bulletin on Invention Designs); *Problemy Postępu Technicznego* (Problems of Technical Progress); *Informator o Zagranicznej Literaturze Techniczno-Handlowej* (Technical and Commercial Firm Literature); *Informator o Polskiej Literaturze Techniczno-Handlowej* (Technical and Commercial Polish Literature).

Politechniki Wrocławska, Biblioteka Główna i Ośrodek Informacji Naukowo-Technicznej   1.58
[Technical University of Wrocław, Main Library and Scientific Information Centre]
Address: 27 Wybrzeze Wyspiańskiego, 50-370 Wrocław, Poland
Telephone: 21 27 07; 20 23 05
Telex: 0715371 bgpw pl
Enquiries to: Library director, Dr Henryk Szarski
Founded: 1946
Subject coverage: Library and information science, science, technology.
Library facilities: Open to all users (nc); loan services (nc); reprographic services (c).
Library holdings: 694 000 volumes; 2778 current periodicals; 34 000 reports.
Information services: Bibliographic services (c); no translation services; literature searches (c); access to on-line information retrieval systems (c); SDI services and retrospective searching.
Courses: Annual course (of five days) on computerized library and information systems functioning in the main library and scientific information centre.
Publications: *Prace Bibliograficzne* (Annual bibliography of the scientific publications of the university); *Nauka-Technika-Przemys* abstracts of scientific, technical and industrial research reports; *Wykazy Nabytków Zagranicznych* list of foreign literature acquisitions; research reports.

PORTUGAL

Biblioteca da Academia das Ciências de Lisboa   1.59
[Library of the Academy of Sciences]
Address: Rua da Academia das Ciências 19, 1000 Lisboa, Portugal
Enquiries to: Director, Joaquim Alberto Iria

Centro de Informação Tecnica para a Indústria *   1.60
– CITI
[Technical Information Centre for Industry]
Address: Azinhaga dos Lameirosà Estrada do Paço do Lumiar, 1699 Lisboa Codex, Portugal
Telephone: 796141; 799181
Telex: 42486 LNETI P
Affiliation: Laboratório Nacional de Engenharia e Tecnologia Industrial LNETI (National Laboratory of Engineering and Industrial Technology), Ministry of Industry, Energy and Exportation
Enquiries to: Director, Dr Ana Maria Ramalho Correia Santos Costa
Founded: 1978
Subject coverage: Industrial technology covering the following fields: industrial studies and analyses; chemical industries; food industries; metallurgy and metalwork; electronics and electrical equipment; energy (conventional energy, renewable energy, nuclear engineering and energy, nuclear techniques and sciences); and radiological protection and safety.
Library facilities: Open to all users; loan services; reprographic services.
Information services: Bibliographic services; literature searches.
The library offers SDI, patent information, on-line subject information searches.
Consultancy: Yes.
Publications: *Documentação Técnico-Científica* (Scientific-Technical Documentation), fortnightly; *Documentação Técnico-Científica/IE/Sacavem* (Scientific-Technical Documentation/IE/Sacavem), fortnightly; *Boletim de Informação Bibliografica/IE/Savacém* (Bibliographic Information Bulletin), monthly.

Direcção Geral da Indústria   1.61
– DGI
[General Directorate for Industry]
Address: Avenida Conselheiro Fernando de Sousa 11, 1092 Lisboa Codex, Portugal
Telephone: 681755
Telex: 13567 MITIL P
Affiliation: Ministry of Industry and Trade
Enquiries to: Information Division head, Maria José Rodrigues Brito

Founded: 1978
Subject coverage: The main objectives of DGI are: to follow industrial development; to undertake action concerning the implementation of industrial policies; to contribute to technical and technological support for industry.
Library facilities: Open to all users; no loan services; reprographic services.
Information services: Bibliographic services; translation services; literature searches; no access to on-line information retrieval systems.
The document section includes more than 21 000 bibliographic references of periodical and non-periodical documents, constituting a manual bibliographic base with files organized by registry number, author, title, editor and keywords.
Consultancy: Yes.
Publications: *Informação*, an information and bibliographic bulletin; *DGI - Sumários*, digest bulletin.

Instituto Nacional de Investigação Científica   1.62
– INIC
[National Institute for Scientific Research]
Address: Avenida Ellias Garcia 137, 1093 Lisboa, Portugal
Telephone: (9) 731300
Telex: 18428 P
Affiliation: Ministry of Education
Enquiries to: Head, Centro de Documentação Científica e Técnica (CDCT)
Founded: 1936
Library facilities: Open to all users; no loan services; reprographic services.
There is no general library of the institute, as each research unit has its own independent collection; the centre maintains a small specialist collection on information science and librarianship.
Information services: Bibliographic services; access to on-line information retrieval systems.
The centre has access to the following on-line information systems: Dialog; SDC Orbit; BRS; Blaise; ESA-IRS; Télésystèmes-Questel.
Consultancy: Yes (on information science).

ROMANIA

Institutul National de Informare si Documentare   1.63
– INID
[National Institute for Information and Documentation]
Address: Cosmonauţilor 27-29, 70141 Sector 1, Bucureşti, Romania
Telephone: 13 40 10
Telex: 11247

Telex: 11247
Affiliation: National Council for Science and Technology
Enquiries to: Director, Gheorghe Anghel
Founded: 1949
Subject coverage: Information on science and technology.
Library facilities: Open to all users (nc); loan services (nc); reprographic services (only for INID documentary stock - c).
Library holdings: 650 000 bound volumes; 5000 current periodicals; 2000 reports.
Information services: Bibliographic services (c); translation services (c); literature searches (c); no access to on-line information retrieval systems.
Consultancy: No consultancy.
Courses: The course held by NIID for non-staff users is entitled 'The Information and Documentation Technique' and comprises both theoretical lectures and practical work.
Publications: *Abstracts of Romanian Scientific and Technical Literature; Information and Documentation Problems;* bulletin listing bibliographic searches made by the INID-CERBIEF system.

SPAIN

Gabinete de Formación y Documentación *
[Training and Documentation Office]
1.64

Address: Alfonso XXI 3, 28014 Madrid, Spain
Telephone: (91) 227 38 95
Enquiries to: Director, Felipe Martínez Martínez
Subject coverage: Documentation concerning civil engineering, construction and environment; technical information and teledocumentation service.
Publications: Bibliography; documents.

Instituto de Información y Documentación en Ciencia y Tecnología*
– ICYT
[Institute for Information and Documentation in Science and Technology]
1.65

Address: Joaquím Costa 22, 28006 Madrid 6, Spain
Telephone: 261 48 08
Telex: 2 26 28
Affiliation: Consejo Superior de Investigaciones Científicas (CSIC)
Enquiries to: Secretary, Milagros Villarreal
Subject coverage: Industrial chemistry; electrotechnology and electronics; metallurgy; agricultural engineering; management.

Library facilities: Open to all users; no loan services; reprographic services (c).
Regular subscription to basic referral repertories such as the CA Engineering Index.
Library holdings: 1800 titles.
Information services: Bibliographic services (c); translation services (c); literature searches (c); access to on-line information retrieval systems (c).
Consultancy: Yes (c).
Publications: A bibliographic service covers Spanish scientific and technical literature, published as *Indice Español de Ciencia y Tecnología.*

SWEDEN

Delegationen för Vetenskaplig och Teknisk Informations Försörjning *
– DFI
[Swedish Delegation for Scientific and Technical Information]
1.66

Address: PO Box 43033, S-100 72 Stockholm, Sweden
Telephone: (08) 7442840
Telex: 17150 teamcon
Affiliation: Ministry of Education; Ministry of Industry
Enquiries to: Secretary, Bjorn Thomasson
Founded: 1979
Subject coverage: DFI plans and coordinates information supply on research and development and related activities.
Library facilities: Not available.

Kungliga Tekniska Högskolans Bibliotek
– KTHB
[Royal Institute of Technology Library]
1.67

Address: Valhallavägen 81, S-100 44 Stockholm, Sweden
Telephone: (8) 7877000
Telex: 10389 kthb-s
Facsimile: (8) 109190
Enquiries to: Head librarian, Dr Stephan Schwarz
Founded: 1827
Subject coverage: Science and technology: engineering; architecture; applied and basic science.
Library facilities: Open to all users (nc); loan services (nc); reprographic services (c).
The library has a microfiche camera and a PDP minicomputer.
Library holdings: 500 000 volumes; about 3500 current periodicals.
Information services: Bibliographic services (c); no translation services; literature searches (c); access to

on-line information retrieval systems (c).

The library is an ESA/QUEST national centre, and also a Euronet Diane centre, and has access to most data bases and systems, including Dialog, SDC, Blaise, Télésystèmes-Questel, DIMDI, INKA, and data bases on Scannet.

The library maintains the MechEn bibliographic data base in the field of mechanical engineering; also DOLDIS data base, on-line data bases and data banks produced in Sweden.

The library also offers current awareness service (SDI - EPOS/VIRA) on 21 bibliographic data bases.

The library participates in the Swedish national union catalogue *LIBRIS* and the associated national network for inter-library resource sharing. The library also participates in the build-up of Scandinavian Union Catalogues for monographs, conference proceedings, etc (SMOT) and for periodicals (SPOT) in the areas of technology and associated basic sciences. They feature on-line searching and ordering, and are based on the ALIS system, developed and operated by the Danish Technological Library, Lyngby, Denmark.

Consultancy: Telephone answers to technical queries (nc); writing of technical reports (c); no compilation of statistical trade data; no market surveys in technical areas.

Courses: Permanent programme of courses, seminars and demonstrations on the following topics: on-line bibliographic search; profiling for SDI systems; use of library resources; research in library and information science; remarketing of various vendors and hosts (including ESA, INKA, DIMDI, ISI and CA).

Publications: *Stockholm papers in Library and Information Science*; *Stockholm Papers in the History and Philosophy of Science and Technology*; Royal Institute of Technology dissertations (microfiche).

SWITZERLAND

Eidgenössische Technische Hochschule Bibliothek, Zürich   1.68
– ETH- Bibliothek
[Swiss Federal Institute of Technology, Zürich]
Address: Rämistrasse 101, CH-8092 Zürich, Switzerland
Telephone: (01) 256 21 35
Telex: 53178 ethbich
Enquiries to: Director, ETH- Bibliothek
Founded: 1855
Subject coverage: Science and technology.
Library facilities: Open to all users; loan services; reprographic services.
Library holdings: 3.5m bound volumes; 10 000 current journals; 21 000 current serials; 1.4m reports.

Information services: Bibliographic services; no translation services; literature searches; access to on-line information retrieval systems.

The library has access to the following on-line information retrieval systems: Edis; SDC; Dialog; Euronet; Datastar.

Consultancy: Telephone answers to technical queries (nc); no writing of technical reports; no compilation of statistical trade data; no market surveys in technical areas.

TURKEY

Türkiye Bilimsel ve Teknik Dokümantasyon Merkezi   1.69
– TÜRDOK
[Turkish Scientific and Technical Documentation Centre]
Address: Atatürk Bulvari 221, Kavaklidere, Ankara, Turkey
Telephone: (41) 258689
Telex: 43186 BTAK TR
Facsimile: 9-41-285629
Affiliation: Scientific and Technical Research Council of Turkey
Enquiries to: Deputy director, Mrs Zerrin Esensoy
Founded: 1966
Subject coverage: Documentation in the fields of science and technology.
Library facilities: Open to all users (nc); no loan services; reprographic services (c).
Library holdings: 10 000 bound volumes; 600 current periodicals.
Information services: Bibliographic services (c); translation services (c); literature searches (c); access to on-line information retrieval systems (c).

The centre acts as a referral centre for translations, and keeps a list of translators in various languages. It also maintains a bibliographic data base (1980-) for Turkish literature on pure and applied sciences.

Consultancy: No consultancy.
Publications: *Turkish Dissertation Index*, annual; *Current titles in Turkish science*, irregular; lists of subject bibliographies and monographs on request.

UNITED KINGDOM

Aberdeen University Research and Industrial Services Limited   1.70
– AURIS
Address: King's College, Old Aberdeen, Aberdeen AB9 1FX, UK
Telephone: (0224) 492681

Telex: 73458
Facsimile: (0224) 491439
Enquiries to: Managing director, W. Jamieson
Founded: 1981
Subject coverage: Marine biology; microbial growth and activity; offshore structures; environmental management and planning; computing services; engineering services.
Library facilities: No library facilities.
Information services: No information services.
Consultancy: Telephone answers to technical queries (some queries charged for); writing of technical reports (c); no compilation of statistical trade data; no market surveys in technical areas.
Courses: Primarily in the area of environmental assessment - in conjunction with United Nations, World Health Organization and other government bodies. Courses are held in many non-European countries.

Aslib, Association for Information Management
1.71

Address: 26/27 Boswell Street, Holborn, London WC1N 3JZ, UK
Telephone: (01) 430 2671
Telex: 23667
Facsimile: (0372) 374496
Enquiries to: Director, Dr D.A. Lewis
Information Resources Centre
Founded: 1924
Subject coverage: All aspects of information management.
Library facilities: Not open to all users; loan services (nc); reprographic services (c).
Aslib holds the most comprehensive collection in the UK on information handling and special librarianship.
Information services: The information centre answers members' enquiries on a wide range of subjects in the field of information handling and management, including advice on on-line information retrieval and library automation technology.

British Library, Document Supply Centre
1.72

– BLDSC
Address: Boston Spa, Wetherby, West Yorkshire LS23 7BQ, UK
Telephone: (0937) 843434
Telex: 557381
Facsimile: (0937) 845520
Enquiries to: Head, user services, Katy King
Founded: 1973
Subject coverage: All subjects, including all areas of science and technology, excluding children's books, fiction and ephemeral works.
Library facilities: Not open to all users (c); loan services (c); reprographic services (c).

The BLDSC is part of the Science Technology and Industry Division of the British Library. It was formed in 1973 by the merger of the National Central Library and the National Lending Library for Science and Technology.
The centre is unique in being planned specifically with the main objective of supplying items on request to other organizations.
Requests are accepted only on the centre's own prepaid request forms, or by telex, data base order systems and on-line direct to the centre's computer. Only registered users may make requests: to register, please write to the above address.
Library holdings: 4.5m books and serials, and over 3m documents in microform; 54 000 current serials; English language, Russian science and other language monographs; UK theses; conference proceedings; official publications; over 75 000 British reports.
Information services: Bibliographic services (c); no translation services; literature searches (c); access to on-line information retrieval systems (c).
Consultancy: Telephone answers to technical queries (consultancy limited to document supply and associated matters - nc); no writing of technical reports; no compilation of statistical trade data; no market surveys in technical areas.
Courses: Courses for staff engaged in interlending (UK) - two one-day courses and three two-day courses every year; Microcomputers in Interlending seminars, international users course - two two-day courses every year.
Publications: Publications list on request.

Department of Trade and Industry Library *
1.73

Address: Ashdown House, 123 Victoria Street, London SW1E 6RB, UK
Telephone: (01) 212 6189
Telex: 8813148 DIHQG
Enquiries to: Librarian
Subject coverage: Industrial development, science policy, technology and industry.
Library facilities: Not open to all users (occasional visits at librarian's discretion); loan services (through national interlending scheme only); reprographic services (restricted to small range of unique departmental material - c).
Information services: Bibliographic services.
Publications: *Information Technology - a Bibliography*, annual; publications list on request.

ERA Technology Limited
1.74

Address: Cleeve Road, Leatherhead, Surrey, KT22 7SA, UK
Telephone: (0372) 374151
Telex: 264045
Facsimile: (0372) 374496

Enquiries to: Manager, P.A. Shepherd
Information and Administrative Services Department
Founded: 1920
Subject coverage: Electrical and electronic engineering; energy conservation; alternative energy; product design and prototype construction; new materials; testing and certification; national and international standardization; failure analysis; safety; explosion hazards; electromagnetic compatibility using electricity under water; electric vehicles; servo components and small motors; contacts; fuses; insulation; switching transients and harmonics; cables; electrostatics; radio frequency technology; microelectronics; microprocessors; creep of steel; plastics and industrial market research.
Library facilities: Not open to all users.
Library holdings: 6000 bound volumes; 350 current periodicals; 2000 reports.
Information services: Information services available to members only.
Consultancy: Writing of technical reports (c); compilation of statistical trade data (c); market surveys in technical areas (c).
On-the-phone answers to technical queries given to members only.
Courses: On-line workshop.

IRS-Dialtech, Department of Trade and Industry   1.75

Address: Room 392, Ashdown House, 123 Victoria Street, London SW1E 6RB, UK
Telephone: (01) 212 5638
Telex: 8813148
Facsimile: (01) 828 3258
Affiliation: European Space Agency
Enquiries to: Head, R. Kitley
Subject coverage: IRS-Dialtech is an on-line information retrieval service providing access to over 80 scientific and technical bibliographic data bases, including CHEMABS, INSPEC, BIOSIS, COMPENDEX, NASA, and CAB.
Library facilities: No library facilities.
Information services: No bibliographic services; no translation services; literature searches (c); access to on-line information retrieval systems (c).
Consultancy: No consultancy.
Courses: On-line information retrieval courses - for beginners and advanced.
Publications: Leaflets; free user manual; newsletters.

Loughborough Consultants Limited   1.76

Address: University of Technology, Loughborough, Leicestershire LE11 3TF, UK
Telephone: (0509) 230426
Telex: 34319
Parent: Loughborough University of Technology

Enquiries to: General manager, Dr M.A. Brown
Founded: 1969
Subject coverage: Science and technology; human and social sciences; consultancy, research and development.
Library facilities: Not open to all users; loan services; reprographic services.
Loughborough Consultants Limited does not have a library separate from that of the University of Technology.
Information services: No information services.
Consultancy: Telephone answers to technical queries (c); writing of technical reports (c); compilation of statistical trade data (c); market surveys in technical areas (c).

National Centre for Information Media and Technology   1.77
– CIMTECH

Address: PO Box 109, Hatfield Polytechnic, College Lane, Hatfield, Hertfordshire AL10 9AB, UK
Telephone: (07072) 79691
Telex: 262413
Affiliation: British Library; Hatfield Polytechnic
Enquiries to: Information officer, Anne Grimshaw
Founded: 1967
Subject coverage: Micrographics, word processing, facsimile, videotex, optical discs, laser printing, photocopiers, etc.
Library facilities: Not open to all users (nc); loan services (members only); reprographic services (free to members).
Library holdings: About 1200 bound volumes; 80 current periodicals; about 200 reports.
Information services: No bibliographic services; no translation services; literature searches (nc); access to on-line information retrieval systems (only for Cimtech staff).
Consultancy: Telephone answers to technical queries (nc); writing of technical reports (c); no compilation of statistical trade data; no market surveys in technical areas.
Courses: Courses are held in spring and autumn on the following topics: micrographics; word and information processing; computer assisted retrieval of microforms; optical digital data discs; laser printing; in-house publishing; document delivery.
Publications: *Reprographics Quarterly*; *Information Media and Technology* (journal); evaluation reports. List of publications available on request.

Science Reference and Information Service

1.78

– SRIS

Address: 25 Southampton Buildings, London WC2A 1AW, UK

Telephone: (01) 405 8721; (01) 636 1544 (life sciences); (01) 379 6488 (European Biotechnology Information Project)

Telex: 266959

Facsimile: (01) 242 8046

Affiliation: British Library

Enquiries to: Head, external relations and liaison section; Head, reader services

Founded: 1855 as Patent Office Library

Subject coverage: Patents; science and technology; commerce and industry.

The Life Sciences Department is at 9 Kean Street, London WC2B 4AT. Other departments of the service include the following: British patents; foreign patents; computer search service; Japanese information service; European Biotechnology Information Project.

Library facilities: Open to all users; reprographic services (c).

Extensive collection of trade, business, and house journals.

Information services: Translation services (linguistic aid service); literature searches.

Prestel, on-line services (c). The SRIS computer search service exploits major on-line information retrieval services; searches carried out by SRIS staff or in person. List of files available and their prices available on request.

Consultancy: Telephone answers to technical queries (nc); no writing of technical reports; no compilation of statistical trade data; no market surveys in technical areas.

Publications: *SRIS News*, 4 per year; *Patents Information Network Bulletin*, 4 per year; *Catalogue on Microfiche*; subject lists of periodicals; list of publications available on request.

Sira Limited

1.79

Address: South Hill, Chislehurst, Kent, UK

Telephone: (01) 467 2636

Telex: 896649

Facsimile: (01) 467 6515

Enquiries to: Head, training and membership group, Patricia West

Founded: 1918

Subject coverage: Instrumentation; measurement; control; optics; expert systems.

Library facilities: No library facilities.

Information services: No bibliographic services; no translation services; no literature searches; access to on-line information retrieval systems (for members of Sira - c).

Consultancy: Telephone answers to technical queries (c - but Technivise provide a limited free service);

writing of technical reports (c); no compilation of statistical trade data; no market surveys in technical areas.

Courses: List of courses available on request. Topics include vibration measurement using laser technology, fibre optics for instrumentation, and testing thermal imagers.

Publications: Publications list available on request.

University of Sheffield Library

1.80

Address: Western Bank, Sheffield S10 2TN, UK

Telephone: (0742) 78555

Enquiries to: Science librarian, Dr Susan Frank; applied science librarian, Helen Workman

Subject coverage: Subjects include pure and life sciences, and applied sciences. The Applied Science Library is at the following address: Mappin Street, Sheffield S1 3JD.

Library facilities: Not open to all users (open for reference purposes to non-university users on application); loan services (via inter-library-loan lending services).

Library holdings: About 800 000 bound volumes; about 5000 current periodicals.

Information services: No bibliographic services; translation services (the library holds a register of university translators, who may be prepared to undertake external work); no literature searches (advice will be given where possible); access to on-line information retrieval systems (usually via university staff undertaking external work - c).

Consultancy: No telephone answers to technical queries (redirection of queries to the appropriate university department may be possible); no writing of technical reports; no compilation of statistical trade data; no market surveys in technical areas.

Courses: Courses on information retrieval and sources are sometimes given in collaboration with university departments for external users.

University of Sussex, Services for Industry *

1.81

Address: Sussex House, Falmer, Brighton BN1 9RF, UK

Telephone: (0273) 606755

Enquiries to: John Golds

Subject coverage: Consultancy services are offered by the university's science schools (engineering and applied science, mathematical and physical sciences, chemistry and molecular sciences, biological sciences), in the special research areas (microprocessors and industrial control, thermo-fluid mechanics, biotechnology, energy research, environmental studies), and in science policy.

Information services: Bibliographic services; translation services; literature searches.

The university offers staff training and test and equipment facilities in the areas mentioned above; translations from French, German, Japanese, Polish and Russian; media services, including preparation of presentations using low-band U-matic television, 16 mm film, tapeslide and audio tape, and television and film studio facilities.

Advice is also given on creation of data banks and technical editing.

Consultancy: Yes.
Publications: *Services for Industry Newsletter.*

YUGOSLAVIA

Centar za Dokumentaciju i Informacije *   1.82
– CDI
[Documentation and Information Centre]
Address: Trg Republike 1/1, 41000 Zagreb, Yugoslavia
Telephone: (041) 449 769
Telex: 21-156
Enquiries to: Manager, Draggoljub Krstić
Subject coverage: Collection and dissemination of information including information about publishers' programmes, conferences, seminars and consultancy institutions.
Library facilities: Not open to all users; no loan services; reprographic services.
Information services: Bibliographic services; translation services; literature searches; access to on-line information retrieval systems.
Consultancy: Yes.
Publications: *Org Registar.*

Institut Rudjer Bošković   1.83
– IRB
[Rudjer Bošković Institute]
Address: PO Box 1016, 41001 Zagreb, Yugoslavia
Telephone: (041) 435 111
Telex: 21383
Affiliation: University of Zagreb; Vladimir Bakarić University in Rijeka
Enquiries to: Director general, Dr Sergije Kveder
Founded: 1950
Subject coverage: Physics; chemistry; biology; biomedical sciences; oceanography; environmental sciences.
Library facilities: Open to all users (nc); loan services (nc); reprographic services (c).
Library holdings: 50 000 bound volumes; 350 current periodicals.
Information services: No information services.
Consultancy: No consultancy.

Courses: Training courses in radiological protection and work with ionizing radiation, about 2-3 times per annum; International Atomic Energy Agency advanced interregional training course on data processing and interpretation in X-ray fluorescence analysis, about one per annum.

Jugoslovenski Centar za Techničku i Naučnu Dokumentaciju   1.84
– JCTND
[Yugoslav Centre for Technical and Scientific Documentation]
Address: Slobodana Penezića-Krcuna br 29-31, PF724, 11000 Beograd, Yugoslavia
Telephone: (011) 644 546
Telex: 12497-YU-JCTND
Enquiries to: Director, Dr Ljiljana Kojić-Bogdanović
Founded: 1952
Subject coverage: Scientific, technological and economic documentation and information.
Library facilities: Open to all users; loan services; reprographic services (c).
Library holdings: 900 000 copies of periodicals (for 30 years in continuity); 60 000 monographs; encyclopaedias; handbooks; technical dictionaries, etc. 1500-2000 foreign journal titles are collected per annum.
Information services: Bibliographic services (c); translation services (c); literature searches (c).
JCTND provides information retrieval services on special topics upon user's request or SDI on a basis of long-term contracts with enterprises using the following kinds of data bases in its own collection: JCTND data file; *Chemical Abstracts Journal;* *Engineering Index;* *Food Abstracts Journal;* *Official Gazette of the US Patent and Trade Mark Office;* *Referativnyi Zhurnal.* Also available is a card file of Yugoslav institutions for patents and standards. In addition, JCTND uses external specialized and computerized data banks such as AGRIS, INIS, etc.
Courses: Regular seminars, consultations and training courses for staff and users on the following topics: information - processing, storage and dissemination, benefit from its use; status and trends in information activities throughout the world. Regular courses on micrographics, COM, microforms in information handling, processing and duplicating methods.
Publications: *Bilten Dokumentacije.* (Bulletin of Documentation); journals and monographs concerning information science and practice.

2 PATENT OFFICES

ALBANIA

Bureau for the Registration of Patents and Trademarks *   2.1

Address: la Chambre de Commerce, Rue Kongresi Permetit, Tiranë, Albania

AUSTRIA

Internationales Patentdokumentations-Zentrum GmbH *   2.2

– INPADOC
[International Patent Documentation Centre]
Address: Möllwaldplatz 4, A-1040 Wien, Austria
Telephone: 0222/658784
Telex: 136337
Enquiries to: Dipl Kfm Norbert Fux
Founded: 1972
Library facilities: Not open to all users; loan services (c); reprographic services (c).
Non-subscribers of the INPADOC services are provided with searches according to technical fields, applicant names, inventor names, patent families (ie patent documents claiming common priority), legal status data; copies of patent specifications; translations of Japanese patent publications; special analysis using the IDB INPADOC data base.
Information services: Consultancy (c); bibliographic (c); translation (c); no literature searches; access to on-line information retrieval systems (c).
INPADOC PFS/PRS on-line for patent families and legal status data; INKA Karlsruhe for bibliographic searches and SDI; Pergamon InfoLine for bibliographic services, patent families and SDI.

Österreichisches Patentamt *   2.3

[Austrian Patent Office]
Address: Kohlmarkt 8-10, A-1014 Wien, Austria
Telephone: 0222/633636
Telex: 136847 OEPA
Parent: Ministry of Trade and Commerce
Enquiries to: Director of Library, Dr Ingrid Weidinger
Founded: 1899
Subject coverage: Twenty-seven million patent documents in twenty-five countries and three international organizations; technical and legal literature.
Library facilities: Open to all users; no loan services; reprographic services.
Information services: Consultancy; bibliographic; no translation; literature searches; no access to on-line information retrieval systems.
Search reports; SDI services.
Publications: Austrian Patent Documents; Austrian Patent Journal; Austrian Trade Marks Journal; anniversaries; annual reports.

BELGIUM

Service de la Propriété Industrielle et Commerciale *   2.4

– SPIC
[Department of Industrial and Commercial Property]
Address: Rue J.A. De Mot 24-26, 1040 Bruxelles, Belgium
Telephone: (02) 233 61 11
Telex: 23658 Verli
Parent: Ministère des Affaires Économiques
Enquiries to: D. Van Der Gheynst; P. Ceuninck; J. Michiels
Subject coverage: Service for registration of patents, models and trade marks.

Library facilities: Open to all users; no loan services; reprographic services (c).
Information services: Consultancy (nc); bibliographic (nc); no translation.
Only technical literature searches (c).
Publications: Monthly list of Belgian patents.

CYPRUS

Ministère du Commerce et de l'Industrie, Service du Receveur Officiel et Directeur de l'Enregistrement *   2.5
[Ministry of Commerce and Industry, Department of the Official Receiver and Registrar]
Address: PO Box 1720, 9 Byron Avenue, Nicosia, Cyprus
Telephone: 40 2317
Enquiries to: Official receiver and registrar, Takis Christodoulides

CZECHOSLOVAKIA

Urad pro Patenty a Vynálezy *   2.6
[Czechoslovak Patent Office]
Address: Václavské nám 19, 1, Nové Mesto, 11346 Praha, Czechoslovakia
Telephone: 226 845
Telex: 12 1948
Enquiries to: President, M. Bélohlávek

DENMARK

Direktoratet for Patent- og Varemaerkevaesenet   2.7
[Danish Patent Office]
Address: Nyropsgede 45, DK-1602 København V, Denmark
Telephone: (01) 128440
Telex: 16046 DPO DK
Enquiries to: Librarian, Peter Havnø Patentdirektoratets Bibliotek (the library); and Serviceafdelingen (the service section)
Founded: 1894

Subject coverage: Issuing of patents; registration of trademarks and designs; providing information on patents to the public.
Library facilities: Open to all users; no loan services, reprographic services (c).
Library holdings: About 22 500 bound volumes; 125 current periodicals; 22.2m patent documents.
Information services: Bibliographic services (c); no translation services; literature searches (c); access to on-line information retrieval systems (c).
Consultancy: No consultancy.
Publications: *Dansk Patenttidende* (The Patent Gazette); *Dansk fremlaeggelsesskrifter* (printed examined applications on paper and microfiche); *Trillaeg til Dansk Pattenttidende* (abstracts of Danish patent applications); *Registreringstidende for Vare- og Faellesmaerker* (trademark journal); *Registreringstidende for Mønstre* (design journal); *Register over danske patenter udstedt* (annual index of Danish patents).

FINLAND

Patentti- ja Rekisterihallitus   2.8
– PRH
[National Board of Patents and Registration]
Address: PO Box 18154, Albertinkatu 25, 00180 Helsinki 18, Finland
Telephone: (90) 69531
Telex: 121394
Parent: Ministry of Commerce and Industry
Enquiries to: Documentation officer, Lolan Kölhi
Founded: 1942
Subject coverage: Patents, trade marks, industrial designs and trade names.
Library facilities: Open to all users; no loan services; reprographic services (c).
The library contains about 19 million patent publications from all over the world.
Information services: Consultancy (c); bibliographic (c); no translation; no literature searches; no access to on-line information retrieval systems.
The library provides INPADOC services such as PFS, PAS, PIS, PCS and IPG.

FRANCE

Societé Française pour la Gestion des Brevêts d'Application Nucléaire *   2.9

[French Nuclear Patents Society]
Address: 25 rue de Ponthieu, 75008 Paris, France
Telex: Spibrev 660028 F
Parent: Commissariat à l'Énergie Atomique
Enquiries to: A. Mongredien
Patent Department
Subject coverage: Patent agent.

GERMAN DEMOCRATIC REPUBLIC

Amt für Erfindungs-und Patentwesen *   2.10

[Inventions and Patents Office]
Address: Mohrenstrasse 37b, 1080 Berlin, German Democratic Republic
Enquiries to: President, Professor Hemmerling

GERMAN FEDERAL REPUBLIC

Deutsches Patentamt   2.11
– DPA
[German Patent Office]
Address: Zweibrückenstrasse 12, D-8000 München 22, German Federal Republic
Telephone: (089) 2195
Telex: 523534
Facsimile: (089) 21952221
Parent: Ministry of Justice
Enquiries to: Regierungsdirektor, Johann Lang
Bibliothek des Deutschen Patentamts
Founded: 1877
Subject coverage: Patent literature of all countries and non-patent literature in the fields of industrial property, technology and other natural sciences.
Library facilities: Open to all users (nc); loan services (nc); reprographic services (c).
Library holdings: 899 697 bound volumes; 1386 current periodicals; 27 million patent documents.
Information services: No bibliographic services; no translation services; no literature searches; access to on-line information retrieval systems (c - but in-house

patent information systems free of charge).
Consultancy: No consultancy services.
Publications: Deutsches Patentamt Bibliothek Benutzer-Information.

European Patent Office/ Europäisches Patentamt/Office Européen des Brevêts   2.12
– EPO
Address: Erhardstrasse 27, D-8000 München 2, German Federal Republic
Telephone: (089) 2399-0
Telex: 5 23 656
Enquiries to: Head of library, Dr G. Kruse
Founded: 1977
Subject coverage: The task of the European Patent Office is to grant patents for member states of the European Patent Convention. European patents can be granted with effect in up to eleven European states, namely: Austria, Belgium, France, the Federal Republic of Germany, Italy, Liechtenstein, Luxembourg, the Netherlands, Sweden, Switzerland, and the United Kingdom.
Library facilities: No library facilities.
Library holdings: 30 000 bound volumes; 600 current periodicals; 170 000 European patent applications.
Information services: No bibliographic services; no translation services; no literature searches; access to on-line information retrieval systems (c - on-line access to register entries on European patent applications and European patents).
Consultancy: No consultancy.
Publications: Publications list available on request; *Official Journal EPO.*

GREECE

Commercial and Industrial Property Directorate *   2.13
Address: Canning Square, Athinai, Greece
Telephone: 36 23 593
Telex: 215 282
Affiliation: Ministry of Commerce
Enquiries to: Director, Emile Tiranas
Founded: 1920
Subject coverage: Patents.
Library facilities: Open to all users; no loan services; reprographic services - photocopies of patents, descriptions and designs.
Information services: Informal consultancy services.

Ministère de la Recherche et de la Technologie, Section des Brevêts *   2.14
[Ministry of Research and Technology Patents Section]
Address: Ermou 2, Athinai, Greece
Telephone: 3239459
Enquiries to: Head, Vasilios Triantafillis

HUNGARY

Országos Találmányi Hivatal, Szabadalmi Informácios Központ *   2.15
– OTH
[National Office for Inventions, Patent Information Centre]
Address: POB 552, Garibaldi útca 2, H-1370 Budapest V, Hungary
Telephone: (01) 124-400
Telex: 224700
Enquiries to: Director, Dr J. Zsiga
Founded: 1895
Subject coverage: Patents and similar documents.
Library facilities: Open to all users; no loan services; reprographic services (c).
Information services: Consultancy (nc); bibliographic (c); no translation; no literature searches; no access to on-line information retrieval systems.
Bibliographic services for patent documents from around 50 countries. Services only available to Hungarian users.

IRELAND

Oifig na bPaitinní   2.16
[Patents Office]
Address: 45 Merrion Square, Dublin 2, Ireland
Telephone: (01) 614144
Telex: 24651
Facsimile: (01) 765776
Parent: Department of Industry and Commerce
Enquiries to: Controller of patents, designs and trade marks
Subject coverage: Patents, designs and trade marks.
Library facilities: Open to all users (c); no loan services; reprographic services (c).
The library holds reference books referring specifically to patents, trade marks, and designs, Irish and British patent specifications, and USA abridgements. Also European applications and granted European patents in paper form, and on microfiche PCT applications.
Information services: No bibliographic services; no translation; no literature searches; no access to on-line information retrieval systems.
Consultancy: No consultancy.

ITALY

Ufficio Centrale Brevetti *   2.17
[Central Patents Office]
Address: Via Molise 19, 00187 Roma, Italy
Telephone: 4751188
Telex: 610154 minindustria
Enquiries to: Director, Professor S. Samperi

NETHERLANDS

Bureau voor de Industriële Eigendom Octrooiraad *   2.18
[Netherlands Patent Office]
Address: Postbus 5820, 2280 HV Rijswijk, Zuid Holland, Netherlands
Telephone: (070) 90 76 16
Telex: 31651
Affiliation: Ministry of Economic Affairs
Enquiries to: Librarian, Library and Documentation Department
Founded: 1911
Subject coverage: Patents and patent literature of 28 countries and organizations.
Library facilities: Open to all users; loan services; reprographic services (c).
Information services: Consultancy; access to on-line information retrieval systems - legal status data base for Dutch patent applications.

Octrooiinformatiedienst *   2.19
[Patent Information Office]
Address: Postbus 309, 2280 AH Rijswijk, Netherlands
Telephone: (070) 90 78 58
Telex: 31622 nider nl
Affiliation: Organisatie voor Toegepast Natuurwetenschappelij Onderzoek
Founded: 1934
Subject coverage: All fields of technology.
Library facilities: Not open to all users; no loan services; reprographic services (c - copies of patent specifications).

Information services: Translation (c); literature searches; access to on-line information retrieval systems.

The office conducts patent searches, manually and on-line via the Central Information Department of TNO, in the following areas: collection searches; novelty searches; validity searches; infringement searches; counterpart searches in foreign countries; name searches; patent application status searches.

NORWAY

Styret for det Industrielle Rettsvern * 2.20
[Norwegian Patent Office]
Address: Post Boks 8160 Dep, N-0033 Oslo 1, Norway
Telephone: (02) 461900
Telex: 19152
Enquiries to: Chief librarian, Turi Stokke
Founded: 1910
Subject coverage: Industrial property; technical literature.
Library facilities: Open to all users; loan services (no c); reprographic services (c).
Information services: Consultancy (c); bibliographic (c); no translation; literature searches (nc); access to on-line information retrieval systems.

POLAND

Urząd Patentowy Polskiej Rzeczpospolitej Ludowej 2.21
[Patent Office of the Polish People's Republic]
Address: Al Niepodległości 188, Skryt poczt 203, 00-950 Warszawa, Poland
Telephone: (022) 25 14 43
Enquiries to: Director, information division, Maria Jurczakowska
Collections Division
Founded: 1919
Subject coverage: Inventions and industrial property.
Library facilities: Open to all users; no loan services; reprographic services (c).
The library collects, produces and provides all users with Polish patent specifications; also available are foreign patent specifications and patent literature from many countries.
Information services: Bibliographic (c); no translation; no literature searches.

The library provides copies on request, subscribes to patent specifications, according to patent classification, and maintains INPADOC magnetic tape EDT and IFD (for patent office use only).
Consultancy: No consultancy.
Publications: *Opisy patentowe* (Patent Specifications); *Biuletyn Urzędu Patentowego* (Bulletin of the Polish Patent Office), biweekly; *Wiadomości Urzędu Patentowego* (News of the Polish Patent Office), monthly; *Wykaz patentów na wynalazki*, annual index of patents granted in Poland; *Wykaz wzorów użytkowych*, annual index of utility models; *Cztery edycje MKP w języku polskim* (Four editions of the IPC in Polish).

PORTUGAL

Instituto Nacional da Propriedade Industrial * 2.22
– INPI
[National Industrial Property Institute]
Address: Campo das Cebolas, 1100 Lisboa, Portugal
Telephone: 87 61 51
Telex: 18 356 INPI P
Parent: Ministério da Indústria e Energia.
Enquiries to: Chief technical officer, Madalena Rios Abreu
Founded: 1896
Subject coverage: Register of patents and trademarks.
Library facilities: Open to all users; no loan services; reprographic services (c).
Information services: Consultancy (c); bibliographic (c); no translation; literature searches (c); access to on-line information retrieval systems (c).

SWEDEN

Patent- och Registreringsverket 2.23
[Swedish Patent Office]
Address: Box 5055, S-102 42 Stockholm, Sweden
Telephone: (08) 22 55 40
Telex: 17978 PATOREG-S
Affiliation: Companies Department, Sundsvall
Enquiries to: Head, documentation division, Kerstin Bergström; head of division, Jan Averdal
Library/information centre: InterPat Sweden
Founded: 1885
Subject coverage: The office is the central industrial property office of Sweden.

Library facilities: Open to all users; no loan services; no reprographic services.

Information services: Bibliographic services (c); no translation services; literature searches (c); access to on-line information retrieval systems (c).

The office provides state-of-the-art searches, special information services, and patent family searches, for which charges are made; access is available to ESA - IRS (Italy), Dialog and SDC (USA), Derwent-SDC and Pergamon-InfoLine (UK).

Consultancy: A consultancy service is available.

Publications: *Svensk patenttidning* (Swedish Patent Gazette); *Svensk varumärkestidning* (Swedish Trademark Gazette); *Registreringstidning för mönster* (Design Gazette); Swedish Patent Documents; Abstracts of Swedish Patent Applications; Annual Index of Granted Patents; Swedish Patent Office Annual Report.

SWITZERLAND

Office Fédéral de la Propriété Intellectuelle   2.24
– OFPI

[Swiss Federal Intellectual Property Office]

Address: Einsteinstrasse 2, CH-3003 Bern, Switzerland

Telephone: (031) 61 48 06

Telex: 912805 bagech

Facsimile: (031) 61 48 95

Enquiries to: Head of TIPAT-Service, Valerio Candolfi

Technical Information Patents (TIPAT)

Founded: 1888

Subject coverage: Patents, industrial designs, trademarks, copyright.

Library facilities: Open to all users (nc); no loan services; reprographic services (c).

In the library, patents organized according to the International Patent Classification system can be consulted as follows: Swiss patents from 1969; patents from German FR, France, UK, Netherlands, Austria and USA from 1975; European and PCT applications from 1978; patents from Canada from 1983.

Information services: Bibliographic services (c); no translation services; literature searches (c); no access to on-line information retrieval systems.

The in-house computerized patents data base is for internal use only; bibliographic services to non-staff users are based on this and operated by OFPI staff, in addition bibliographic patents on-line searches and subject-oriented technical and patent on-line searches can be performed in external data bases and in

connection with these searches full text documents can be delivered.

Consultancy: Telephone answers to technical queries (nc - consultancy is restricted to assistance in the drawing up of patents specifications); writing of technical reports (reports and statistical data related to OFPI activities, are published in annual reports and in the Swiss Patents, Designs and Trademarks Bulletin); compilation of statistical trade data; no market surveys in technical areas.

Verband Schweizerischer Patentanwalte *   2.25

[Association of Swiss Patent Attorneys]

Address: c/o Patentabteilung Ciba-Geigy AG, CH-4002 Basel, Switzerland

Telephone: (061) 36 50 65

UNITED KINGDOM

Patent Office   2.26

Address: State House, 66-71 High Holborn, WC1R 4TP, UK

Telephone: (01) 831 2525

Parent: Department of Trade

Enquiries to: Classification Section T3

Founded: 1852

Subject coverage: Patent applications cover all areas of technology; trademark registration; design registration.

Library facilities: Open to all users.

Information services: Search facilities for registered and pending trade marks, soon to be provided on-line; off-line computer search for patents classified on UK classification key.

Consultancy: Telephone answers to technical queries (nc); no writing of technical reports; no compilation of statistical trade data; no market surveys in technical areas.

Search and advice service on registering specific marks.

Publications: List of publications with their price available on request. There are also several free publications.

Science Reference and Information Service   2.27
– SRIS

Address: 25 Southampton Buildings, London WC2A 1AW, UK

Telephone: (01) 405 8721; (01) 636 1544 (life sciences); (01) 379 6488 (European Biotechnology Information Project)

Telex: 266959

Facsimile: (01) 242 8046
Affiliation: British Library
Enquiries to: Head, external relations and liaison section; Head, reader services
Founded: 1855 as Patent Office Library
Subject coverage: Patents; science and technology; commerce and industry.
The Life Sciences Department is at 9 Kean Street, London WC2B 4AT. Other departments of the service include the following: British patents; foreign patents; computer search service; Japanese information service; European Biotechnology Information Project.
Library facilities: Open to all users; reprographic services (c).
Extensive collection of trade, business, and house journals.
Information services: Translation services (linguistic aid service); literature searches.
Prestel, on-line services (c). The SRIS computer search service exploits major on-line information retrieval services; searches carried out by SRIS staff or in person. List of files available and their prices available on request.
Consultancy: Telephone answers to technical queries (nc); no writing of technical reports; no compilation of statistical trade data; no market surveys in technical areas.
Publications: *SRIS News*, 4 per year; *Patents Information Network Bulletin*, 4 per year; *Catalogue on Microfiche*; subject lists of periodicals; list of publications available on request.

YUGOSLAVIA

Savezni Zavod za Patente   2.28
[Federal Patent Office]
Address: Uzum Mirkova 1, 11000 Beograd, Yugoslavia
Telephone: (011) 637 787
Telex: 12761
Affiliation: Federal Executive Council
Enquiries to: Director
Library/information centre: Centar za Patentnu Dokumentaciju i Rěseřse
Founded: 1920
Subject coverage: Protection of inventions, technical improvements, trade marks, trade names, designs and models.
Library facilities: Open to all users; no loan services; reprographic services.
Information services: Consultancy; bibliographic.
Publications: *Patentni Glasnik*, bimonthly; patent specifications.

3 STANDARDS OFFICES

ALBANIA

Albanian Standards Office * 3.1
Address: c/o Prane Komisionit te Planit te Shetit, Tiranné, Albania
Telephone: 75 00
Affiliation: International Organization for Standardization

AUSTRIA

Österreichisches 3.2
Normungsinstitut *
[Austrian Standards Institute]
Address: Postfach 130, Leopoldgasse 4, A-1021 Wien 2, Austria
Telephone: (0222) 33 55 19-0
Telex: 7-5960
Affiliation: International Organization for Standardization
Enquiries to: Chief librarian, M. Tastel
Subject coverage: Responsible for National Standards. Conducts research, organizes conferences, and has a library with over 500 volumes.
Publications: Standards catalogue.

BELGIUM

Institut Belge de Normalisation 3.3
– IBN
[Belgian Standardization Institute]
Address: 29 Avenue de la Brabanconne, B-1040 Bruxelles, Belgium

Telephone: (02) 734 9205
Telex: 23877 BENOR
Affiliation: International Organization for Standardization
Subject coverage: Drafting and publication of Belgian standards.
Library facilities: Open to all users.
Library holdings: The library has a collection of national and international standardization documents.
Publications: *Revue IBN*, monthly; annual report; *Catalogue des Normes Belges*, annual.

BULGARIA

State Committee for 3.4
Standardization at the Council of
Ministers *
Address: 6 Septemvri 21, Sofia, Bulgaria
Telephone: 85 91
Telex: 22570 dks bg
Affiliation: International Organization for Standardization

CYPRUS

Cyprus Organization for 3.5
Standards and Control of
Quality
Address: c/o Ministry of Commerce and Industry, Nicosia, Cyprus
Telephone: 40 34 41
Telex: 2283 minco mind
Affiliation: International Organization for Standardization

CZECHOSLOVAKIA

Úřad pro normalizaci a Měření 3.6
– ÚNM
[Standards and Measurements Office]
Address: Václavské nám 19, 113 47 Praha 1, Czechoslovakia
Telephone: 26 22 51
Telex: 121948 unm
Parent: State Commission for Scientific and Technological Development
Enquiries to: Head, Dr Ivo Prokop
Information Department, Na příkopě 17, 113 47 Praha 1, Czechoslovakia
Founded: 1951
Subject coverage: National, foreign and international standards and standard type documents.
Library facilities: Open to all users (nc); loan services (nc); reprographic services (c).
Information services: No information services.
Consultancy: No consultancy.

DENMARK

Dansk Standardiseringsråd 3.7
– DS
[Danish Standards Association]
Address: 12 Aurehoejvej, DK-2900 Hellerup, Denmark
Telephone: (01) 62 93 15
Telex: 15616 dansta dk
Affiliation: International Organization for Standardization
Enquiries to: Commercial and Finance Manager
Founded: 1926
Subject coverage: Standardization in all fields except electrical technology.
Library facilities: Open to all users; no loan services; no reprographic services.
Information services: Bibliographic services (c).
Consultancy: Yes (c).
Publications: Biennial catalogue of standards.

FINLAND

Suomen Standardisoimisliitto 3.8
Informaatiopalvelu
– SFS Information Service
[Finnish Standards Association Information Service]
Address: PO Box 205, Bulevardi 5 A7, SF-00121 Helsinki 12, Finland
Telephone: (90) 645 601
Telex: 122303 stand sf
Facsimile: (90) 64 31 47
Enquiries to: Manager, Marjatta Aarniala
Founded: 1978
Subject coverage: Standards and technical regulations from all over the world and in any technical field.
Library facilities: Open to all users (c); no loan services; reprographic services (c).
Library holdings: 220 000 standards.
Information services: Bibliographic services (c); no translation services; literature searches (c); access to on-line information retrieval systems (c).
SFS data base was founded by the Finnish Standards Association, which is a member of ISO information network ISONET and acts as the Finnish GATT enquiry point as defined by the GATT Agreement on Technical Barriers to Trade. The SFS data base is operated on the information storage and retrieval system, MINTTU, using the computers at the Finnish State Computer Centre. SFS data base is accessible through the Finnish PTT's packet switched network Datapak which is interconnected with the Nordic countries, Europe, USA, Canada, Japan, Hong Kong, Singapore, Australia etc, through direct dial to the Finnish host computer in Espoo or through the Finnish Videotex network.
Publications: *SFS Standards*, national SFS standards numbering to 3000, 350 translations available in Swedish and 400 in English; *SFS Tiedotus*, bimonthly standardization news magazine; *SFS Catalogue*, annual bilingual (Finnish and English) list and eight cumulative supplements of valid SFS standards and other SFS publications, information on translations available and SFS standards to which governmental bodies refer; *SFS Handbooks*, information packages on standards for specific fields of technology together with other relevant information.

FRANCE

Association Française de 3.9
Normalisation *
– AFNOR
[French Standardization Association]
Address: Tour Europe, 92080 Paris La Défense, Cedex 7, France
Telephone: 47 88 13 26
Telex: 611 974 F
Subject coverage: The association coordinates all work relating to standardization in France, and represents French standardization organizations at international meetings.
Publications: *Catalogue des Normes Françaises*, annual, with monthly supplement; *Courier de la Normalisation*, bimonthly; *Bulletin Mensuel de la Normalisation Française*, monthly.

Bureau National de Métrologie * 3.10
– BNM
[National Bureau of Metrology]
Address: 21 rue Casimir Perier, 75007 Paris, France
Telephone: 47 05 11 95
Telex: Dica Cri 680233 F
Parent: Ministère de l'Industrie
Enquiries to: Secrétaire général
Founded: 1969
Subject coverage: Metrology; standardization.
Library facilities: Loan services and reprographic services only in special cases.
Information services: Bibliographic services; literature searches.
Consultancy: Yes.

Centre Technique de l'Industrie 3.11
Horlogère *
– CETEHOR
[French Technical Centre for Watch- and Clock-making]
Address: BP 1145, 39 avenue de l'Observatoire, 25003 Besançon Cedex, France
Telephone: 81 50 38 88
Telex: 360293 FO88
Founded: 1949
Subject coverage: Horology; microtechnology; biomedicine; establishing standards; technical assistance and inspection tests; horological and mechanical/electronic precision products. The centre acts as a regional standards office.
Library facilities: Open to all users; no loan services; reprographic services (c).
Information services: Bibliographic services (c); no translation services; literature searches (c); access to on-line information retrieval systems (c).

Access is available to ESA, Télésystème, Sligos, and INKA.
Consultancy: Yes (c).

International Bureau of Weights 3.12
and Measures *
Address: Pavillon de Breteuil, 92310 Sèvres, France
Enquiries to: Director, Pierre Giacomo
Founded: 1875
Subject coverage: The bureau is responsible for preservation of the International System of Units (SI); determines national standards; and advises on international standards of measurement.

GERMAN DEMOCRATIC REPUBLIC

Amt für Standardisierung, 3.13
Messwesen und Warenprüfung *
[Office for Standardization, Measurement and Testing]
Address: Fürstenwalder Damm 388, 1162 Berlin, German Democratic Republic
Enquiries to: President, Professor Lilie

GERMAN FEDERAL REPUBLIC

Deutsches Informationszentrum 3.14
für Technische Regeln
– DITR
[German Information Centre for Technical Rules]
Address: Postfach 1107, Burggrafenstrasse 6, D-1000 Berlin 30, German Federal Republic
Telephone: (030) 2601 600
Telex: 185269 ditr d
Parent: Deutsches Institut für Normung eV (DIN)
Enquiries to: Head, sales department, Jürgen Kölling
Founded: 1979
Subject coverage: All German technical rules including the legislative provisions and administrative regulations pertaining to technical matters published by the Federal German Government, the Länder parliaments and the European Communities.
Library facilities: Open to all users (nc); loan services (for foreign, regional and international standards - c); no reprographic services.
The DIN library is an integral part of DITR serving as a central reference point for: German (DIN) Standards including English translations; other

voluntary technical rules - such as codes of practice and recommendations; official gazettes containing technical laws, regulations and ordinances; foreign national, regional and international standards; other material pertinent to standardization in general (bibliographies, journals, and yearbooks).

For foreign national, regional and international standards the library offers a loan service restricted to national users. Users in other countries should apply to their national standards organization.

Library holdings: 8200 bound volumes; 370 current periodicals; 330 000 standards.

Information services: Bibliographic services (c); no translation services; literature searches (c); access to on-line information retrieval systems (c).

The services offered by the DITR include: lists which are provided either as high-speed printer output or on magnetic tape, where subject and volume of the information supplied can either be defined by DITR (standard profiles) or chosen by the user (individual profiles); direct on-line access to the DITR bibliographic data base via data display terminal, giving access to 47 000 valid and 30 000 historical German technical rules.

Consultancy: Telephone answers to technical queries (c); no writing of technical reports; no compilation of statistical trade data; no market surveys in technical areas.

Information can be given by telex or letter, in addition to the telephone service mentioned above.

Publications: *Liste der Regelwerke*, list of technical rules and standards contained in the DITR data base; *Standardprofile*, list of subject areas in which the DITR publishes quarterly technical rules information, in print and on magnetic tape; leaflets; *DIN Catalogue of Technical Rules*, annual, with monthly supplements.

Kerntechnischer Ausschuss 3.15
– KTA
[Nuclear Safety Standards Commission]
Address: Postfach 10 16 50, Schwertnergasse 1, D-5000 Köln, German Federal Republic
Telephone: (0221) 2068-0
Telex: 8881807 GRSD
Enquiries to: Executive director, W. Schwarzer
Founded: 1972
Subject coverage: Development of nuclear safety standards.
Library facilities: No library facilities.
Information services: No information services.
Consultancy: No consultancy.
Publications: Bibliographic list.

Physikalisch-Technische Bundesanstalt Braunschweig und Berlin 3.16
– PTB
[Federal Institute of Metrology in Brunswick and Berlin]
Address: Postfach 3345, Bundesallee 100, D-3300 Braunschweig, German Federal Republic
Telephone: Braunschweig: (0531) 5920; Berlin: (030) 34811
Telex: Braunschweig 9-52822 ptb d
Parent: Bundesminister für Wirtschaft
Enquiries to: Deputy head, library, E. Bretnütz
Alternative address for PTB: Abbestrasse 2-12, D-1000 Berlin 10, German FR
Founded: 1887
Subject coverage: The PTB is the national institute for science and technology and the highest technical authority for metrology in the Federal Republic of Germany.

Research into: physics and technology, realization and dissemination of SI units, realization and propagation of legal time, realization of the international practical temperature scale, precision determination of physical constants, type testing and approval of measuring devices, slot machines and civilian firearms, type testing in the field of safety technology, radiation protection, medicine and traffic control, technical testing and storage of nuclear fuel, long-term storage and final disposal of radioactive waste. Cooperation in national and international technical committees, elaboration of technical regulations and directives, test work on commission and scientific and technological consultation, aid in metrological training and development overseas.

Library facilities: Open to all users (nc); loan services (nc); reprographic services (nc).

Special library for selected areas in science and technology, mainly for providing staff members with desired literature and information from external information centres. Reference library with restricted loans. The PTB is incorporated in the national library loan service.

Library holdings: 100 800 bound volumes; 860 current periodicals; 5000 reports.

Information services: No bibliographic services; no translation services; no literature searches; access to on-line information retrieval systems (for internal use only).

Consultancy: No consultancy.

Publications: *PTB-Mitteilungen Forschen und Prüfen*, bimonthly; *Jahresbericht*, annual; *PTB-Berichte*, irregular series of reports in the various research areas of the PTB; *PTB-Prüfregeln* etc, various directives and instructions on calibration and technical standards.

GREECE

Hellenic Organization for Standardization *   3.17

Address: Didodou 15, 106 80 Athinai, Greece
Telephone: 360 99 47
Affiliation: International Organization for Standardization
Founded: 1976
Subject coverage: Preparation and control of national standards.
Publications: Catalogue of Hellenic standards.

HUNGARY

Magyar Szabványügyi Hivatal *   3.18
– MSZH
[Hungarian Office for Standardization]
Address: Ülloi utca 25, H-1450 Budapest, Hungary
Telephone: (361) 183011
Telex: 225723
Enquiries to: Director, Gyula Róka
Documentation Centre and Technical Library
Founded: 1951
Subject coverage: Directing and controlling authority of standardization on national level.
Library facilities: Open to all users; loan services; reprographic services (c).
The main fields of interest of the Documentation Centre and Technical Library are as follows: Hungarian national and branch standards; standards of 30 countries; standards of ISO, IEC and CMEA; complete collection of national literature in connection with standardization and technical development; collecting selected foreign technical literature.
Information services: Bibliographic services (c); translation services (c); literature searches (c); no access to on-line information retrieval systems.
Consultancy: Yes.
Publications: List of standardization publications.

ICELAND

Idntaeknistofnun Islands   3.19
– ITI
[Technological Institute of Iceland]
Address: Keldnaholt, IS-112 Reykjavik, Iceland
Telephone: (91) 687000

Telex: 3020 Istech is
Enquiries to: Librarian, E. Arnvĭdardóttir
Subject coverage: Industrial research and development; technical assistance; standardization.
Library facilities: Open to all users (nc); loan services (nc); reprographic services (nc).
Library holdings: 4700 bound volumes; 230 current periodicals; 1550 reports.
Information services: Bibliographic services (c); no translation services; literature searches (c); access to on-line information retrieval systems (c).
Consultancy: Telephone answers to technical queries (nc); no writing of technical reports; no compilation of statistical trade data; market surveys in technical areas (c).
Courses: Various courses are run through the year for non-staff users.

IRELAND

Institute for Industrial Research and Standards *   3.20
– IIRS
Address: Ballymun Road, Dublin 9, Ireland
Telephone: (01) 370101
Telex: 25449
Enquiries to: Head, information services department, J. McClusky
Founded: 1947
Subject coverage: Science and technology (excluding agriculture, medicine and nuclear).
Library facilities: Open to all users; loan services (nc); reprographic services (c).
The library has a reference collection of national, ISO and other international standards.
Information services: Bibliographic services (c); no translation services; literature searches (c); access to on-line information retrieval systems.
The library has access to the following: Blaise, ESA/IRS, InfoLine, Dialog, SOC. It also maintains Biomass, an international information data base for the IEA.
Consultancy: Telephone answers to technical queries.
Publications: *Technology Ireland*, monthly; *Irish Construction Directory; Irish Engineering Directory; Survey of Scientific and Technical Information in Ireland;* full list on request.

ITALY

Ente Nazionale Italiano di Unificazione *

3.21

– UNI

[Italian National Standards Institute]
Address: Piazza Armando Diaz 2, 20123 Milano, Italy
Telephone: (02) 876914
Telex: 312481 uni
Enquiries to: Standards Information Service
Founded: 1921
Subject coverage: Italian standardization.
Library facilities: Open to all users; no loan services; no reprographic services.
Information services: Bibliographic services (c); access to on-line information retrieval systems.
Consultancy: Yes (c).
Publications: List on request; *Unificazione* (quarterly).

Istituto di Metrologia 'G.Colonnetti'

3.22

– IMGC

['G.Colonnetti'Metrology Institute]
Address: Strada delle Cacce 73, 10135 Torino, Italy
Telephone: (011) 34 878 34
Telex: 212209 IMGCTO I
Parent: Consiglio Nazionale delle Ricerche
Enquiries to: Director, Professor Anthos Bray
Founded: 1968, in its present form; evolved from the Istituto Dinamometrico Italiano, founded 1956.
Subject coverage: Realization, updating and dissemination of the standards of length, mass, temperature, angle, volume, density, force, pressure, hardness, flow-rate; determination of material properties; vibration measurements; absolute gravimetry; absolute determination of silicon lattice constant.
Library facilities: Library facilities are normally available only to members of IMGC or other CNR bodies, but direct individual arrangements can be made for external users.
Library holdings: 4800 bound volumes, including 600 volumes of bound journals; 130 current periodicals; about 760 internal technical reports by IMGC staff.
Information services: Access to information services is limited to the staff of CNR organizations.
Consultancy: Consultancy on measurement methods and procedures, instrumentation.
Publications: *IMGC Publications*, biannual, a booklet with summaries (in English) of books and papers by IMGC personnel.

UNIPREA

3.23

Address: Via Montevecchio 29, 10128 Torino, Italy
Telephone: (011) 531712
Telex: 216825 UNIPTO I
Affiliation: Ente Nazionale Italiano di Unificazione
Enquiries to: Secretary general, Giorgio Jannuzzi
Founded: 1947
Subject coverage: Standardization in the following fields: dimensional metrology, cinematography, photography, documentation, metrology and properties of surfaces, banking, computers, credit cards, data processing, identification cards, office machines, metallic and other non-organic coatings, vacuum technology, horology, plain bearings, sewing machines, documents in administration, commerce and industry, micrographics, optics and optical instruments, industrial automation.
Library facilities: No loan services; reprographic services.
UNIPREA does not have a library; only a collection of the working documents produced by International Standardization Organization TCs.
Consultancy: Yes.

NETHERLANDS

Nederlands Normalisatie-instituut *

3.24

– NNI

[Standardization Institute of the Netherlands]
Address: Postbus 5059, Kalfjeslaan 2, 2600 GB Delft, Netherlands
Telephone: (015) 906800
Telex: 32123
Enquiries to: Director, N.J.F. Zandstra
NNI infotheek
Founded: 1916
Library facilities: Open to all users; no loan services; reprographic services.
Library holdings: The library holds complete sets of European standards, international standards and national standards from the USA.
Information services: Bibliographic services; no translation services; literature searches; no access to on-line information retrieval systems.
Consultancy: Yes (c).

NORWAY

Norges Standardiseringsforbund   3.25
– NSF
[Norwegian Standards Association]
Address: Postboks 7020, Homansbyen, N-0306 Oslo 3, Norway
Telephone: (02) 466094
Telex: 19050 nsf n
Enquiries to: Sales manager, Pal Eddie
Founded: 1923
Subject coverage: The association has the following functions: editor and supplier of Norwegian Standards; exclusive representative for members of the International Organization for Standardization (ISO) in 89 countries around the world; enquiry point according to the GATT agreement.
Library facilities: Not open to all users; no loan services; reprographic services (c).
Information services: Bibliographic services (c); no translation services; no literature searches; access to on-line information retrieval systems.
The NSF maintains a data base covering Norwegian standards, Norwegian electrotechnical norms, and Norwegian technical regulations of importance to international trade, which is updated six times per year. The data base is open to all ISO member bodies, and can be reached by telephone via data terminal each working day. It is also connected to the Nordic data net, Scannet, which comprises over 20 data bases, for which a special application is required.
Consultancy: Yes (nc).

POLAND

Centralny Ośrodek Informacji   3.26
Normalizacyjnej i
Metrologicznej *
– COINiM
[Standardization and Metrology Information Centre]
Address: Pl Dzierz. yńskiego 1, 00-139 Warszawa, Poland
Telephone: (022) 20 96 06
Telex: 814855
Parent: Polish Committee for Standardization, Measures and Quality Control
Enquiries to: Director, Danuta Planer-Górska
Founded: 1972
Subject coverage: Standardization, metrology, and quality control.

Library facilities: Open to all users; loan services (nc); no reprographic services.
The library lends out literature and periodicals.
Information services: Bibliographic services (nc); no translation services; no literature searches; no access to on-line information retrieval systems.
COINiM gives information and advice, formulates documentary comparison, lends out international, foreign and Polish standards publications, and issues guides to scientific and technical terminology for general publications use.
Consultancy: Yes (nc).
Publications: The centre prepares bibliographic lists of Polish standards in Polish, English, French and Russian with the help of a computer.

Polski Komitet Normalizacji   3.27
Miar i Jakości *
[Polish Committee for Standardization, Measures and Quality Control]
Address: ul. Elektoralna 2, 00-139 Warszawa, Poland
Telex: 813642 pkn pl
Affiliation: International Organization for Standardization
Publications: *Normalizacja.*

PORTUGAL

Direcção-Geral da Qualidade   3.28
– DGQ
[General-Directorate for Quality]
Address: Rua José Estevão 83-A, 1199 Lisboa Codex, Portugal
Telephone: 35119-539891
Telex: 13042 QUALIT P
Parent: Ministério da Industria e Comércio
Enquiries to: Head, information and education division, M.O. Fernandes
Founded: 1971
Subject coverage: Standardization and metrology; quality control and certification; technical regulations; industrial environment.
Library facilities: Open to all users (nc); loan services (nc - except for standards); reprographic services (c).
Library holdings: 5000 bound volumes; 300 current periodicals; 120 000 standards.
Information services: Bibliographic services (nc); translation services (c); literature searches (nc); access to on-line information retrieval systems (c).
SDI services on national standards.
Consultancy: Telephone answers to technical queries (nc); no writing of technical reports; no compilation of statistical trade data; no market surveys in technical areas.

SPAIN

Instituto Español de Normalización * 3.29

[Spanish Standards Institute]
Address: Zurbano 46, Madrid 10, Spain
Telephone: 91 410 46 76
Telex: 46545 unor e
Affiliation: International Organization for Standardization
Founded: 1946

SWEDEN

Standardiseringskommissionen I Sverige * 3.30

– SIS
[Swedish Standards Institution]
Address: Box 3295, S-103 66 Stockholm, Sweden
Telephone: (08) 23 04 00
Telex: 17453 sis s
Enquiries to: Head, information and public affairs, Folke Hermanson-Snickars
Founded: 1922
Subject coverage: SIS, the central Swedish standards body and a member of CEN and ISO, is a private organization with its charter approved by the government. SIS is the Swedish member of ISONET, and together with the National Swedish Board of Commerce is a GATT enquiry point.
In 1984 there were approximately 6000 voluntary Swedish standards, prepared by technical committees, and administered by SIS or the following technical divisions: BST (building standards), Drottning Kristinas väg 73, S-114 28 Stockholm; MNC (metals standards), Åsögatan 122, S-116 24 Stockholm; SEK (electrotechnical standards), Box 5177, S-102 44 Stockholm; SMS (mechanical standards), Box 5393, S-102 46 Stockholm.
Library facilities: Open to all users (visiting in person); reprographic services.
Information services: Bibliographic services (c); translation services (c); literature searches (c); access to on-line information retrieval systems (c).
Access is available to the REGIS data base (registry of Swedish Technical rules) within the TESS search service. The SIS also sells its own publications and foreign standards.
Consultancy: Yes (c).

SWITZERLAND

Eidgenössisches Amt für Messwesen * 3.31

– EAM
[Swiss Office of Metrology]
Address: Lindenweg 50, CH-3084 Wabern, Bern, Switzerland
Telephone: (031) 54 10 61
Telex: 33385 LATOP
Parent: Swiss Federal Government Department of Justice and Police
Enquiries to: Librarian, P. Tröhler
Founded: 1864
Subject coverage: Legal metrology; metrological standards.
Library facilities: Open to all users; loan services; no reprographic services.
Information services: No bibliographic services; no translation services; no literature searches; no access to on-line information retrieval systems.
Consultancy: Yes (c).

International Organization for Standardization 3.32

– ISO
Address: 1 rue de Varembé, Case postale 56, CH-1211 Genève 20, Switzerland
Telephone: (022) 34 12 40
Telex: 23 887 ISO CH
Enquiries to: Information Centre
Founded: 1947
Library facilities: Open to all users; no loan services; no reprographic services.
Information services: Bibliographic services (nc - bibliographies of International Standards); no translation services; literature searches (nc); no access to on-line information retrieval systems.
National users should contact their National Standards Organization (ISO Member Body) in the first instance. International organizations should contact the ISO Information Centre direct. An internal data base is maintained, which contains bibliographic information regarding the entire ISO documentation - published International Standards, drafts under study and working items. This data base serves when establishing various listings and catalogues.
Publications: *International Standards, Catalogue ISO,* annual; *Bibliographies ISO; ISO Handbooks,* collected standards in technical fields; *Memento ISO,* annual; *Bulletin ISO,* monthly; annual ISO activities report, and general information brochure.

Schweizerische Normen-Vereinigung

3.33

– SNV

[Swiss Standardization Association]

Address: Kirchenweg 4, CH-8032 Zürich, Switzerland
Telephone: (01) 47 69 70
Telex: 815 036 SNV
Facsimile: (01) 47 08 80
Enquiries to: Documentation officer, M. Freimann
Founded: 1919
Subject coverage: National and foreign standards.
Library facilities: Open to all users; no loan services; reprographic services (c).
Information services: Bibliographic services; no translation services; literature searches; access to on-line information retrieval systems (in the near future).
Consultancy: Yes (charge dependent on working time).
Publications: Mostly in German and French.

TURKEY

Türk Standardlari Enstitüsü *

3.34

– TSE

[Turkish Standards Institution]

Address: Necatibey Caddesi 112, Bakanliklar, Ankara, Turkey
Telephone: (41) 18 72 40
Telex: 42047 TSE TR
Enquiries to: Director, external relations and information, T. Türker
Library, Documentation and Information Centre
Founded: 1960
Subject coverage: Preparation and publication of Turkish standards; cooperation with sectors of industry and universities.
Library facilities: Open to all users; loan services; no reprographic services.
Sales of national and international standards.
Information services: Bibliographic services; literature searches.
The centre provides information on Turkish, foreign and international standards.
Consultancy: Yes (c).
Publications: Turkish Standards Catalogue.

UNITED KINGDOM

British Standards Institution, Milton Keynes Information Centre

3.35

– BSI

Address: Linford Wood, Milton Keynes MK14 6LE, UK
Telephone: (0908) 320066
Telex: 825777 (BSIMK G)
Facsimile: (0908) 320856
Enquiries to: Enquiry section
Founded: 1901
Subject coverage: Standards and technical requirements in the UK and overseas.
Library facilities: Open to all users (nc); loan services (only available to BSI subscribing members); no reprographic services.
The information centre houses the following: BSI's Central Enquiry Service for all non-specific enquiries and telephone orders; the Data Base Section responsible for the BSI STANDARDLINE data base and other manual and mechanized information systems on standards and technical requirements; the BSI Library, with the world's largest collection of standards and technical requirements; and Technical Help to Exporters consultant engineering service. Privileged access to the services is available to BSI members.
Library holdings: Over half a million standards, laws, regulations and technical specifications.
Information services: Bibliographic services (c); translation services (c); literature searches (c); access to on-line information retrieval systems (c).
Consultancy: Telephone answers to technical queries (c); writing of technical reports (c); no compilation of statistical trade data; no market surveys in technical areas.
BSI has a current awareness service covering standards worldwide.
Courses: Occasional courses on development of library and information services for standards organizations; occasional open days.
Publications: BSI ROOT Thesaurus; catalogue of publications; bibliographies of standards; leaflets.

National Physical Laboratory *

3.36

– NPL

Address: Queens Road, Teddington, Middlesex TW11 0LW, UK
Telephone: (01) 977 3222
Telex: 262344
Affiliation: Department of Trade and Industry
Enquiries to: Librarian, P.M. Udy
Founded: 1900

Subject coverage: The laboratory acts as the UK National Standards Laboratory, responsible for the establishment of internationally acceptable basic standards of measurement; development of new techniques of measurement and the promotion of their use by government, industry and universities.

It has the following divisions: electrical science; information technology and computing; materials applications; mechanical and optical metrology; quantum metrology; and radiation science and acoustics.

Library facilities: Not open to all users; no loan services (unless material is unavailable from NLL); no reprographic services.

Information services: The NPL houses the British Calibration Service, the National Testing Laboratory Accreditation Scheme and the National Corrosion Service. The Metallurgical and Thermochemical Data Service provides thermodynamic data from a bank. Results of research investigations are available (c).

Consultancy: Yes (c).

Publications: *NPL Publications*, available on request.

Sheffield Assay Office 3.37

Address: PO Box 187, 137 Portobello Street, Sheffield S1 4DS, UK

Telephone: (0742) 755111

Telex: 54659

Enquiries to: The Assay Master

Founded: 1773

Subject coverage: Hallmarking of platinum, gold and silver; assaying of all precious metal alloys.

Library facilities: Not open to all users (approved users only - nc); no loan services; reprographic services (c).

Library holdings: 1600 bound volumes; 16 current periodicals.

Information services: No bibliographic services; no translation services; literature searches (in certain cases); no access to on-line information retrieval systems.

Consultancy: Telephone answers to technical queries (straightforward enquiries only - nc); writing of technical reports (c); no compilation of statistical trade data; no market surveys in technical areas.

Courses: Specialist courses arranged individually by agreement.

YUGOSLAVIA

Yugoslovenski Zavod za 3.38
Standardizaciju *
[Yugoslavian Standardization Institution]

Address: Slobodana Penezića-Krcuna 35, 11 000 Beograd, Yugoslavia

Telephone: (011) 644066

Telex: 12089 YU JUS

Enquiries to: Librarian, Pivić Miroslava

Founded: 1946

Subject coverage: Library and standardization department.

Library facilities: Open to all users; no loan services; reprographic services.

Information services: Bibliographic services; no translation services; no literature searches; no access to on-line information retrieval systems.

Consultancy: Yes.

4 AGRICULTURE, HORTICULTURE, FISHERIES AND FORESTRY

AUSTRIA

Bundesanstalt für Agrarwirtschaft * 4.1
[Federal Institute of Agricultural Economics]
Address: Schweizertalstrasse 36, A-1133 Wien, Austria
Telephone: (0222) 823651
Affiliation: Bundesministerium für Land- und Forstwirtschaft
Enquiries to: Librarian, W. Pevetz
Bibliothek und Dokumentationsstelle
Founded: 1960
Subject coverage: Agricultural economics; regional economics; rural sociology.
Library facilities: Open to all users; loan services (nationally); no reprographic services.
Library holdings: The library holds nominal and subject catalogues, documentation files, collections of abstracts.
Information services: Bibliographic services; literature searches; access to on-line information retrieval systems (AGRIS).
Consultancy: Yes.
Publications: *Monatsberichte über die Österreichische Landwirtschaft*, monthly economic report; *Schrifttum der Agrarwirtschaft*, bimonthly; *Zugangsverzeichnis der Bibliothek*, semiannual; *Gesamtverzeichnis Österreichischer Landwirtschaftlicher Zeitschriften* (General Catalogue of Austrian Agricultural Economics Periodicals), every five years. Publications list on request.

Bundesanstalt für Fischereiwirtschaft 4.2
[Federal Institute of Fisheries]
Address: Scharfling 18, A-5310 Mondsee, Austria
Telephone: (06232) 245618
Affiliation: Ministry of Agriculture
Enquiries to: Information officer, Dr Ilse Butz
Founded: 1953
Subject coverage: Freshwater research, eutrophication of lakes, water quality control, trout and carp fishery, fish parasitology.
Library facilities: Open to all users; no loan services; reprographic services (xerox, c - or exchange).
Library holdings: 2070 bound volumes; 116 journals; 9058 reprints.
Consultancy: No consultancy.
Publications: Publications list on request.

Forstliche Bundesversuchsanstalt 4.3
– FBVA
[Federal Forest Research Station]
Address: A-1131 Wien, Austria
Telephone: (0222) 823638
Affiliation: Bundesministerium für Land- und Forstwirtschaft
Enquiries to: Director, F. Ruhm
Founded: 1874
Subject coverage: Forestry.
Library facilities: Not open to all users (office library); loan services; reprographic services.
Library holdings: 32 980 bound volumes; 700 current periodicals.
Information services: No bibliographic services; no translation services; literature searches; access to on-line information retrieval systems.
Consultancy: Telephone answers to technical queries; no writing of technical reports; no compilation of statistical trade data; no market surveys in technical areas.

BELGIUM

Rijksstation voor Zeevisserij 4.4
– RvZ
[Fisheries Research Station]
Address: Ankerstraat 1, B-8400 Ostend, Belgium
Telephone: (059) 320 805
Affiliation: Ministry of Agriculture
Enquiries to: Director, P. Hovart
Founded: 1962
Subject coverage: Marine biology; fishing techniques; food chemistry; food technology.
Library facilities: Open to all users (nc); no loan services; reprographic services (c).
Library holdings: 1350 bound volumes; 33 current periodicals.
Information services: No information services.
Consultancy: Yes (nc).

BULGARIA

Centre for Scientific, Technical and Economic Information in Agriculture and Forestry 4.5
Address: 6 Boulevard Dragan Tsankov, Sofia, Bulgaria
Telephone: 66 55 11
Affiliation: Georgi Dimitrov Agricultural Academy
Library facilities: Open to all users (scientists, teachers and specialists); reprographic services.
Information services: Bibliographic services; translation services (from all European languages into Bulgarian, and vica versa).
Abstracts are published in Russian and English.
Publications: Annual catalogue available.

Fisheries Institute and Aquarium 4.6
Address: Bul Tchervenoarmejski 4, 9000 Varna, Bulgaria
Telephone: 2 41 93; 3 18 52
Affiliation: National Agricultural and Industrial Association
Enquiries to: Director, Dr Ivanov
Founded: 1954
Information services: Bibliographic services; translation services; literature searches (bibliographic and thematic searches).
Consultancy: Yes.
Publications: Reports of proceedings.

CYPRUS

Agricultural Research Institute 4.7
Address: Box 2016, Nicosia, Cyprus
Telephone: (02) 403431
Telex: 4660
Affiliation: Ministry of Agriculture and Natural Resources
Enquiries to: Senior agricultural research officer i/c library, P.I. Orphanos
Founded: 1962
Subject coverage: Applied research in crop and animal production.
Library facilities: Open to all users (nc); loan services (nc); reprographic services (nc).
The library acts as the national input centre for FAO AGRIS.
Library holdings: 3600 bound volumes; 110 current periodicals.
Information services: No information services.
Consultancy: No consultancy.
Publications: List of publications available on request.

CZECHOSLOVAKIA

Laboratórium Rybárstva a Hydrobiológie * 4.8
– LRH
[Laboratory of Fishery Research and Hydrobiology]
Address: Drieňová 3, 826 24 Bratislava, Czechoslovakia
Telephone: 22 87 57
Enquiries to: Information officer, Štefan Ziegler
Founded: 1956
Library facilities: Open to all users; loan services; reprographic services.
Information services: Bibliographic services; no translation services; literature searches; no access to on-line information retrieval systems.
Consultancy: Yes.

Ústřední Zemědělská a Lesnická Knihovna 4.9
[Central Agricultural and Forestry Library]
Address: 7 Slezská, 120 56 Praha 2, Czechoslovakia
Telephone: 25 75 41
Parent: Ústav Vedeckotechnickych Informaci, Československá Akademie Zemědělská
Enquiries to: Director, Josef Bočan

Výskumný Ústav Lesného Hospodárstva

4.10

– VÚLH

[Forest Research Institute]

Address: K. Marxa 22, 960 92 Zvolen, Czechoslovakia

Telephone: (0855) 273 11

Telex: 72284 VÚLHz

Affiliation: Ministry of Forest and Water Management

Enquiries to: Librarian, J. Sobocká

Founded: 1898

Subject coverage: Tree breeding, silviculture, forest nurseries, forest pathology, game management, forest inventory, integrated use, forestry economics, logging.

Library facilities: Open to all users (nc); loan services (nc); no reprographic services.

Library holdings: 60 000 bound volumes; 584 current periodicals per annum; 1000 reports.

Information services: No information services.

Consultancy: No consultancy.

Publications: *Acta Instituti forestalis Zvolenenzis*; *Vedecké práce Výskumného ústavu lesného hospodárstva vo Zvolene*; *Lesnícke štúdie*; *Lesnícky časopis*; *Folia venatoria*; *Polovníckeštúdie.*

Výskumný Ústav Pôdoznalectva a Výživy Rastlín

4.11

[Soil Science and Plant Nutrition Research Institute]

Address: Rožňavská 23, 823 69 Bratislava, Czechoslovakia

Telephone: 227 264

Enquiries to: Information officer

Founded: 1960

Subject coverage: Soil science; plant nutrition.

Library facilities: Open to all users; loan services; reprographic services.

The institute undertakes exchange of scientific publications dealing with soil science and plant nutrition.

Library holdings: 10 100 bound volumes; 140 current periodicals; 350 reports.

Information services: Bibliographic services; translation services; literature searches; access to on-line information retrieval systems.

Consultancy: No telephone answers to technical queries; writing of technical reports (within the ČSSR); no compilation of statistical trade data; no market surveys in technical areas.

DENMARK

Bioteknisk Institut *

4.12

– BI

[Biotechnical Institute]

Address: Holbergsvej 10, DK-6000 Kolding, Denmark

Telephone: (05) 520433

Telex: 16600 fotex dk, att. Biotech Kolding

Affiliation: Danish Academy of Technical Sciences

Enquiries to: Deputy director, C. Skov Larsen

Founded: 1959

Subject coverage: Utilization of agricultural plant products for feed, food and technical purposes; optimization of horticultural produce usage.

Library facilities: No library facilities.

Information services: No information services.

Consultancy: Yes (c).

Danmarks Fiskeri- og Havundersøgelsers Bibliotek

4.13

– DF&H

[Danish Institute for Fisheries and Marine Research Library]

Address: Charlottenlund Castle, DK-2920 Charlottenlund, Denmark

Telephone: (02) 628550

Telex: 19960 dfh dk

Affiliation: Danish Ministry of Fisheries

Enquiries to: Chief librarian, Mogens Sandfair Librarian, Bent Gaardestrup

Founded: 1889

Subject coverage: Fishery biology; marine biology; marine ecology; aquaculture; oceanography; limnology.

Library facilities: Open to all users (nc); loan services (nc); reprographic services (c).

Library holdings: 30 000 bound volumes; 700 current periodicals and reports.

Information services: Bibliographic services (c); no translation services; literature searches (c); access to on-line information retrieval systems (c).

Consultancy: Telephone answers to technical queries; writing of technical reports (c); compilation of statistical trade data (c); no market surveys in technical areas.

Danmarks Veterinaer- og Jordbrugsbibliotek *

4.14

– DVJB

[Danish Veterinary and Agricultural Library]

Address: Bülowsvej 13, DK-1870 København V, Denmark

Telephone: (01) 35 17 88

Telex: 15061 dvj bib dk

Affiliation: Den Kongelige Veterinaer- og Landbohøjskole (Royal Veterinary and Agricultural University)
Enquiries to: Chief librarian, Inge Berg Hansen
Det veterinaer- og jordbrugsfaglige Dokumentationscenter
Founded: 1783
Subject coverage: Agriculture, horticulture, fisheries, food science, veterinary medicine.
Library facilities: Open to all users; loan services (nc); reprographic services (c).
Information services: Bibliographic services; translation services; literature searches (nc for manual searches); access to on-line information retrieval systems (c).
The library has access to Euronet, Lockheed Dialog, AGRIS (Vienna), and IRS (Frascati) for documentation services. For document location there is access to Blaise, Libris (Sweden), and Alis (Denmark). DVJB provides on-line access to its own monograph collection (1979-84) as part of Alis at I/S Datacentralen.
Consultancy: Yes.

Dansk Fiskeriteknologisk Institut *   4.15
[Danish Institute of Fisheries Technology]
Address: Postbox 93, Nordsøcentret, DK-9850 Hirtshals, Denmark
Telephone: (08) 94 43 00
Enquiries to: Information officer, Gunnar Petersen Dokumentationscenter
Founded: 1981
Library facilities: Open to all users; loan services; reprographic services (c).
Information services: Bibliographic services (c); no translation services; literature searches (c); access to on-line information retrieval systems (c).
The institute has access to the following data hosts: Blaise; BNDO; Datacentralen CDCJ; Dialog; DIMDI; ESA; INKA; Libris; Polydoc; Pergamon InfoLine; Samson; SDC; and TED.
Consultancy: Yes (c).

Jordbrugsteknisk Institut *   4.16
[Agricultural Engineering Institute]
Address: Rolighedsvej 23, DK-1958 København V, Denmark
Telephone: (01) 351788
Parent: Royal Veterinary and Agricultural University
Enquiries to: Head, Professor T. Tougaard Pedersen
Subject coverage: Agricultural machinery; tractors; processing on the farm; alternative energy (wind and biomass).
Library facilities: Not open to all users; no loan services; reprographic services (c).
Information services: Bibliographic services (c); translation services (c); no literature searches; no access to on-line information retrieval systems.

Consultancy: Yes (nc).

Statens Husdyrbrugsforsøg   4.17
Biblioteket *
[National Institute of Animal Science Library]
Address: Forsøgsanlaeg Foulum, Postbox 39, DK-8833 Ørum Sønderlyng, Denmark
Telephone: (06) 652500
Enquiries to: Librarian, Anne Mette Emdal
Founded: 1984
Subject coverage: Animal husbandry.
Library facilities: The library is not open to the public, and its primary purpose is to assist research workers with reference questions and with on-line literature searches.
Information services: No information services.

Statens jordbrugstekniske Forsøg   4.18
[Danish Agricultural Engineering Institute]
Address: Bygholm, DK-8700 Horsens, Denmark
Telephone: (05) 623199
Enquiries to: Head, F. Guul-Simonsen
Founded: 1978
Subject coverage: Machinery, buildings, and work studies for farms.
Library facilities: Open to all users (nc); no loan services; no reprographic services.
Library holdings: 550 bound volumes; 90 current periodicals; 75 reports.
Consultancy: Telephone answers to technical queries (nc); writing of technical reports (c); no compilation of statistical trade data; market surveys in technical areas (c).

FINLAND

Helsingin Yliopiston   4.19
Maatalouskirjasto
[Helsinki University Library of Agriculture]
Address: Viikki, SF-00710 Helsinki 71, Finland
Telephone: (90) 378011
Telex: 122352
Enquiries to: Librarian, Annikki Kaivosoja
Founded: 1930
Subject coverage: Agricultural, food and nutrition, household, and environmental sciences.
Library facilities: Open to all users; loan services; reprographic services.
Library holdings: 250 000 bound volumes; 3500 current periodicals.
Information services: Bibliographic services (c); no translation services; literature searches (c); access to on-line information retrieval systems (c).
Consultancy: No consultancy.

Metsäntutkimuslaitos *   4.20
– METLA
[Forest Research Institute of Finland]
Address: Postbox 18, Rillitie 8-12, SF-01301 Vantaa,
Finland
Telephone: (90) 831941
Enquiries to: Librarian, Liisa Ikävalko-Ahvonen
Founded: 1917
Subject coverage: Forestry, forest resources and
development.
Library facilities: Not open to all users; loan
services; reprographic services.
Information services: Bibliographic services; no
translation services; literature searches; no access to
on-line information retrieval systems.
Consultancy: Yes.
Publications: Communicationes Instituti Forestalis
Fenniae; Folio Forestalia.

Valtion Maatalous-Metsätieteellinen   4.21
Toimikunta *
[Research Council for Agriculture and Forestry]
Address: Ratamestarinkatu 12, SF-00520 Helsinki
52, Finland
Telephone: (90) 141 611
Telex: 123416 acad sf
Affiliation: Academy of Finland
Enquiries to: Secretary, Mirja Suurnäkki
Science Policy Library of the Academy of Finland
Founded: 1969
Subject coverage: Agriculture, forestry, horticulture,
fisheries, food and drink.
Library facilities: Open to all users; loan services;
no reprographic services.
Information services: Bibliographic services; no
translation services; literature searches; access to on-
line information retrieval systems.
Consultancy: No consultancy.

Valtion Maatalousteknologian   4.22
Tutkimuslaitos
– VAKOLA
**[State Research Institute of Engineering in Agriculture
and Forestry]**
Address: PPA 1, SF-03400 Vihti, Finland
Telephone: (913) 46 211
Parent: Ministry of Agriculture and Forestry
Enquiries to: Inspector, P. Olkinuora
Founded: 1949
Subject coverage: Machinery and equipment used in
agriculture, forestry, horticulture, dairies, home
economics and industry; standardization and safety;
promotion of research cooperation with regard to
agricultural buildings.
Library facilities: Open to all users; no loan services;
no reprographic services.
Information services: No information services.

Consultancy: Telephone answers to technical queries
(nc); writing of technical reports; compilation of
statistical trade data (nc); no market surveys in
technical areas.
Publications: Test reports; study reports; test
bulletins.

FRANCE

Centre National d'Études et   4.23
d'Expérimentation de
Machinisme Agricole *
– CNEEMA
[National Institute of Agricultural Engineering]
Address: Parc de Tourvoie, 92160 Antony, Hauts de
Seine, France
Telephone: (1) 46 66 21 09
Telex: 204565 CNEEMA ANTY
Affiliation: Ministry of Agriculture
Enquiries to: Head of documentation, M Ganneau;
M Dao
Founded: 1957
Subject coverage: Farm machinery; agricultural
mechanization.
Library facilities: Open to all users; no loan services;
reprographic services (c).
Information services: Bibliographic services;
translation services; literature searches.
Consultancy: Yes.

Centre National de Recherches   4.24
Forestières *
[National Centre for Forestry Research]
Address: Champenoux, 54280 Seichamps, France

Institut d'Élevage et de Médecine   4.25
Vétérinaire des Pays Tropicaux *
– IEMVT
**[Animal Production and Veterinary Science of Tropical
Countries Institute]**
Address: 10 rue Pierre-Curie, 94704 Maisons-Alfort
Cedex, France
Telephone: (1) 43 68 88 73
Affiliation: Ministère des Relations Extérieures,
Coopération et Développement
Enquiries to: Information officer, Dr J.F.
Giovannetti
Founded: 1921
Subject coverage: Research into tropical animal
husbandry, development and improvement of livestock
and animal production, grazing and nutrition, fodder
production and pathology; training; documentation.

Library facilities: Open to all users; no loan services; reprographic services.
Information services: Bibliographic services; literature searches.
Consultancy: Yes.
Publications: Revue d'Élevage et de Médecine Vétérinaire des Pays Tropicaux.

Institut National de la Recherche Agronomique 4.26
– INRA
[National Agronomic Research Institute]
Address: route de St Cyr, 78000 Versailles, France
Telephone: 39 50 75 22
Affiliation: Ministère de l'Agriculture
Enquiries to: Head, Documentation Service
Founded: 1951
Subject coverage: Bioagronomy.
Library facilities: Loan services; reprographic services.
The library is open to non-staff users at the director's discretion; it participates in interlibrary loan services.
Library holdings: 2500 periodicals.
Information services: Bibliographic services (c); translation services; literature searches (c); access to on-line information retrieval systems.
Translations are available from German, Russian, Bulgarian, Polish. The library has on-line access to over 100 bibliographic data bases, and provides SDI services.
Publications: Bulletin Sciences Agronomiques - Productions Végétales, 10 issues a year; *Bulletin Protozoaires et Invertébrés - Zoologie,* 10 issues a year.

Institut de Recherches sur les Fruits et Agrumes 4.27
– IRFA-CIRAD
[Fruit and Citrus Research Institute]
Address: 6 rue du Général Clergerie, 75116 Paris, France
Telephone: (1) 45 53 16 92
Telex: ELITA A 641155F
Affiliation: CIRAD, 42 rue Scheffer, 75116 Paris
Enquiries to: Documentation officer, B. Moreau
Founded: 1942
Subject coverage: Tropical/subtropical fruits; agronomy; economics; food science technology; culture of temperate fruit trees in warm zones.
Library facilities: Not open to all users; no loan services; reprographic services.
Library holdings: 4000 bound volumes.
Information services: Bibliographic services; no translation services; literature searches; access to on-line information retrieval systems (FAIREC).
Consultancy: No consultancy.

GERMAN FEDERAL REPUBLIC

Bundesforschungsanstalt für Fischerei 4.28
[Federal Research Centre for Fisheries]
Address: Palmaille 9, D-2000 Hamburg 50, German Federal Republic
Telephone: (040) 389050
Telex: 215716 bfafi
Enquiries to: Information officer, Dr W.P. Kirchner Informations- und Dokumentationsstelle mit Bibliothek - (040) 389051 13
Founded: 1948
Subject coverage: Fisheries science.
Library facilities: Open to all users; no loan services; reprographic services (c).
Information services: Bibliographic services (c); no translation services; literature searches (c); access to on-line information retrieval systems (c).
The information office provides and supervises input to the multinational Aquatic Science and Fisheries Information System; this data base is accessible on-line or through published abstracts.
Consultancy: No consultancy.
Publications: Archiv für Fischereiwissenschaft; Informationen für die Fischwirtschaft.

Bundesforschungsanstalt für Forst- und Holzwirtschaft * 4.29
– BFH
[Federal Research Centre for Forestry and Forest Products]
Address: Leuschnerstrasse 91, D-2050 Hamburg 80, German Federal Republic
Telephone: (040) 739621
Affiliation: Bundesministerium für Ernährung, Landwirtschaft und Forsten
Enquiries to: Librarian
Founded: 1939
Library facilities: Open to all users; loan services (nc); reprographic services (c).
Information services: No bibliographic services; no translation services; literature searches (c); no access to on-line information retrieval systems.
Consultancy: No consultancy.
Publications: Mitteilungen der Bundesforschungsanstalt für Forst- und Holzwirtschaft, irregular; *BFH-Nachrichten,* quarterly; *Jahresbericht der Bundesforschungsanstalt für Forst- und Holzwirtschaft,* (annual report).

Zentralstelle für Agrardokumentation 4.30 und -Information

– ZADI

[Centre for Agricultural Documentation and Information]

Address: Postfach 20 05 69, Villichgasse 17, D-5300 Bonn 2, German Federal Republic
Telephone: (0228) 357097
Affiliation: Bundesministerium für Ernährung, Landwirtschaft und Forsten
Enquiries to: Director, Dr Eugen Müller
Founded: 1969
Subject coverage: Food, agriculture, forestry, including viticulture, plant protection, animal nutrition and foodstuffs, animal production and husbandry, fisheries, food hygiene.
Library facilities: No library facilities. The centre uses the facilities of the Zentralbibliothek der Landbauwissenschaft und Abteilungsbibliothek für Naturwissenschaft und vorklinische Medizin, Postfach 2460, Nussallee 15a, D-5300 Bonn 1, German FR.
Information services: Bibliographic services (c); no translation services; literature searches; access to on-line information retrieval systems.
The centre has access to the following:
CAB; AGRIS; AGRICOLA; ASFA; FSTA; PSTA; Biosis; Agrep; PHYTOMED; VITIS.
Consultancy: No consultancy.
Publications: *Kongresse und Tagungen der Ernährungswissenschaften, Landbauwissenschaften, Forstwissenschaften, Holzwirtschaftswissenschaften, Veterinärmedizin* (Conferences in Nutrition, Agricultural, Forestry and Veterinary Sciences), annual; *Forschungsvorhaben im Bereich der Landbau- Ernährungs- Forst- und Holzwirtschaftswissenschaften sowie der Veterinärmedizin; Teil 1: Landbauwissenschaften; Teil 2: Tierische Produktion/ Veterinärmedizin; Teil 3: Ernährungswissenschaften; Teil 4: Forst- und Holzwirtschaftswissenschaften,* research reports in agriculture, animal production and veterinary science, nutrition and forestry; *Nachweise von Literatur und Forschungsvorhaben; Informationsbereitstellung und Datenverarbeitung im Agrarbereich,* quarterly review of literature and research, information science in agriculture; *Nachweise von Literatur und Forschungsvorhaben: Alternativen im Landbau,* quarterly review of literature and research, alternatives in agriculture.
Publications list available on request.

GREECE

Institute of Oceanographic and Fisheries Research * 4.31

Address: Aghios Kosmas, Helleniko, Athinai 166 04, Greece
Telephone: (01) 9820214
Affiliation: Ministry of Research and Technology
Enquiries to: Director, Dr A. Boussoulengas
Founded: 1965
Subject coverage: Marine biology; marine chemistry; physical oceanography, marine geology, fisheries; aquaculture; marine pollution; freshwater biology.
Library facilities: Open to all users; no loan services; reprographic services (c).
Library holdings: The library houses a collection of scientific papers referring to Greek seas, lakes, and rivers on the subjects covered above.
Information services: No bibliographic services; no translation services; literature searches; access to on-line information retrieval systems (via Euronet).
Consultancy: Yes.

HUNGARY

Agrártudományi Egyetem, Gödöllő * 4.32

– GATE

[University of Agricultural Sciences, Gödöllő]

Address: Páter Károly utca 1, H-2103 Gödöllő, Hungary
Telephone: (Gödöllő) 1
Telex: 224892 gateh
Enquiries to: Director, Dr Gyula Lőrincz
Central Library
Founded: 1945
Subject coverage: Agriculture, agronomy, agricultural engineering, agricultural economics.
Library facilities: Open to all users; loan services (nc); reprographic services (c).
Information services: Bibliographic services (c); no translation services; literature searches (c); no access to on-line information retrieval systems.
Consultancy: Yes (nc).
Publications: *Bibliography of Scientific Publications.*

Erdészeti és Faipari Egyetem Központi Könyvtára * 4.33

[University of Forestry and Timber Industry, Central Library]

Address: POB 132, Bajcsy-Zsilinszky utca 4, H-9401 Sopron, Hungary
Telephone: 111-00

Telex: 249126
Affiliation: Ministry of Agriculture and Food
Enquiries to: Director general, Dr István Hiller
Founded: 1735
Subject coverage: Forestry, timber, and woodworking industry.
Library facilities: Open to all users; loan services (nc); reprographic services (c).
Exchange of publications; forestry history research.
Information services: Bibliographic services (nc); translation services (c); literature searches; no access to on-line information retrieval systems.
Consultancy: Yes (nc).

Erdészeti Tudományos Intézet 4.34
– ERTI
[Forest Research Institute]
Address: Frankel Leó utca 42-44, H-1023 Budapest, Hungary
Telephone: (01) 150 624
Telex: 226914
Affiliation: Ministry of Agriculture and Food
Enquiries to: Director general
Founded: 1949
Subject coverage: Forestry site conditions, tree-breeding, seed and seedling production, afforestation, tending of young stands, clearing, thinning, wood production, stand structure, protection, utilization, mechanization, economics, forest management, ergonomics, environmental protection.
Library facilities: Open to all users (nc); no loan services; reprographic services (nc).
Library holdings: 25 000 bound volumes; 81 current periodicals; 945 reports.
Information services: No information services.
Consultancy: Telephone answers to technical queries (nc); no writing of technical reports; compilation of statistical trade data (nc); no market surveys in technical areas.
Publications: *Erdészeti Kutatások* (Forest Research).

Mezógazasági és Élelmeześugyi Minisztérium Informacios Központja * 4.35
– AGROINFORM
[Information Centre of the Ministry of Agriculture and Food]
Address: Akadémia utca 1, H-1034 Budapest, Hungary
Telephone: 118 218; 160 020
Enquiries to: Attila utca 93, H-1012 Budapest, Hungary
Library facilities: Open to all users.
Information services: Bibliographic services; translation services; literature searches.

ICELAND

Búnaðarfélag Íslands 4.36
– BI
[Agricultural Society of Iceland]
Address: Pósthólf 7080, 127 Reykjavík, Iceland
Telephone: (91) 19200
Enquiries to: Director, Jónas Jónsson
Library
Founded: 1837
Subject coverage: Advisory services for the agricultural community.
Library facilities: Open to all users; loan services (nc); reprographic services (c).
Information services: Bibliographic services (nc); translation services (c); literature searches (nc); access to on-line information retrieval systems (c - Dialog).
Publications: *Búnadarrit; Freyr; Handbok baenda; Rádunautafundir; Saudfjárraektin; Nautgriparaektin Fraedslurit Búnadarfélags Íslands.*

Rannsóknastofnun Fiskiðnaðarins 4.37
[Icelandic Fisheries Laboratories]
Address: PO Box 1390, Skúlagata 4, 121 Reykjavík, Iceland
Telephone: (91) 20240
Telex: Simtex IS 3000 - Fishlab
Enquiries to: Librarian, Eiríkur T. Einarsson
Founded: 1965
Subject coverage: Sea food research; bacteriology; microbiology; biotechnology; mechanical engineering concerning the fish industry.
Library facilities: Open to all users (nc); loan services (nc); reprographic services.
Books only, not periodicals, may be borrowed.
Library holdings: About 5000 bound volumes; 145 current periodicals.
Information services: Bibliographic services (c); no translation services; literature searches (c); access to on-line information retrieval systems (c).
Data banks used are Dialog, Pergamon InfoLine, and the data bases of Food RA in Leatherhead, United Kingdom.
Consultancy: Telephone answers to technical queries (nc); no writing of technical reports; no compilation of statistical trade data; no market surveys in technical areas.
Publications: Annual report; technical reports.

Rannsóknastofnun Landbúnaðarins 4.38
– RALA
[Agricultural Research Institute]
Address: Keldnaholti v/Vesturlandsveg, 112 Reykjavík, Iceland
Telephone: (91) 82230
Telex: 2307 Isinfo IS

Affiliation: Ministry of Agriculture
Enquiries to: Librarian, Guyyrún Pálsdóttir
Founded: 1965
Subject coverage: Agricultural research: feeds, soils, livestock breeding and production, foods, grassland utilization, plant production, farm technology.
Library facilities: Open to all users (nc); loan services (nc); reprographic services (nc).
Information services: Bibliographic services (nc); no translation services; literature searches (nc); access to on-line information retrieval systems (for staff only - Dialog).
Publications: *Íslenskar landbúnaðarrabbsóknir* (Journal of Agricultural Research in Iceland); *Fjölrit RALA.*

IRELAND

An Foras Taluntais * 4.39
[Agricultural Institute]
Address: 19 Sandymount Avenue, Dublin 4, Ireland
Telephone: (01) 688188
Telex: 30459
Enquiries to: Librarian, Bonnie Keane
Founded: 1958
Library facilities: Open to all users; loan services (interlibrary loan service); reprographic services (c).
Information services: Bibliographic services (c); translation services (c); literature searches (c); access to on-line information retrieval systems (c).
The library has access to Lockheed, Euronet, and Esanet.
Consultancy: Yes.
Publications: List on request.

Department of Agriculture, Ireland * 4.40
Address: Agriculture House, Kildare Street, Dublin 2, Ireland
Telephone: (01) 789011
Telex: 4280
Enquiries to: Librarian
Founded: 1899
Library facilities: Open to researchers and postgraduate students; interlibrary loan and reprographic services.
The library holds FAO and USDA publications.
Consultancy: Yes.

Department of Fisheries and Forestry 4.41
Address: Leeson Lane, Dublin 2, Ireland
Telephone: (01) 60044
Telex: 90253 FFWS
Enquiries to: Librarian, Mary Moore

Subject coverage: Fisheries: freshwater and marine ecology; forestry: timber, plant and animal ecology.
Library facilities: Open to all users (nc); loan services (to libraries only - nc); reprographic services (nc).
Information services: Bibliographic services (nc); no translation services; literature searches (nc for manual searches).
Limited access to on-line information retrieval systems (c).
Consultancy: Telephone answers to technical queries (nc); no writing of technical reports; compilation of statistical trade data (nc); no market surveys in technical areas.
Market surveys in technical areas are undertaken by subsidiary bodies of the department.
Publications: List on request.

ITALY

Food and Agriculture Organization 4.42 of the United Nations/ Organisation des Nations Unies pour l'Alimentation et l'Agriculture *
– FAO
Address: Via delle Terme di Caracalla, 00100 Roma, Italy
Telephone: (06) 57971
Telex: 610181 FAOI
Enquiries to: Library and Documentation Systems Division
Subject coverage: The FAO exists to provide technical advice, to serve as a worldwide source of information, to sponsor and service international consultation and cooperation, and to render technical assistance.

Istituto Ricerche sulla Pesca 4.43 Marittima *
– IRPEM
[Institute for Marine Fisheries Research]
Address: Molo Mandracchio, 60100 Ancona, Italy
Telephone: (071) 55314
Affiliation: Consiglio Nazionale delle Ricerche
Enquiries to: Information officer, Dr Carlo Froglia
Founded: 1969
Subject coverage: Biology of marine living resources; statistics and fish population dynamics; echo surveys for evaluation of pelagic resources; fishing gear technology; net materials testing; fishing vessels technology; oceanography.

Library facilities: Open to all users; no loan services; no reprographic services.
Consultancy: Yes.

Istituto Sperimentale per la Cerealicoltura *

4.44

– ISC
[Experimental Institute for Cereal Research]
Address: Via Cassia 176, 00191 Roma, Italy
Telephone: (06) 3285705
Parent: Ministero per l'Agricoltura e Foreste
Enquiries to: Director, Professor Angelo Bianchi
Founded: 1968
Subject coverage: Improvement of cereal crops from genetic, agronomic, and technological standpoints.
Library facilities: No library facilities.
Information services: No information services.

Istituto Sperimentale per la Floriculture *

4.45

– ISF
[Experimental Institute for Floriculture]
Address: Corso Inglesi 508, 18038 San Remo, Italy
Telephone: (0184) 884944
Parent: Ministero per l'Agricoltura e Foreste
Enquiries to: Information officer, Dr Luigi Volpi
Founded: 1967
Subject coverage: Floriculture.
Library facilities: Open to all users; no loan services; reprographic services (c).
Information services: No information services.

Istituto Sperimentale per la Selvicoltura *

4.46

– ISS
[Experimental Institute for Silviculture]
Address: Viale S. Margherita 80, 52100 Arezzo, Italy
Telephone: (0575) 353021
Affiliation: Ministero per l'Agricoltura e Foreste
Enquiries to: Researcher, Emilio Amorini
Founded: 1922
Subject coverage: Forestry and related subjects.
Library facilities: Open to all users; no loan services; reprographic services (c).
Information services: No information services.

Istituto Sperimentale per la Viticoltura

4.47

[Experimental Institute for Viticulture]
Address: Via 28 Aprile 26, 31015 Conegliano, Treviso, Italy
Telephone: (0438) 61635
Affiliation: Ministero per l'Agricoltura e Foreste
Enquiries to: Director, Professor Antonio Calò
Founded: 1923

Library facilities: Open to all users (nc); no loan services; no reprographic services.
Library holdings: 10 000 bound volumes; 100 current periodicals.
Information services: No information services.
Consultancy: Yes.

LUXEMBOURG

Administration des Eaux et Forêts *

4.48

[Water and Forestry Department]
Address: PO Box 411, 34 Avenue de la Porte-Neuve, L-2014 Luxembourg
Telephone: 293 54
Enquiries to: Head, A. Krier
Service de la Statistique Forestière
Founded: 1669
Subject coverage: Administration of public woodlands; wood and timber; hunting; fisheries; nature conservation.
Library facilities: Open to all users; no loan services; reprographic services (nc).
Information services: No information services.
Consultancy: Yes (nc).
Publications: *Statistiques Forestières.*

NETHERLANDS

Centrum voor Boomteelt en Stedlijk Groen *

4.49

– CBSG
[Centre for Nursery Stocks and Urban Horticulture]
Address: Valkenburgerlaan 3, Postbus 118, 2770 AC Boskoop, Zuid-Holland, Netherlands
Telephone: (01727) 3220
Affiliation: Ministerie van Landbouw en Visserij
Enquiries to: Information officer, G.P.M. Faaij-Groenen
Founded: 1900
Subject coverage: Research on nursery stock production and urban horticulture.
Library facilities: Open to all users; no loan services; reprographic services (c).
Library holdings: The library has an extensive collection of books and periodicals, covering horticulture, phytopathology, propagation and growing of horticultural plants. It also maintains a register of nursery specialists.
Information services: No information services.

Centrum voor Landbouwpublikaties en Landbouwdocumentatie   4.50
– PUDOC
[Agricultural Publishing and Documentation Centre]
Address: Postbus 4, 6700 AA Wageningen, Netherlands
Telephone: (08370) 89222
Telex: 45015 BLHWG NL
Affiliation: Ministerie van Landbouw en Visserij
Enquiries to: Head, Documentation and Information Department
Founded: 1957
Subject coverage: Agriculture.
Library facilities: No library facilities.
Information services: Bibliographic services (c); no translation services; literature searches (c); access to on-line information retrieval systems (c).
Consultancy: Telephone answers to technical queries (c); writing of technical reports (c); no compilation of statistical trade data; no market surveys in technical areas.
Courses: Quarterly course entitled 'Searching the Agralin-Database'; annual post-academic course entitled 'Searching Agricultural Information'.
Publications: List of publications available on request.

Instituut voor Mechanisatie, Arbeid en Gebouwen *   4.51
– IMAG
[Agricultural Engineering Institute]
Address: Postbus 43, 6700 AA Wageningen, Netherlands
Telephone: (08370) 19119
Telex: 45330 CTWAG
Enquiries to: Assistant librarian, A.F.C. Bonenberg
Founded: 1974
Subject coverage: Agricultural mechanization; horticultural mechanization; agricultural and horticultural buildings; labour.
Library facilities: Open to all users; loan services; reprographic services.
Information services: No information services.
Consultancy: Yes (nc).
Publications: *Research Reports*; publications list on request.

Instituut voor Plantenziektenkundig Onderzoek   4.52
– IPO
[Plant Protection Research Institute]
Address: Postbus 9060, Binnenhaven 12, 6700 GW Wageningen, Netherlands
Telephone: (08370) 19151
Enquiries to: Librarian
Library of the Phytopathology and Entomology Centre, Postbus 8122, Binnenhaven 8, 6700 ER Wageningen, Netherlands
Founded: 1949
Subject coverage: Plant protection; entomology, mycology and bacteriology; virology; nematology; phytotoxicology of air pollution.
Library facilities: Open to all users; loan services (nc); reprographic services (c).
Library holdings: 26 500 bound volumes; 480 current periodicals and reports.
Information services: Bibliographic services (c); no translation services; literature searches (c); access to on-line information retrieval systems (c).
Consultancy: No consultancy.

Instituut voor Veredeling van Tuinbouwgewassen *   4.53
– IVT
[Horticultural Plant Breeding Institute]
Address: Mansholtlaan 15, Postbus 16, 6700 AA Wageningen, Netherlands
Telephone: (08370) 19123
Enquiries to: Information officer, B.P. van der Post Documentation Department
Founded: 1943
Subject coverage: Horticulture (including vegetable and fruit growing, floriculture, flower-bulb culture, and arboriculture); disease resistance, energy efficiency, species crosses and cellular techniques for genetic improvements.
Library facilities: Open to all users; reprographic services.
Information services: No information services.

Internationaal Bodemreferentie en Informatie Centrum   4.54
– ISRIC
[International Soil Reference and Information Centre]
Address: Postbus 353, 6700 AJ Wageningen, Netherlands
Telephone: (08370) 19063
Enquiries to: Director, Dr W.G. Sombroek
Subject coverage: Soil science.
Library facilities: Open to all users (nc); no loan services; reprographic services (c).
Library holdings: 10 500 bound volumes; 50 current periodicals.
Information services: Bibliographic services (usually nc); no translation services; no literature searches; no access to on-line information retrieval systems.
Consultancy: Telephone answers to technical queries (nc); no writing of technical reports; no compilation of statistical trade data; no market surveys in technical areas.

Laboratorium voor Bloembollen Onderzoek
4.55

– LBO

[Bulb Research Laboratory]

Address: Vennestraat 22, Postbus 85, 2100 AB Lisse, Netherlands

Telephone: (02521) 19104

Enquiries to: Assistant librarian, Selma Gallacher, E van Breda

Founded: 1920

Subject coverage: Growing of ornamental bulbous crops.

Library facilities: Open to all users (nc); loan services (nc); reprographic services (c).

Information services: Bibliographic services (c); no translation services; no literature searches; access to on-line information retrieval systems (c - Agralin, for agricultural information).

Consultancy: No consultancy.

Landbouwhogeschool Wageningen, Centrale Bibliotheek *
4.56

[Wageningen Agricultural University, Central Library]

Address: Salverdaplein, Postbus 9101, 6701 DB Wageningen, Netherlands

Telephone: (08370) 84472

Telex: 45015

Enquiries to: Head, P. Aben

Information Department

Founded: 1918

Subject coverage: Agricultural sciences; biological sciences; environmental sciences; economics; sociology; soil sciences.

Library facilities: Open to all users; loan services (nc - interlibrary loans with Europe); reprographic services (c).

Library holdings: The central library, including some 80 branch libraries, holds 850 000 volumes and 2000 current periodicals.

Information services: Bibliographic services (c); no translation services; literature searches (c); access to on-line information retrieval systems (c).

Consultancy: Yes (c).

Ministerie van Landbouw en Visserij *
4.57

– MLV

[Ministry of Agriculture and Fisheries]

Address: Bezuidenhoutseweg 73, Postbus 20401, 2500 EK 's-Gravenhage, Netherlands

Telephone: (070) 792080/86

Telex: 32040

Enquiries to: Head, P. Rolteveel

Department of Documentary Information

Founded: 1932

Subject coverage: Nutrition, agriculture, nature conservation, agricultural policy, agricultural trade, agricultural industry, animal production, law, fisheries.

Library facilities: Open to all users; loan services (nc); reprographic services (nc).

The library's catalogue is issued on COM (computer output on microfiche), and distributed to other major libraries.

Information services: Bibliographic services (nc); no translation services; literature searches (nc); access to on-line information retrieval systems (nc).

The library has access to the national data bases on periodicals (CCP), on agricultural literature (Agralin), and on government publications (PARAC).

Consultancy: Yes (nc).

Publications: *Recent*, weekly review of scientific literature; *MLV-literatuur rapporten*.

Nederlands Instituut voor Zuivelonderzoek *
4.58

– NIZO

[Netherlands Institute for Dairy Research]

Address: Postbus 20, 6710 BA Ede, Netherlands

Telephone: (08380) 19013

Telex: 37205 NL

Enquiries to: Head, E. Otter

Founded: 1948

Subject coverage: Dairy technology; dairy microbiology; analytical chemistry; process technology; biochemistry; nutrition.

Library facilities: Not open to all users; loan services (within the Netherlands); reprographic services (c).

Information services: No information services.

Proefstation voor de Bloemisterij
4.59

– PBN

[Research Station for Floriculture]

Address: Linnaeuslaan 2a, 1431 JV Aalsmeer, Netherlands

Telephone: (02977) 26151

Enquiries to: Librarian, C.M.M. de Boer

Founded: 1899

Subject coverage: Floriculture.

Library facilities: Open to all users (nc); loan services (nc); reprographic services (c).

Library holdings: 5000 bound volumes; 100 current periodicals; 40 000 slides.

Information services: No bibliographic services; no translation services; no literature searches; access to on-line information retrieval systems.

The library has on-line access to the bibliographic automation system (LH Wageningen/Pudoc).

Consultancy: No consultancy.

Proefstation voor de Fruitteelt *
4.60

– PFW

[Fruit Growing Research Station]

Address: Brugstraat 51, 4475 AN Wilhelminadorp, Netherlands

Telephone: (01100) 16390
Enquiries to: Information officer, M.M. Crembyers-van Scherpenzeel
Founded: 1902
Subject coverage: Research on growing top fruit and small fruit (including strawberries), and on raising planting material for these crops (temperate zones only).
Library facilities: Open to all users; no loan services; reprographic services (c).
Library holdings: The station's library holds a collection of textbooks on fruit growing, scientific horticultural magazines, fruit growers journals, abstract periodicals and reports in Dutch, English, German and French.
Information services: No information services.
Publications: List on request.

Proefstation voor Tuinbouw onder Glas 4.61
[Glasshouse Crops Research and Experiment Station]
Address: Zuidweg 38, Postbus 8, 2670 AA Naaldwijk, Zuid-Holland, Netherlands
Telephone: (01740) 26541
Affiliation: Ministerie van Landbouw en Visserij
Enquiries to: Librarian, W.A. van Winden
Founded: 1900
Subject coverage: Glasshouse horticulture (vegetable and some floricultural crops).
Library facilities: Open to all users (nc); loan services (nc); reprographic services (nc).
Library holdings: 6000 bound volumes; 145 current periodicals; 3000 reports.
Information services: Bibliographic services (c); no translation services; literature searches (c); access to on-line information retrieval systems (c).
Together with other agricultural libraries in the Netherlands, the library is being incorporated in Agralin (Agrarisch Literatuur Informatiesysteem Nederland), the national agricultural data base.
Consultancy: Telephone answers to technical queries (nc); no writing of technical reports; no compilation of statistical trade data; no market surveys in technical areas.

Ryksinstituut voor Visserijonderzoek * 4.62
– RIVO
[Netherlands Institute for Fisheries Research]
Address: Haringkade 1, Postbus 68, 1970 AB IJmuiden, Netherlands
Telephone: (2550) 31614
Telex: 71044 RIVO
Affiliation: Ministerie van Landbouw en Visserij
Enquiries to: Librarian, S.E. Kondenburg
Founded: 1887

Subject coverage: Marine and freshwater fisheries worldwide - fish, crustacea and molluscs; chemistry related to fisheries; oceanography; fishing gear and fish capture methods.
Library facilities: Open to all users; loan services (c); reprographic services (c).
Information services: Bibliographic services; no translation services; literature searches; no access to on-line information retrieval systems.
Consultancy: No consultancy.

NORWAY

Fiskeridirektoratet, Havforskningsinstituttet * 4.63
[Fisheries Directorate, Institute of Marine Research]
Address: Nordnesparken 2, Box 1870-72, N-5011 Bergen, Norway
Telephone: (05) 327760
Telex: 42 297 ocean n
Enquiries to: Librarian, Leif Takvam
Founded: 1900
Subject coverage: Marine research (oceanography); hydrography; marine biology, fisheries.
Library facilities: Open to all users; loan services (nc); reprographic services (limited - nc).
Copying machine; reading instrument for microfiche.
Information services: Bibliographic services (nc); no translation services; literature searches (nc); no access to on-line information retrieval systems.
Literature searching within the subjects covered, by means of the library's subject catalogue.
Consultancy: Yes (nc).

Kontoret for Informasjon og Rettleing i Landbruk 4.64
[Governmental Information and Advisory Service on Agriculture]
Address: Moorveien 12, N-1430 Ås, Norway
Telephone: (02) 941365
Founded: 1948
Library facilities: No library facilities.
Information services: Various kinds of agricultural information are made available to farmers and agricultural information specialists; the service prepares a weekly 5-minute radio broadcast on agricultural problems; and organizes conferences on agricultural information.
Consultancy: Yes.
Publications: Information pamphlets for farmers and agriculture advisors.

Landbruksteknisk Institutt   4.65
– LTI
[Norwegian Institute of Agricultural Engineering]
Address: Postboks 65, N-1432 Ås-NLH, Norway
Telephone: (02) 949370
Affiliation: Royal Ministry of Agriculture
Enquiries to: Professor Kristian Aas
Founded: 1947
Subject coverage: Research and testing activities in agricultural and horticultural engineering, extension work.
Library facilities: Not open to all users; no loan services; no reprographic services.
The library is available only to the staff at the institute and to students at the Agricultural University of Norway. The library is a special library consisting mainly of text books, periodicals, and magazines, concerned with agricultural and horticultural engineering. The library cooperates with the main library of the Agricultural University of Norway.
Information services: Bibliographic services (c); no translation services; no literature searches; no access to on-line information retrieval systems.
The International Association on Mechanization of Field Experiments (IAMFE) has its temporary secretariat and Information Centre at the Institute.
Consultancy: Yes (nc).
Publications: *Test and Research Reports; Orientation*; these publications are distributed abroad on an exchange basis, as is the annual report.

Norges Landbrukshøgskoles   4.66
Bibliotek *
[Agricultural University of Norway Library]
Address: Postboks 12, N-1432 Ås-NLH, Norway
Telephone: (02) 949060
Telex: 17125 n
Enquiries to: Head librarian, Jon Hjeltnes
Founded: 1859
Subject coverage: Agriculture; horticulture; forestry; animal husbandry; dairy science; food technology.
Library facilities: Open to all users; loan services (nc); reprographic services (nc).
Information services: Bibliographic services; no translation services; literature searches; access to on-line information retrieval systems.
The library has access to the following: Lockheed; ESA/IRS; AGRIS.
Consultancy: No consultancy.

Norsk Institutt for Skogforskning   4.67
– NISK
[Norwegian Forest Research Institute]
Address: Postboks 61, N-1432 Ås-NLH, Norway
Telephone: (02) 949642
Affiliation: Department of Agriculture
Enquiries to: Librarian, Mari Nordang
Founded: 1917

Subject coverage: Forestry.
Library facilities: Open to all users (nc); loan services (nc); reprographic services (nc).
Library holdings: 35 000 bound volumes; 900 reports.
Information services: No information services.
Consultancy: No consultancy.
Publications: Reports (in English); research papers (in Norwegian, with a summary in English).

POLAND

Centralna Biblioteka Rolnicza *   4.68
– CBR
[Central Agricultural Library]
Address: Krakowskie Przedmieście 66, PO Box 360, 00-950 Warszawa, Poland
Telephone: 26 60 41
Telex: 81 64 81 cbrol pl
Affiliation: Ministry of Agriculture
Enquiries to: General director, Dr Jerzy Rasiński
Founded: 1956
Subject coverage: Agriculture and related sciences.
Library facilities: Open to all users; loan services; reprographic services.
Library holdings: The library with its two branches at Puławy and Bydgoszcz holds 388 000 volumes, and conducts exchange of scientific material with foreign institutions.
Information services: Bibliographic services; literature searches; access to on-line information retrieval systems.
The library has access to FAO AGRIS and MS Agroinform; it maintains the National Information System on Agriculture and Food Management, from which SDI services are offered.
Consultancy: Yes.

Centralny Ośrodek Badawczo-   4.69
Rozwojowy Ogrodnictwa *
– COBRO
[Horticultural Research and Development Centre]
Address: ul Chodakowska 53/57, 03-816 Warszawa, Poland
Telephone: 10 56 87; 10 62 88
Affiliation: Central Union of Peasant Self-Aid Cooperatives
Enquiries to: Director, Dr Stanisław Rosowski
Founded: 1971
Subject coverage: Production, technology, conservation, refrigeration, packaging and transportation of fruits and vegetables; economics of horticulture; horticultural products marketing; costs and prices.

Library facilities: Not open to all users; loan services; reprographic services.
Facilities are available for staff of COBRO and Central Union of Peasant Self-Aid Cooperatives only.
Information services: Bibliographic services; translation services; literature searches; access to on-line information retrieval systems (System Informacji Naukowej Technicznej i Organizacyjnej - SINTO).
Consultancy: Yes.

Instytut Badawczy Leśnictwa 4.70
– IBL
[Forestry Research Institute]
Address: ul Wery Kostrzewy 3, skr poczt 61, 00-973 Warszawa, Poland
Telephone: 22 32 01
Telex: 812476
Enquiries to: Head, Dr Leopold Rossakiewicz
Scientific, Technical and Economic Forestry Information Centre (ZINTE)
Founded: 1933
Subject coverage: Forest science, game management.
Library facilities: Open to all users (nc); loan services (nc); reprographic services (c).
Library holdings: 58 650 bound volumes; 384 current periodicals; 2400 reports.
Information services: Bibliographic services (c); no translation services; literature searches (nc); no access to on-line information retrieval systems.
Consultancy: Telephone answers to technical queries (nc); writing of technical reports (nc); no compilation of statistical trade data; no market surveys in technical areas.
Courses: Annual courses for research information users.
Publications: *Nowości Piśmiennictwa Leśnego* (Forest Literature News); *Komunikaty Informacyjne* (Information News); *Prace Instytutu Badawczego Leśnictwa* (Proceedings of the Forestry Research Institute).

Instytut Rybactwa Śródladowégo * 4.71
– IRS
[Inland Fisheries Institute]
Address: Olsztyn-Kortowo bl.5, 10-957 Olsztyn 5, Poland
Telephone: 259 81
Telex: 0512316
Enquiries to: Head, Krystyna Korpacz
Branch Information Centre
Founded: 1951
Subject coverage: Fisheries, hydrobiology.
Library facilities: Open to all users; loan services; reprographic services.
National and international interlibrary loan services.
Information services: Bibliographic services.
Consultancy: Yes.

Publications: *Bibliografia Polskiej Literatury Rybackiej* (Bibliography of Polish Literature in the Field of Fisheries), semiannual; *Informacja o Pracach Zamieszczonych w Czasopismach Polskich i Zagranicznych* (Information on Papers Published in Polish and Foreign Periodicals), bimonthly; *Instytut Rybactwa Śródlądowégo Problematyka - Bibliografia* (Inland Fisheries Institute - Bibliography), every 5 years, in Polish and English.

Morski Instytut Rybacki 4.72
– MIR
[Sea Fisheries Institute]
Address: Al Zjednoczenia 1, 81-345 Gdynia, Poland
Telephone: 21 70 21
Telex: 051 543
Affiliation: Board of Maritime Economy
Enquiries to: Head, Henryk Ganowiak
Scientific and Information Centre with Library
Founded: 1921
Subject coverage: Oceanography, marine biology, fishing techniques, fish processing technology, fish processing mechanization, fishery economics and statistics, aquarium and oceanographic museum.
Library facilities: Open to all users; loan services (nc); reprographic services (c).
Library holdings: The library receives 250 Polish and foreign current periodicals.
Information services: Bibliographic services (nc); translation services (c); literature searches (nc); no access to on-line information retrieval systems.
Consultancy: Yes (c).
Publications: *Polska Bibliografia Rybotówstwa Morskiego*, annual bibliography of fisheries and related literature (in Polish), available on request; bulletin.

PORTUGAL

Instituto de Investigação Científica 4.73
Tropical, Centro de
Documentação e Informação
[Institute for Tropical Scientific Research, Documentation and Information Centre]
Address: Rua Jau 47, 1300 Lisboa, Portugal
Telephone: 645321
Affiliation: Ministério da Educação
Enquiries to: Director, F. Almeida Ribeiro
Founded: 1957
Subject coverage: Earth sciences; geographic engineering science; biology; agronomy; ethnological sciences.
Library facilities: Open to all users; loan services (only to staff); reprographic services (c).

Library holdings: 33 200 bound volumes; 980 current periodicals.
Information services: Bibliographic services (nc); no translation services; literature searches (nc); no access to on-line information retrieval systems.
Consultancy: Telephone answers to technical queries; no writing of technical reports; no compilation of statistical trade data; no market surveys in technical areas.
Publications: List of publications available on request.

Instituto Nacional de Investigação das Pescas *   4.74
– INIP
[National Institute of Fisheries Research]
Address: Avenida Brasília, 1400 Lisboa, Portugal
Telephone: 610814; 616361/9
Telex: 15857 INIP P
Affiliation: Ministerio da Agricultura e Pescas
Enquiries to: Chief, Lidia Nunes
Information and Documentation Division
Founded: 1976
Subject coverage: Research and technology in living marine resources; marine and freshwater biology.
Library facilities: Open to all users; loan services; reprographic services (nc).
Information services: Bibliographic services; no translation services; literature searches.
The institute acts as the Portuguese input centre for Aquatic Sciences and Fisheries Abstracts (ASFA).

Ministerio da Agricultura e Pescas   4.75
[Ministry of Agriculture and Fisheries]
Address: Praça do Comércio, 1194 Lisboa Codex, Portugal
Telephone: 368371
Telex: 18 803 DGHEA P
Enquiries to: Head, Documentation Division
Founded: 1977
Subject coverage: Agriculture, horticulture, fisheries and forestry.
Library facilities: Open to all users; loan services; reprographic services.
Information services: Bibliographic services; translation services; literature searches; access to on-line information retrieval systems (FAO AGRIS).
Consultancy: Yes.

ROMANIA

Biblioteca Centrală a Academiei de   4.76 Ştiinţe Agricole şi Silvice
[Central Library of the Academy of Agricultural and Forestry Sciences]
Address: Bulevardul Mărăşti 61, Bucureşti, Romania
Library holdings: 92 000 volumes.

Institutul de Cercetare şi Producţie   4.77 pentru Creşterea Bovinelor
– ICPCB
[Bovine Breeding Research and Production Institute]
Address: 8113, Baloteşti, Sectorul Agricol Ilfov, Romania
Telephone: 17 77 43
Affiliation: Academy of Agricultural and Forestry Sciences; Ministry of Agriculture
Enquiries to: Director, Mircea Pătraşcu; Librarian, Lucia Cureu
Founded: 1981
Subject coverage: Cattle and buffalo breeding and genetics; embryo transfer; climatic chamber-behaviour; bovine feeding, physiology, blood groups; management, biochemistry, economics.
Library facilities: Open to all users; loan services; reprographic services.
Library holdings: 11 332 bound volumes; 58 current periodicals.
Information services: Translation services.
Publications: *Taurine* (scientific papers), annual.

Institutul de Cercetare şi Producţie   4.78 Pomicolă
– ICPP
[Fruit Growing Research Institute]
Address: 0300 Piteşti Mărăcineni, Romania
Telephone: (976) 32066
Enquiries to: Librarian, Mariana Morlova
Founded: 1967
Library facilities: Open to all users; loan services.
Library holdings: 9677 bound volumes.

Institutul de Cercetări pentru   4.79 Protectia Plantelor
– ICPP
[Plant Protection Research Institute]
Address: Bulevardul Ion Ionescu de la Brad 8, 71592 Bucureşti, Romania
Telephone: 33 58 50
Affiliation: Academy of Agricultural and Forestry Sciences
Founded: 1967
Subject coverage: Plant protection: phytopathology, applied entomology, pesticides, forecasting and

monitoring.
Library facilities: Open to all users; loan services; reprographic services.

Institutul de Cercetări pentru Viticultură şi Vinificaţie *   4.80
– ICVV
[Vine-growing and Wine-making Research Institute]
Address: Valea Călugărească, Prahova, Romania
Telephone: (971) 20200
Affiliation: Economic Production Trust of Viticulture
Enquiries to: Scientific manager, M. Macici
Technical Library
Founded: 1967
Subject coverage: Research in the field of vine genetics and amelioration; production of vine planting material; technology of vine crops; oenology; implementation of research results in production.
Library facilities: Open to all users; loan services (nc - scientific papers only); no reprographic services. The library holds a collection of specialized papers in the field of viticulture and oenology; it exchanges scientific literature with foreign institutions.
Information services: Bibliographic services (nc); translation services (nc); no literature searches; access to on-line information retrieval systems (c).
Consultancy: Yes (nc).
Publications: Annual report.

SPAIN

Instituto Español de Oceanografía   4.81
– IEO
[Spanish Institute of Oceanography]
Address: Calle Alcalá 27, Madrid 14, Spain
Telephone: 91-222 74 93
Telex: 44460
Affiliation: Ministerio de Agricultura y Pesca
Enquiries to: Head, oceanographic data centre, Dr F. Fernandez
Subject coverage: Marine biology; fisheries biology and technology; marine pollution; physical oceanography; marine geology.

Instituto Nacional de Investigaciones Agrarias   4.82
– INIA
[National Institute for Agricultural Research]
Address: José Abascal 56, Madrid 3, Spain
Telephone: 442 3199
Telex: 23425 AGRIME
Affiliation: Ministerio de Agricultura, Pesca y Alimentación
Enquiries to: Head, José Ramón Cadahía

Library and Documentation Department
Founded: 1971
Subject coverage: Agriculture and allied sciences.
Library facilities: Open to all users; no loan services; reprographic services.
Information services: Bibliographic services; no translation services; literature searches; access to on-line information retrieval systems.
This institute is the AGRIS liaison centre, providing input to this international information system, and using output tapes on request.
Consultancy: Yes.

Instituto de Investigaciones Pesqueras de Barcelona *   4.83
[Fishery Research Institute of Barcelona]
Address: Paseo Nacional s/n, Barcelona 3, Spain
Telephone: 310 64 50
Telex: 59367 impb e
Affiliation: Consejo Superior de Investigaciones Científicas
Enquiries to: Director, Juan José López Gómez; Librarian, Angeles Moreno
Founded: 1949
Subject coverage: Marine sciences and oceanography.
Library facilities: Open to all users; no loan services; reprographic services (c).
Information services: Bibliographic services; no translation services; no literature searches; no access to on-line information retrieval systems.
Consultancy: Yes (nc).
Publications: *Investigación Pesquera; Resultados Expediciones Científicas; Informes Técnicos; Datos Informativos.*

SWEDEN

Fiskeristyrelsen *   4.84
[National Board of Fisheries]
Address: PO Box 2565, S-403 17 Göteborg, Sweden
Telephone: (031) 17 63 80
Telex: 27108 Natfish S
Affiliation: Ministry of Agriculture
Enquiries to: Information secretary
Founded: 1948
Subject coverage: Professional and recreational fishing; fishery conservation.
Library facilities: Not open to all users; loan services; reprographic services.
Consultancy: Yes.
Publications: List on request.

Sveriges Lantbruksuniversitet 4.85
[Swedish University of Agricultural Sciences]
Address: Ultuna, S-750 07 Uppsala 7, Sweden
Telephone: (018) 17 10 00
Telex: 76062 Ult Bibl S
Enquiries to: Librarian
Ultunabiblioteket (Ultuna Library)
Founded: 1932
Subject coverage: Agriculture, forestry and veterinary science.
Library facilities: Open to all users; loan services; reprographic services.
Information services: Bibliographic services; literature searches; access to on-line information retrieval systems.
Consultancy: Yes.

SWITZERLAND

Eidgenössische Anstalt für das 4.86
Forstliche Versuchswesen
– EAFV
[Swiss Federal Institute of Forestry Research]
Address: Zürcherstrasse 111, CH-8903 Birmensdorf, Zürich, Switzerland
Telephone: (01) 737 1411; (01) 739 2207 (library)
Telex: 829 803 eafv ch
Affiliation: Schweizerischer Schulrat
Enquiries to: Librarian, Regina Schenker
Founded: 1885
Subject coverage: Forestry and related subjects.
Library facilities: Open to all users; loan services (nc); reprographic services (c).
Library holdings: The library holds a bibliography in catalogue form of Swiss forestry since 1760, containing 35 000 references; also a bibliography in catalogue form of worldwide forestry since 1953, containing 280 000 references. The library contains 20 000 bound volumes and 850 current periodicals.
Information services: Bibliographic services (nc); no translation services; literature searches (nc); access to on-line information retrieval systems (only to researchers - nc).
Consultancy: Yes (nc).
Publications: List on request.

Eidgenössische Forschungsanstalt für 4.87
Obst-, Wein- und Gartenbau *
[Swiss Federal Research Station for Fruit-growing, Viticulture and Horticulture]
Address: Schloss, CH-8820 Wädenswil, Switzerland
Telephone: (01) 75 13 33
Affiliation: Department of Public Economy, Division of Agriculture
Enquiries to: Chief librarian; Information officer

Founded: 1890
Subject coverage: Fruit-growing, viticulture, horticulture; biochemistry, agricultural chemistry; entomology, nematology, and phytopathology; weeds; biology and chemistry of wines and fruit juices; storage and processing of fruits and vegetables.
Library facilities: Not open to all users; loan services (c); reprographic services (c).
Information services: Bibliographic services (c); translation services (c).
Consultancy: Yes (c).

Eidgenössische Forschungsanstalt für 4.88
Viehwirtschaftliche Produktion,
Grangeneuve *
– FAG
[Swiss Federal Research Station for Animal Production, Grangeneuve]
Address: CH-1725 Posieux, Switzerland
Telephone: (037) 82 11 81
Enquiries to: Librarian, G. Mangold
Founded: 1974
Subject coverage: Animal nutrition, veterinary medicine, forage conservation, feedstuffs.
Library facilities: Open to all users; loan services (nc); reprographic services (nc).
Information services: Bibliographic services (nc); no translation services; no literature searches; no access to on-line information retrieval systems.
Consultancy: Yes (nc).

TURKEY

Ormancilik Araştirma Enstitüsü, 4.89
Kütüphane Müdür
[Forest Research Institute, Library Directorate]
Address: PK 24, Bahçelievler, Ankara, Turkey
Telephone: 225390 2024
Enquiries to: Director, Nihat Yilmaz
Founded: 1956
Subject coverage: General forestry; statistics; forest soils; forest entomology; systematic botany; plant ecology; silviculture; nursery practice; watershed management; soil erosion; forest engineering; forest economy; forest transportation; logging; forest inventory; forest management; forest products; pulp and paper; minor forest products; research studies.
Library facilities: Open to all users (nc); no loan services; reprographic services (photocopying only - nc).
Library holdings: 10 500 bound volumes; 140 current periodicals; 500 reports.

Information services: No bibliographic services; translation services (nc); literature searches (nc); no access to on-line information retrieval systems.
Consultancy: Telephone answers to technical queries (nc); no writing of technical reports; no compilation of statistical trade data.
Publications: Technical bulletin, irregular, with English, German or French summary; technical reports, irregular; *Ormancilik Araştirma Enstitüsü Dergi Serisi* (Forest Research Institute Journal), biannual; *Ormancilik Araştirma Enstitüsü Muhtelif Yayinlar Serisi* (Miscellaneous Publications of the Forest Research Institute), irregular.

Su Urünleri Dairesi Başkanliği * 4.90
[Department of Fisheries]
Address: Olgunlar Sokak 10, Bakanliklar, Ankara, Turkey
Telephone: 25 43 53
Affiliation: Tarim Orman ve Köyişleri Bakanliği (Ministry of Agriculture, Forestry and Rural Affairs)
Enquiries to: General director, Irfan Sahin
Founded: 1971
Subject coverage: Fisheries; water pollution control; access to control and policy-making.
Library facilities: No library facilities.
Consultancy: The department offers consultancy services to fishermen, and carries out research on their behalf; it also advises in legal matters.

UNITED KINGDOM

Animal and Grassland Research Institute 4.91
– AGRI
Address: Hurley, Maidenhead, Berkshire SL6 5LR, UK
Telephone: (082) 882 3631
Affiliation: Agricultural and Food Research Council
Enquiries to: Head, Liaison, Information and Library Department, Dr O.R. Jewiss
Founded: 1985
Subject coverage: Agricultural research in ruminant animal production from forage, and pig production.
Library facilities: Open to all users (by appointment only - occasional c); loan services (to other libraries - occasional c); reprographic services (to other libraries - occasional c).
The institute maintains a joint library with the Commonwealth Bureau of Pastures and Field Crops. The library is a back-up library to the British Library.
Library holdings: Over 12 000 books excluding bound journals; 1590 journals; 700 annual reports; about 485 000 reports.

Information services: No information services.
Consultancy: Telephone answers to technical queries; writing of technical reports; no compilation of statistical trade data; market surveys in technical areas.
Consultancies and collaborative research.

CAB International 4.92
– CAB
Address: Farnham House, Farnham Royal, Slough SL2 3BN, UK
Telephone: (02814) 2281
Telex: 847964 (COMAGG G)
Affiliation: CAB International, formerly called Commonwealth Agricultural Bureaux, is an international non-profit organization sponsored by 30 Commonwealth governments.
Founded: 1929
Subject coverage: Worldwide information service on agriculture and biology, covering agricultural science; forestry; horticulture; applied biology; pest and disease identification; biological control; agricultural economics; dairy technology; veterinary science; rural development; land use; rural extension, education and training; soil science; leisure, recreation and training.
Library facilities: Reprographic services.
CAB International does not have a library as such; it maintains a data base, accessible on-line via Dialog, ESA/IRS, ESA, CAN/OLE, BRS, and DIMDI. The data base contained around 2m records as of March 1986, with around 12 000 added monthly. Photocopies of original articles can be provided subject to copyright restrictions.
Information services: Bibliographic services (c); no translation services; literature searches (c); access to on-line information retrieval systems (c).
The following publications are now included in the CAB Abstracts data base: all the main and specialist in-house abstracts journals, plus *Animal Disease Occurrence, Biocontrol News and Information*, and *International Biodeterioration*.
Consultancy: Yes (c). Consultancy services on biocontrol and identification are offered free of charge to Commonwealth member countries.
Publications: *Journals, Back Volumes and Serial Publications List 1986* available on request, listing CAB's 25 main abstracts journals, 17 specialist abstracts journals, and other serial publications.

Forestry Commission * 4.93
Address: Forest Research Station, Alice Holt Lodge, Wrecclesham, Farnham, Surrey GU10 4LH, UK
Telephone: (0420) 22255
Enquiries to: Librarian, E.M. Harland
Founded: 1919
Subject coverage: Temperate forestry; arboriculture.
Library facilities: Open to all users (by appointment with the librarian); loan services (nc); reprographic

services (c).
Information services: Bibliographic services (nc); no translation services; literature searches; no access to on-line information retrieval systems.
Consultancy: Yes (nc).
Publications: Catalogue of Publications.

International Bee Research Association   4.94
– IBRA
Address: Hill House, Gerrards Cross, Buckinghamshire SL9 0NR, UK
Telephone: (0753) 885011
Telex: 23152 monref G 8390
Enquiries to: Director, Dr Margaret Adey
Founded: 1949
Subject coverage: Bees and beekeeping; pollination; honey and other hive products.
Library facilities: Not open to all users; loan services (c); reprographic services (c).
The library is open to members of the association; loan services to members only.
Information services: Bibliographic services (c); no translation services; literature searches (c); no access to on-line information retrieval systems.
An abstracting service is provided, based on the IBRA journal *Apicultural Abstracts.* Sets of cards on specific subjects can be supplied. IBRA provides COM microfiche supplement (1973-83) to *Index to Apicultural Abstracts 1950-72.*
Consultancy: Yes (c).
Publications: List on request.

Land Resources Development Centre   4.95
– LRDC
Address: Tolworth Tower, Surbiton, Surrey KT6 7DY, UK
Telephone: (01) 399 5281
Affiliation: Overseas Development Administration, Foreign and Commonwealth Office
Enquiries to: Librarian and information officer, Philip Reilly
Library and Information Services
Founded: 1964
Subject coverage: Assessment of tropical natural resources, and evaluation of potential land uses for agriculture and forestry.
Library facilities: Open to all users (by appointment only - nc); loan services (c); reprographic services (c).
Library holdings: 27 000 bound volumes, including reports; 500 current periodicals.
Information services: Bibliographic services (c on application); no translation services; no literature searches; access to on-line information retrieval systems (c).

The library data base is TRADIS (Tropical Resources for Agricultural Development Information System) which contains over 40 000 records. Retrospective searching covers TRADIS, external data bases and other libraries.
Consultancy: No consultancy.
Publications: Biennial progress report; publications list; leaflets on how to obtain consultancy services on topics covered by LRDC.

Ministry of Agriculture, Fisheries and Food, Main Library *   4.96
– MAFF
Address: 3 Whitehall Place, London SW1A 2HH, UK
Telephone: (01) 233 5092
Telex: 889351
Enquiries to: Librarian-in-Charge
Branch libraries: Great Westminster House, Horseferry Road, London SW1P 2AE, United Kingdom; Government Buildings, Hook Rise South, Surbiton, Surrey KT6 7NF, United Kingdom
Subject coverage: Agriculture, horticulture, fisheries, food; particularly temperate agriculture. At Horseferry Road: human nutrition; food science, food standards; food engineering and processing; dietetics. At Surbiton: animal health; vertebrate pest biology and control.
Library facilities: Open to all users (to research workers); loan services (by direct application, or through BLLD); reprographic services (c).
Information services: Bibliographic services.
Publications: MAFF publications are listed in HMSO sectional Lists Nos 1, 2 and 3, and in *MAFF Publications Catalogue.*

Ministry of Agriculture, Fisheries and Food, Slough Laboratory *   4.97
Address: London Road, Slough, Berkshire SL3 7HJ, UK
Telephone: (0753) 34626
Telex: 889078
Enquiries to: Librarian, T. Cullen
Founded: The laboratory became part of MAFF in 1970.
Subject coverage: Stored product and household pests (insect and mite); pesticides (including physiological means of control); stored products (eg grain, pulses, dried fruit, but not frozen or tinned); microbial associations with stored products, and with stored product pests.
Library facilities: Open to all users (scientific researchers by appointment only); loan services (interlibrary, through BLLD); reprographic services (through BLLD).
Library holdings: The library holds a card index of some 250 000 references on storage and associated

subjects.

Information services: Bibliographic services (nc); no translation services; literature searches (nc); no access to on-line information retrieval systems.

Consultancy: No consultancy (but information advice may be offered).

Publications: *SETA (Storage Entomology Title Alert)*, a current awareness bulletin issued 10 times a year, available on request.

National Institute of Agricultural Engineering 4.98

– NIAE

Address: Wrest Park, Silsoe, Bedford MK45 4HS, UK

Telephone: (0525) 60000

Telex: 825808 NIAE-WP-G

Affiliation: British Society for Research in Agricultural Engineering

Enquiries to: Head, scientific information department

Founded: 1924

Subject coverage: Agricultural and horticultural engineering research and development.

Library facilities: Open to all users (by appointment); loan services; reprographic services (c - plus postage).

Information services: Limited literature searches.

National Vegetable Research Station 4.99

– NVRS

Address: Wellesbourne, Warwick CV35 9EF, UK

Telephone: (0789) 840382

Affiliation: Agricultural and Food Research Council

Enquiries to: Librarian, D.A. Woodroffe

Founded: 1949

Subject coverage: Research on growing vegetables in the open ground.

Library facilities: Not open to all users; loan services (interlibrary); reprographic services (c).

Members of the public may use the library and its facilities by arrangement with the librarian.

Library holdings: 10 000 bound volumes; 400 current periodicals.

Information services: No information services.

Consultancy: Telephone answers to technical queries (nc - ask for Liaison Officer); no writing of technical reports; no compilation of statistical trade data; no market surveys in technical areas.

Oxford Forestry Institute 4.100

– OFI

Address: Department of Plant Sciences, University of Oxford, South Parks Road, Oxford OX1 3RB, UK

Telephone: (0865) 511431

Telex: 83147 VIAORG attn FOROX

Affiliation: University of Oxford

Enquiries to: Head, Library and Information Service, R.A. Mills

OFI Library

Founded: 1905

Subject coverage: Forestry and forest products.

Library facilities: Open to all users (nc); loan services (interlibrary loans in UK only); reprographic services (c).

Library holdings: The library holds 120 000 items, including about 1100 current periodicals, giving worldwide coverage, and has comprehensive catalogues containing 1 250 000 references; also available for sale on microfilm.

Information services: Bibliographic services (c); no translation services; literature searches (c); access to on-line information retrieval systems (c).

The institute is the host library for *Forestry Abstracts* and *Forest Products Abstracts*, published by CAB International and available on-line as CAB Abstracts. Document delivery service available for most material listed.

Consultancy: Telephone answers to technical queries (nc); no writing of technical reports; no compilation of statistical trade data; no market surveys in technical areas.

Publications: Lists of library and forestry information services, publications list and list of microfilm materials available on request.

Tropical Development and Research 4.101 Institute

Address: 56-62 Gray's Inn Road, London WC1X 8LU, UK

Telephone: (01) 242 5412

Affiliation: Overseas Development Administration, Foreign and Commonwealth Office

Enquiries to: Head, J.A. Wright

Library and information services

Founded: 1895

Subject coverage: Agricultural and health problems caused by crop pests and disease vectors; post-harvest agricultural problems of plant and animal products in developing countries.

Library facilities: Open to all users (by appointment only); loan services (interlibrary loans only); reprographic services (c).

Information services are generally available free of charge within the aims of British government overseas aid policy (ie for official bodies in developing countries and other bodies connected with the British aid programme).

Information services: Bibliographic services (nc - selfhelp, by visiting the library); no translation services; literature searches (nc - self help, by visiting the library); no access to on-line information retrieval

systems.
Consultancy: Yes (c).
Publications: Publications list on request.

Welsh Plant Breeding Station *   4.102

Address: Plas Gogerddan, near Aberystwyth, Dyfed
SY23 3EB, UK
Telephone: (0970) 828255
Affiliation: University College of Wales, Aberystwyth;
Agricultural and Food Research Council
Enquiries to: Librarian, D. Collins
Founded: 1919
Subject coverage: Plant breeding of grasses, clovers
and cereals, including related topics of genetics,
cytology, plant physiology, biochemistry, agronomy,
and seed production.
Library facilities: Not open to all users; loan
services; no reprographic services.
Information services: No information services.

Subject coverage: Oceanography, marine biology,
fisheries.
Library facilities: Not open to all users; loan services
(nc); reprographic services (c).
Information services: Bibliographic services (nc);
translation services (nc); literature searches (c); no
access to on-line information retrieval systems.
Consultancy: Telephone answers to technical queries
(c); writing of technical reports (c); no compilation of
statistical trade data; no market surveys in technical
areas.

YUGOSLAVIA

Institut za Mehanizaciju   4.103
Poljoprivrede *

– IMP
[Agricultural Mechanization Institute]
Address: PO Box 41, 11080 Zemun, Yugoslavia
Telephone: (011) 212 403
Enquiries to: Librarian, Voja Mišić
Founded: 1946
Subject coverage: Design and testing of agricultural
machines; projects concerning farms, dairies,
slaughterhouses, cereal driers, etc; mechanization of
various processes in agricultural production and
introduction of new machines, implements and
technologies.
Library facilities: Open to all users; no loan services;
no reprographic services.
Information services: No bibliographic services; no
translation services; literature searches; no access to
on-line information retrieval systems.
Consultancy: Yes.
Publications: *Poljoprivredna Tehnika.*

Institut za Oceanografiju i   4.104
Ribarstvo

– IOR
[Institute of Oceanography and Fisheries]
Address: PO Box 114, Šetalište M. Pijade 63, 58000
Split, SR Hrvatska, Yugoslavia
Telephone: (058) 76 688
Enquiries to: Librarian, Professor Anka Bokšić
Founded: 1931

5 BIOLOGICAL SCIENCES

AUSTRIA

Naturhistorisches Museum Wien * 5.1
[Natural History Museum of Vienna]
Address: Burgring 7, Postfach 417, A-1014 Wien, Austria
Telephone: (0222) 93 45 41
Affiliation: Ministry of Science and Research
Enquiries to: Head librarian, A.-Ch. Hilgers
Founded: 1748
Subject coverage: Anthropology; botany; geology; mineralogy; palaeontology; prehistory; zoology.
Library facilities: Open to all users; loan services (nc); reprographic services (c).
Information services: Bibliographic services (nc); translation services (nc); literature searches (nc); no access to on-line information retrieval systems. Zoological record.
Consultancy: Yes (nc - scientific staff of the museum may be consulted).

BELGIUM

Institut Royal des Sciences 5.2
Naturelles de Belgique/
Koninklijk Belgisch Instituut voor
Naturwetenschappen *
– IRSc Nat/KBIN
[Royal Belgian Institute for Natural Sciences]
Address: Rue Vautier 29, B-1040 Bruxelles, Belgium
Telephone: (02) 648 04 75
Affiliation: Ministère de l'Éducation Nationale
Enquiries to: Department chief, W. De Smet
Founded: 1842
Subject coverage: Natural history.

Library facilities: Open to all users; loan services (nc); reprographic services (c - only cost).
Information services: No information services.

Instituut voor 5.3
Zeewetenschappelijk Onderzoek
– IZWO
[Institute for Marine Scientific Research]
Address: Prinses Elisabethlaan 69, B-8401 Bredene, Belgium
Telephone: (059) 32 37 15
Enquiries to: Director, Dr E. Jaspers
Founded: 1970
Subject coverage: Marine sciences.
Library facilities: Open to all users (by prior telephone arrangement - nc); no loan services; reprographic services (photocopies of articles for personal use - c).
Library holdings: Approximate holdings all pertaining to marine sciences: 750 journals (including 310 current); 300 books; 200 bibliographies; 170 proceedings; 110 newsletters; 15 atlases; 80 reports; 40 publications on museums and institutes; 300 miscellaneous publications; 7000 reprints.
Information services: No information services.
Consultancy: No consultancy.
Publications: *IZWO Collected Reprints* annual, comprising the marine scientific publications (in English or French) of IZWO's 120 members.

Koninklijke Maatschappij voor 5.4
Dierkunde van Antwerpen
[Royal Zoological Society of Antwerp]
Address: Koningin Astridplein 26, B-2018 Antwerpen, Belgium
Telephone: (03) 231 16 40
Enquiries to: Director, F.J. Daman
Founded: 1843
Subject coverage: Zoology, botany, general biology; current research is particularly concerned with pathology of animals in captivity.

Library facilities: Open to all users; no loan services; reprographic services (c).
Loan services - interlibrary only.
Information services: Bibliographic services; no translation services; literature searches; no access to on-line information retrieval systems.
Consultancy: Yes.
Publications: *Zoo-Antwerpen; Zoo-Anvers* (includes annual report); *Acta Zoologica et Pathlogica Antverpiensia.*

Nationale Plantentuin van Belgie/Jardin Botanique National de Belgique *   5.5
[National Botanical Garden of Belgium]
Address: Domein van Bouchout, B-1860 Meise, Belgium
Telephone: (02) 2693905
Affiliation: Ministry of Agriculture
Enquiries to: Curator, R. Clarysse
Founded: 1870
Subject coverage: Botany.
Library facilities: Open to all users; loan services; reprographic services (c).
Information services: Bibliographic services (limited); no translation services; literature searches (limited); access to on-line information retrieval systems (in preparation).
Consultancy: Yes.
Publications: Bibliographic list.

BULGARIA

National Natural History Museum   5.6
Address: Boulevard Russki 1, Sofia, Bulgaria
Telephone: 88 51 15
Affiliation: Bulgarian Academy of Sciences Central Library
Enquiries to: Librarian, Maria Staneva, National History Museum Library
Founded: 1889
Library facilities: Open to all users (nc); loan services (nc); no reprographic services.
Library holdings: About 6000 bound volumes.
Information services: Bibliographic services (nc); no translation services; literature searches; no access to on-line information retrieval systems.
The Bulgarian Academy of Sciences Central Library provides an information service with publications in foreign languages.
Consultancy: Telephone answers to technical queries; writing of technical reports; compilation of

statistical trade data; no market surveys in technical areas.

CZECHOSLOVAKIA

Přírodovědné Muzeum *   5.7
[Natural Science Museum]
Address: Vítězného Února 74, 110 01 Praha 1, Czechoslovakia
Telephone: 26 94 50
Enquiries to: Director, R. Horný

DENMARK

Gensplejsningsgruppen   5.8
[Genetic Engineering Group]
Address: Technical University of Denmark, Building 227, DK-2800 Lyngby, Denmark
Telephone: (02) 87 66 99
Telex: 37529 DTHDIA
Enquiries to: Director, Claus Christiansen
Technical Library of Denmark
Founded: 1982
Subject coverage: Genetic engineering industry, fermentation and downstream.
Information services: No bibliographic services; no translation services; no literature searches; no access to on-line information retrieval systems.
Nucleic acids data bases (c).
Consultancy: Writing of technical reports (c); market surveys in technical areas (c).
On-the-phone answers to technical queries may on occasion be given.

Københavns Universitet, Botaniske Centralbibliotek   5.9
[Copenhagen University, Central Botanical Library]
Address: Sølvgade 83, opg.S, DK-1307 København K, Denmark
Telephone: (01) 123146
Enquiries to: Head librarian, Annelise Hartmann
Founded: 1752
Subject coverage: Botany.
Library facilities: Open to all users; loan services; reprographic services (c).
The library is the chief botanical library in Denmark.
Information services: Bibliographic services (nc); no translation services; literature searches (nc); no access to on-line information retrieval systems. (All mainly for students of botany at the University of

Copenhagen).
Consultancy: Yes (nc).

Københavns Universitet, Zoologisk Museum   5.10
[Copenhagen University, Zoological Museum]
Address: Universitetsparken 15, DK-2100 København Ø, Denmark
Telephone: (01) 354111
Enquiries to: Librarian, Bent Hansen
Founded: 1770
Subject coverage: Systematic zoology, including scientific collections and popular exhibitions.
Library facilities: Open to all users (nc); loan services (nc); reprographic services (c).
Library holdings: 25 000 bound volumes; 700 current periodicals.
Information services: No information services.
Consultancy: No consultancy.
Publications: Steenstrupia (periodical).

Naturhistorisk Museum, Århus   5.11
Address: Bygning 210, Universitetsparken, DK-8000 Århus C, Denmark
Telephone: (06) 12 97 77
Founded: 1921
Library facilities: Open to all users (through the university library, Århus); no reprographic services.
Information services: Natural History Musuem, Århus.
Consultancy: Yes.

Statens Skadedyrlaboratorium *   5.12
[Danish Pest Infestation Laboratory]
Address: Skovbrynet 14, DK-2800 Lyngby, Denmark
Telephone: (02) 87 80 55
Enquiries to: The Laboratory
Founded: 1948
Subject coverage: Pest control, concerning stored products, textiles, wood, and pests of hygienic importance including rats, mice, and moles.
Library facilities: Open to all users; no loan services; reprographic services (nc).
The library is very small.
Information services: No information services.
Consultancy: Yes (nc).
Publications: Annual report; bibliographic list.

Zoologisk Have   5.13
[Zoological Gardens]
Address: Sdr. Fasanvej 79, DK-2000 København F, Denmark
Telephone: (01) 302555
Enquiries to: Curator, Bengt Holst
Founded: 1859
Subject coverage: Zoo animals, animal behaviour, animal diseases.

Library facilities: Open to all users; no loan services; no reprographic services.
Concentrating on zoo literature (zoo animals, zoo management, zoo periodicals).
Information services: No bibliographic services; no translation services; literature searches (nc); no access to on-line information retrieval systems.
Consultancy: Yes (nc).

FINLAND

Helsingin Yliopiston Kirjasto, Luonnontieteidenkirjasto   5.14
[Helsinki University Library, Science Library]
Address: Tukholmankatu 2, SF-00250 Helsinki 25, Finland
Telephone: (90) 410566
Telex: 122785 tsk sf
Enquiries to: Head of information services, Marita Rosengren
Founded: 1979
Subject coverage: Natural sciences, especially biological sciences.
Library facilities: Open to all users (nc); loan services (nc); reprographic services (c).
The library participates in the interlibrary lending scheme.
Library holdings: The library has 6000 current serials, acquired on an exchange basis. In addition to its own collection, the library keeps a card index of titles held by the science institutes of the university.
Information services: Bibliographic services (c); no translation services; literature searches (c, but no charge for searches taking less than an hour); access to on-line information retrieval systems (c - Dialog, ESA/IRS, STN-Intenational, Pergamon InfoLine, Libris, Alis Recodex, DIMDI, KDOK/MINTTU, etc).
Consultancy: Telephone answers to technical queries; no writing of technical reports; no compilation of statistical trade data; no market surveys in technical areas.
Courses: Courses on information retrieval and library use (for various institutions of the Faculty of Natural and Mathematical Sciences of the university, once a year for each institution).
Publications: Current microfiche publications (list on request) include: *Fennica,* Finnish national bibliography 1978-, microfiche edition; *Fennica ISBN; Fennica ISDS,* Finnish serials with ISSN; *Finuc-S, Union Catalogue of Foreign Serials in the Research Libraries of Finland; Union Catalogue of Foreign Literature in the Research Libraries of Finland, (A:books).*

FRANCE

Institut Pasteur, Bibliothèque 5.15
[Pasteur Institute Library]
Address: 25 rue du Docteur Roux, 75724 Paris Cedex 15, France
Telephone: (1) 45 68 82 80
Telex: PASTEUR 250 609 F
Enquiries to: Head librarian, Nicole Dubois
Founded: 1888
Subject coverage: Biology; medical sciences.
Library facilities: Open to all users (c); loan services (nc); reprographic services (c).
Library holdings: 300 000 bound volumes; 600 current periodicals.
Information services: Access to on-line information retrieval systems (c).
Consultancy: No consultancy.

International Organization for Biological Control of Noxious Animals and Plants/Organisation Internationale de Lutte Biologique contre les Animaux et les Plantes Nuisibles 5.16
– IOBC/OILB
Address: 335 avenue Paul Parguel, 34100 Montpellier, France
Telephone: 67 54 51 40
Telex: 490304
Affiliation: International Union of Biological Sciences
Enquiries to: Secretary-general, Dr J.P. Aeschlimann
Founded: 1971
Subject coverage: Promotion of use of living organisms or their products to prevent or reduce the harm or losses caused by pests; dissemination of information on biological control.

International Union of Biological Sciences/Union Internationale des Sciences Biologiques * 5.17
– IUBS
Address: 51 boulevard de Montmorency, 75016 Paris, France
Telephone: (1) 45 25 00 09
Telex: 630553
Enquiries to: Executive secretary, Dr Talal Younès
Founded: 1919
Subject coverage: Biological sciences.
Library facilities: Open to all users; loan services (c); reprographic services (c).
Library services consist of journal/publications available, which may be mailed to individuals at a cost per issue. Library per se, does not exist.
Information services: Bibliographic services; no translation services; no literature searches; no access to on-line information retrieval systems.
Consultancy: No consultancy.
Publications: *Biology International*, regular issues twice a year, plus special issues.

Muséum National d'Histoire Naturelle 5.18
– MNHN
[National Natural History Museum]
Address: 38 rue Geoffroy Saint-Hilaire, 75005 Paris, France
Telephone: (1) 43 31 71 24
Affiliation: Ministère de l'Éducation Nationale
Enquiries to: Central library director, Yves Laissus
Founded: 1635
Subject coverage: Natural history, earth sciences.
Library facilities: Open to all users; loan services (nc); reprographic services (c).
Information services: Bibliographic services (c); no translation services; no literature searches; access to on-line information retrieval systems (c).
Consultancy: Yes (nc).

Secrétariat Faune-Flora * 5.19
[Fauna and Flora Secretariat]
Address: Muséum National d'Histoire Naturelle, 57 rue Cuvier, 75231 Paris Cedex 05, France
Telephone; (1) 43 36 54 32
Enquiries: Director, F de Beaufort
Founded: 1979
Subject coverage: Ecology, fauna, flora, natural habitats: inventories, mapping scientific bibliography.
Library facilities: Not open to all users; loan services; reprographic services.
Information services; Bibliographic services (c); no translation services; no literature searches; no access to on-line information retrieval systems.
The secretariat has a data base, FAUNA-FLORA.
Concultancy: Yes (c)
Publications: List of publications available on request.

GERMAN FEDERAL REPUBLIC

Biologische Bundesanstalt für Land- und Forstwirtschaft
5.20

– BBA

[Federal Biological Research Centre for Agriculture and Forestry]

Address: Königin-Luise-Strasse 19, D-1000 Berlin 33, German Federal Republic

Enquiries to: Professor W. Laux; Dr W. Sicker Documentation Centre for Phytomedicine

Founded: 1898

Subject coverage: Topics include phytomedicine, phytopathology, plant protection, storage product research, weeds, pesticides.

Library facilities: Open to all users; loan services (nc); no reprographic services.

There is a special library for phytomedicine (phytopathology and plant protection) housing 55,000 volumes.

Information services: Bibliographic services (c); no translation services; literature searches (c); access to on-line information retrieval systems.

Data base, PHYTOMED (298 000 data entries, input 15 000 per year).

Consultancy: No consultancy.

Publications: *Bibliography of Plant Protection*; *PHYTOMED Thesaurus*; list of serials in the centre's libraries.

Botanischer Garten und Botanisches Museum Berlin-Dahlem *
5.21

[Botanical Garden and Botanical Museum Berlin-Dahlem]

Address: Königin-Luise-Strasse 6-8, D-1000 Berlin 35, German Federal Republic

Telephone: (030) 8314041

Enquiries to: Dr H.W. Lack

Founded: 1679

Subject coverage: Systematic botany.

Library facilities: Open to all users; loan services; reprographic services.

The establishment's library is one of the leading libraries for systematic botany in Europe.

Information services: Bibliographic services; no translation services; no literature searches; no access to on-line information retrieval systems.

Consultancy: Yes.

Bundesforschungsanstalt für Getreide- und Kartoffelverarbeitung
5.22

[Federal Research Centre for Grain and Potato Processing]

Address: Postfach 23, Schützenberg 12, D-4930 Detmold, German Federal Republic

Telephone: (05231) 28042

Telex: 09-35851

Affiliation: Senate of Berlin

Enquiries to: Head, Magda Klüver Information/Documentation Department

Founded: 1907

Subject coverage: Research in grain and potato processing; baking technology; milling technology; starch and potato technology; biochemistry and analysis.

Library facilities: Open to all users; no loan services; reprographic services (c).

Library holdings: 45 000 bound volumes; 480 current periodicals.

Information services: Bibliographic services (c); no translation services; literature searches (c); access to on-line information retrieval systems (c).

The centre provides an information service on cereal science and processing (printed version issued fornightly).

Consultancy: Yes (nc).

Publications: *Bibliography of Cereals Processing*, annual; *Bibliography of Publications of the Federal Research Centre for Grain and Potato Processing*, every five years.

Forschungsinstitut und Naturmuseum Senckenberg
5.23

[Senckenberg Research Institute and Nature Museum]

Address: Senckenberganlage 25, D-6000 Frankfurt am Main 1, German Federal Republic

Telephone: (0611) 75421

Telex: 413129 sng

Affiliation: Seckenbergische Naturforschende Gesellschaft, Frankfurt am Main

Publications: *Seckenbergiana biologica*; *Seckenbergiana lethaea*; *Seckenbergiana maritima*; *Natur und Museum, Kleine Seckenberg-Reihe*; *Courier Forschung Institut Seckenberg*; *Abhandlungen der Seckenburgische Naturforschenden Gesellschaft*; *Aufsätze und Reden*; *Archiv für Molluskenkunde*.

Gesellschaft für Biotechnologische Forschung mbH *
5.24

– GBF

[Biotechnology Research Society]

Address: Mascheroder Weg 1, D-3300 Braunschweig - Stockheim, German Federal Republic

Telephone: (0531) 6181 539

Enquiries to: Dr Irene Wagner-Doebler
Subject coverage: An information centre for the German biotechnology institute is expected to be established in late 1986. It will be known as Biotechnologie Informations Knotten für Europa (BIKE) and will collect and distribute information on research and industry to enquirers both inside and outside Germany.

A contact point will be provided for those searching for collaborative partnerships and for visitors wishing to meet German biotechnologists. Services provided will include online searching, training seminars, a telephone enquiry point and a free newsletter. GBF is well placed to provide a home for BIKE as it is one of the foremost of the federal research centres with an interest in most aspects of biotechnology as well as being host to the German Type Culture Collection.

Institut für Meeresforschung Bremerhaven *   5.25

– IMB
[Marine Research Institute, Bremerhaven]
Address: Am Handelshafen 12, D-2850 Bremerhaven 1, German Federal Republic
Telephone: (0471) 181 205
Enquiries to: Librarian, Michael J. Gómez
Founded: 1948
Subject coverage: North Sea and German Bight ecosystems; marine biology; chemistry; bacteriology; oceanography; diatoms and nematodes.
Library facilities: Open to all users; loan services (nc); reprographic services.
Library holdings: The library contains approximately 5000 monographs, 23 000 periodical volumes, and 19 000 reprints of reports and dissertations.
Information services: Bibliographic services; no translation services; no literature searches; no access to on-line information retrieval systems.
Consultancy: Yes (nc).
Publications: Veröffentlichungen des Instituts für Meeresforschung Bremerhaven, biannual list of publications.

HUNGARY

Természettudományi Muzeum Könyvtára   5.26

[Hungarian Natural History Museum Library]
Address: Baross utca 13, H-1088 Budapest, Hungary
Telephone: (01) 139-490
Enquiries to: Director, Dr Elisabeth Szajáni
Founded: 1870

Subject coverage: Biology; systematic zoology-botany; palaeontology, mineralogy, anthropology.
Library facilities: Not open to all users; loan services; reprographic services.
The library is open only to researchers and university students, and possesses a collection of documents concerning the history of biological sciences.
Information services: Bibliographic services; no translation services; literature searches; no access to on-line information retrieval systems.
Consultancy: Yes.

ICELAND

Hafrannsóknastofnunin   5.27

[Marine Research Institute]
Address: Skúlagata 4, PO Box 1390, 121 Reykjavík, Iceland
Telephone: (91) 20240
Telex: ˙ ISRADO IS 2066
Enquiries to: Librarian, Eiríkur Th. Einarsson
Founded: 1937; name changed in 1965
Subject coverage: Marine biology; physical and chemical oceanography; marine geology; fish-farming.
Library facilities: Open to all users (nc); no loan services; reprographic services (c).
Library holdings: About 7500 bound volumes; 560 current periodicals.
Information services: Bibliographic services (nc); no translation services; literature searches (c); access to on-line information retrieval systems (Dialog, DIMDI and Pergamon InfoLine - c).
Consultancy: Telephone answers to technical queries (nc); no writing of technical reports; no compilation of statistical trade data; no market surveys in technical areas.
Publications: Rit Fiskideildar (occasional, in English or German with English summary); Hafrannsóknir (occasional, in Icelandic, with English summary; includes annual report of the institute and fishing prospects reports); Fjölrit Hafrannsóknastofnunar (mimeographed papers published occasionally, in Icelandic).

IRELAND

National Botanic Gardens 5.28

Address: Glasnevin, Dublin 9, Ireland
Telephone: (01) 377596; 374388
Affiliation: Department of Agriculture
Enquiries to: Director, Aidan Brady
Founded: 1795
Subject coverage: Botany, horticulture, biohistory, with emphasis on Ireland.
Library facilities: Open to all users (by appointemnt only - nc); loan services (interlibrary loans - nc); reprographic services (no charge unless large quantities are involved).
The library is open by appointment only. A catalogue is under preparation.
Library holdings: 330 current periodicals; 40 000 reports.
Information services: Bibliographic services (nc); no translation services; no literature searches; no access to on-line information retrieval systems.
Bibliographic services are limited to research interests of current members of staff, and only within historical bibliography of Irish botany and horticulture.
Consultancy: Telephone answers to technical queries (nc); no writing of technical reports; no compilation of statistical trade data; no market surveys in technical areas.
Plant identification in all genera; identification of diseased specimens and pests is included. No charge is made for this service.
Courses: Three-year course in amenity horticulture.
Publications: *A short guide to the National Botanic Gardens; Glasra* (occasional journal); *Index Seminum* (annually); *Census catalogue of the Irish flora, Trees for your garden.*

National Museum of Ireland 5.29

Address: Kildare Street, Dublin 2, Ireland
Telephone: (01) 765521
Parent: Department of the Taoiseach
Enquiries to: Librarian, F. Devlin
Founded: 1877
Subject coverage: Archaeology; ethnography; fine arts; zoology; geology.
Library facilities: Open to all users (nc); no loan services; reprographic services (c).
The library provides facilities for consultation of material not available elsewhere, and has a collection of aerial photographs of archaeological and geographical features available for consultation.
Library holdings: 20 000 bound volumes; 100 current periodicals.
Information services: No bibliographic services; no translation services; literature searches (based on

museum holdings only - nc); no access to on-line information retrieval systems.
Consultancy: Telephone answers to technical queries (nc); no writing of technical reports; no compilation of statistical trade data; no market surveys in technical areas.

ITALY

Società Botanica Italiana * 5.30
– SBI
[Italian Botanical Society]
Address: Via Giorgio la Pira 4, 50121 Firenze, Italy
Telephone: (055) 210755
Enquiries to: Librarian
Founded: 1888
Subject coverage: Botany - plant cytology, morphology, anatomy, taxonomy.
Library facilities: Open to all users; no loan services; no reprographic services.
Information services: No information services.
Consultancy: No consultancy.
Publications: *Giornale Botanico Italiano; Informatore Botanico Italiano.*

Stazione Zoologica di Napoli 5.31
– SZN
[Naples Zoological Society]
Address: Villa Comunale 1, 80121 Napoli, Italy
Telephone: (081) 416703
Affiliation: Laboratorio Ecologia Marina Ischia Porto
Enquiries to: Head librarian
Founded: 1872
Subject coverage: Marine biology.
Library facilities: Open to all users (with scientific qualifications); no loan services; reprographic services (c).
The historical section of the library permits access to documents and other material concerning the history of the institute to scientists from all over the world who have been at any time connected with the institute.
Information services: Bibliographic services.
Consultancy: Yes.

LUXEMBOURG

Société des Naturalistes Luxembourgeois   5.32
[Naturalists of Luxembourg Society]
Address: BP 327, L-2013 Luxembourg
Enquiries to: Librarian, H. van Wersch

NETHERLANDS

Artis-Bibliotheek *   5.33
[Artis Library]
Address: Plantage Middenlaan 45A, 1018 DC Amsterdam, Netherlands
Telephone: (020) 5223614
Affiliation: University of Amsterdam/Royal Zoological Society
Enquiries to: Librarian, Florence F.J.M. Pieters
Founded: 1938
Subject coverage: Zoology; biology, especially history of biology.
Library facilities: Not open to all users; loan services (nc); reprographic services (c).
The library is open to visitors to Artis Zoo, and to persons over 18 years old who have a real scientific interest in the large historical part of the collection. Loan services are only available to students and workers of the University of Amsterdam, and after recommendation.
Information services: Bibliographic services (nc); no translation services; no literature searches; no access to on-line information retrieval systems.
Consultancy: Yes (usually no charge).
Publications: Bijdragen tot de Dierkunde (contributions to Zoology - Amsterdam) biannually.

Rijksmuseum van Natuurlijke Historie   5.34
– RMNH
[National Museum of Natural History]
Address: Postbus 9517, Ramsteeg 2, 2300 RA Leiden, Netherlands
Enquiries to: Librarian, A.P.W.M. Kosten
Founded: 1820
Subject coverage: Systematic zoology; entomology; marine biology; oceanography.
Library facilities: Open to all users (nc); loan services (nc); reprographic services (c).
Library holdings: 73 000 books; 1600 current periodicals.

Information services: Bibliographic services (nc); no translation services; no literature searches; no access to on-line information retrieval systems.
Consultancy: Telephone answers to technical queries (nc); no writing of technical reports; no compilation of statistical trade data; no market surveys in technical areas.

NORWAY

Universitetet i Oslo, Botanisk Hage og Museum   5.35
[Oslo University, Botanical Garden and Museum]
Address: Trondheimsveien 23B, N-0562 Oslo 5, Norway
Telephone: (02) 686960
Enquiries to: Librarian or Director
Founded: 1811
Subject coverage: Taxonomy; biosystematics; plant geography; palynology and history of vegetation; plant sociology; plant ecology; dispersal of plants.
Library facilities: Not open to all users; loan services; reprographic services (photocopying).
Information services: Bibliographic services.
Consultancy: Yes.
Publications: Nordic Journal of Botany.

POLAND

Instytut Botaniki Polska Akademia Nauk/Instytut Botaniki Uniwersytetu Jagiellońskiego *   5.36
[Botanical Institute of the Polish Academy of Sciences and of the Jagiellonian University]
Address: Lubicz 46, 31-512 Kraków, Poland
Telephone: 21 51 44
Enquiries to: Chief librarian, Janina Oleszakowa
Founded: Jagiellonian University 1798; Polish Academy of Sciences 1953
Library facilities: Open to all users; loan services (nc); no reprographic services.
The library is common to the botanical institutes of both the Polish Academy of Sciences and the Jagiellonian University.
Library holdings: The library contains 120 000 volumes of publications, periodicals, books, and reprints and is the largest botanical library in Poland. It maintains a publications exchange with 600

exchange partners in 63 countries.
Information services: Bibliographic services; no translation services; no literature searches; no access to on-line information retrieval systems.
Consultancy: Yes.

Ogród Botaniczny PAN * 5.37
[Botanical Gardens PAN]
Address: Ulica Prawdziwka 2, 02-973 Warszawa 34, Poland
Telephone: 022-42 29 14
Parent: Polska Akademia Nauk
Publications: *Gardens*, botanical journal.

Polska Akademia Nauk, Instytut 5.38
Parazytologi im.W.Stefańskiego,
Biblioteka
[W. Stefański Institute of Parasitology of the Polish Academy of Sciences, Library]
Address: Skr poczt 153, Ulica L. Pasteura 3, 00-973 Warszawa, Poland
Telephone: 22 25 62
Enquiries to: Head librarian, M. Radziejewski
Founded: 1952
Subject coverage: Parasitology, including animal parasitism, its origin, prevalence, manifestations, and effects in natural and experimental parasite-host systems; phylogeny and ontogeny, physiology, protozoology, immunology, environmental parasitology, parasitic zoonoses.
Library facilities: Open to all users (nc); loan services (nc); reprographic services (c).
Library holdings: 26 720 bound volumes; 646 current periodicals.
Information services: No information services.
Consultancy: No consultancy.
Publications: *Acta Parasitologica Polonica* in English, quarterly; *Polska Bibliografia Parazytologiczna*, annually.

PORTUGAL

Biblioteca do Instituto Botânico 5.39
[Botanical Institute Library]
Address: Arcos do Jardim, 3049 Coimbra, Portugal
Telephone: (039) 22897
Affiliation: Universidade de Coimbra
Enquiries to: Librarian, Dr Joaquim Tomás Miguel Pereira
Founded: 1772
Subject coverage: Botany.
Library facilities: Open to all users (nc); loan services (c - if by post); reprographic services (c).

Library holdings: 110 059 bound volumes; 1500 current periodicals; 3145 periodicals.
Information services: Bibliographic services (nc); no translation services; no literature searches; no access to on-line information retrieval systems.
Consultancy: Telephone answers to technical queries (nc - for Portuguese territory only); no writing of technical reports; no compilation of statistical trade data; no market surveys in technical areas.

Museu Nacional de História 5.40
Natural *
[National Natural History Museum]
Address: Faculdade de Ciencias, Rua de Escole Politécnica, Lisboa, Portugal
Enquiries to: Director
Library holdings: 27 000 volumes.

SPAIN

Museo Nacional de Ciencias 5.41
Naturales
[Natural Science Museum]
Address: Paseo de la Castellana 84, 28046 Madrid, Spain
Enquiries to: Director, Francisco Hernández-Pacheco

Real Jardin Botánico de Madrid * 5.42
[Royal Botanic Garden of Madrid]
Address: Calle de Moyano no 1, Madrid 7, Spain
Telephone: (91) 4682025
Affiliation: Consejo Superior de Investigaciones Científicas
Enquiries to: Director, Dr F.D. Calonge
Founded: 1781
Subject coverage: Botany.
Library facilities: Open to all users; loan services (nc); reprographic services (c).
Information services: Bibliographic services (nc); no translation services; no literature searches; no access to on-line information retrieval systems.
Consultancy: Yes (nc).

SWEDEN

Lunds Universitet, Teknisk Mikrobiologi

5.43

[Lund University, Technical Microbiology Division]
Address: Chemical Centre, PO Box 740, S-220 07 Lund 7, Sweden
Telephone: 046-10 83 25
Enquiries to: Librarian
Founded: 1974
Subject coverage: Biochemistry; biotechnology; technical microbiology; microbiology.
Library facilities: Open to all users; loan services; no reprographic services.
Consultancy: Yes.

SWITZERLAND

Conservatoire et Jardin Botaniques de la Ville de Genève

5.44

[Botanical Garden and Herbarium of the City of Geneva]
Address: Case Postale 60, CH-1292 Chambésy/Genève, Switzerland
Telephone: (022) 32 69 69
Affiliation: City of Geneva; University of Geneva
Enquiries to: Keeper of the Library, Hervé M. Burdet
Founded: 1817
Subject coverage: Botany - systematics, taxonomy, horticulture, and related fields.
Library facilities: Open to all users; loan services (to all persons living in Switzerland; documents published before 1900, manuscripts, and drawings excluded); reprographic services.
Loans to persons living outside Switzerland are available only through the reprographic service.
The library is affiliated to the Swiss Inter-libary Loan Service and to the International Loan Service; it holds one of the foremost sets of botanical books in the world.
Library holdings: About 180 000 volumes.
Information services: Bibliographic services; no translation services; literature searches (minor enquiries free of charge, others charged at the discretion of the keeper); access to on-line information retrieval systems.
Access to the major European and American interconnected data base networks (charges according to the fees of the relevant bases).
Consultancy: Yes.
Publications: *Candollea*; *Boissiera*; *Saussurea*.

Muséum d'Histoire Naturelle, Bibliothèque

5.45

[Natural History Museum, Library]
Address: Case Postale 434, Route de Malagnou, CH-1211 Genève 6, Switzerland
Telephone: (022) 35 91 30
Enquiries to: Head librarian, C. Favarger
Founded: 1820
Subject coverage: Zoology; entomology; geology; mineralogy; palaeontology.
Library facilities: Open to all users (nc); loan services (nc); reprographic services (c).
Library holdings: 50 000 bound volumes; 1100 current periodicals.
Information services: Bibliographic services; no translation services; literature searches; no access to on-line information retrieval systems.

UNITED KINGDOM

Bioquest Limited *

5.46

Address: 200 Aston Brook Street, Birmingham B8 3TE, UK
Telephone: (021) 359 4647
Telex: 336997
Affiliation: University of Aston
Enquiries to: Technical director, Dr J.L. Hurst
Founded: 1965
Subject coverage: Biodeterioration of materials of economic importance, mycology, microbiology, and general problems of biological origin.
Library facilities: Not open to all users; loan services (c); reprographic services (c).
Information services: No information services.
Consultancy: Yes.

British Industrial Biological Research Association

5.47

– BIBRA
Address: Woodmansterne Road, Carshalton, Surrey SM5 4DS, UK
Telephone: (01) 643 4411
Telex: 25438
Enquiries to: Information Officer
Information and Advisory Section
Founded: 1961
Subject coverage: Toxicology.
Library facilities: Not open to all users; loan services (nc); reprographic services (normally no charge).
Available to members only.
Library holdings: The library consists of a comprehensive collection of scientific papers on the toxicology of chemicals (well in excess of 100 000 documents) and related worldwide legislation. It holds

most of the toxicological journals and there is a small special library of books on toxicology.

Information services: Bibliographic services (available only to members - usually no charge); translation services (translations of German food packaging recommendations and other foreign legislation - nc); literature searches (available only to members - usually no charge); access to on-line information retrieval systems (available only to members: Blaise, Datastar, Dialog, DIMDI, Pergamon InfoLine, ESA-IRS - usually no charge).

Ready prepared authoritative up-to-date summaries of the most relevant data on industrial and other chemicals (updated as new significant information becomes available - c).

Consultancy: Telephone answers to technical queries (nc); writing of technical reports (often nc); no compilation of statistical trade data; no market surveys in technical areas.

Information and advice given to member companies on the toxicology, potential hazards, and safe handling and use of chemicals, including food additives, food contaminants such as packaging migrants, heavy metals, mycotoxins and agricultural chemicals, cosmetics and toiletry ingredients, drug excipients, surgical products, tobacco additives, industrial chemicals and water, air and soil pollutants. International legislation governing these topics is also covered. In addition advice is provided on the design and interpretation of toxicological tests, and on the need for further specific tests on any chemicals.

Written answers are given to technical queries on safety-in-use including assessment of the acceptability of food packaging materials in specific applications.

Publications: BIBRA *Information Bulletin* (10 times a year to members); *Food and Chemical Toxicology.*

British Museum (Natural History), Department of Library Services 5.48

Address: Cromwell Road, London SW7 5BD, UK
Telephone: (01) 589 6323
Enquiries to: Head, A.P. Harvey; Deputy head, R.E.R. Banks
Founded: 1881
Subject coverage: The department is responsible for six subject libraries, including one in earth sciences. The five holding material in the biological sciences are listed below.
Library facilities: Open to all users (if bonafide research workers); no loan services; reprographic services (c).
Library holdings: The museum libraries contain approximately 750 000 volumes, 18 000 serial titles (of which some 10 500 are currently received), 70 000 maps, and tens of thousands of prints, drawings and manuscripts. The Museum Library is the repository

for the archives of the museum under the Public Records Act 1958 and 1967.
Information services: Bibliographic services (c); literature searches (c); access to on-line information retrieval systems (c).
Dialog.
Publications: Catalogue of the Books, Manuscripts, Maps and Drawings in the British Museum (Natural History) 8 volumes; Serial Publications in the British Museum (Natural History) Library three volumes and a microfiche edition updated annually; Bulletin of the British Museum (Natural History); extensive publishing programme; list of publications available.

Botany Library UK 5.49

Telephone: (01) 589 6323, extension 628
Enquiries to: Keeper of Botany.
Department of Botany
Subject coverage: General botany; regional botany; systematic botany; plant physiology; evolution; plant ecology; medical botany; poisonous plants; plant pathology; bacteria; economic plants; botanic gardens; botanical illustration and the history of botany; biographies of botanists, and botanical reference works.
Consultancy: Telephone answers to technical queries; writing of technical reports; no compilation of statistical trade data; no market surveys in technical areas.
The library will identify plants on request. Charges for all consultancy services are made to all commercial users.

Entomology Library UK 5.50

Telephone: (01) 589 6323, extension 306
Enquiries to: Keeper of Entomology.
Department of Entomology
Subject coverage: Entomology including medical and economic aspects, history of entomology, biographies of entomologists and entomological reference works.
Consultancy: Telephone answers to technical queries; writing of technical reports; no compilation of statistical trade data; no market surveys in technical areas.
Charges for consultancy services are made to all commercial users.
Courses: The annual international course in applied taxonomy of insects of agricultural importance is run by the Commonwealth Institute of Entomology with support from the staff of the Department of Entomology.

General Library UK 5.51

Telephone: (01) 589 6323, extension 382
Enquiries to: Librarian, S.L. Goodman
Subject coverage: The literature of natural history apart from that limited to one particular aspect of the subject; evolution, genetics, ecology, microbi-

ology, taxonomy, expeditions and descriptive natural history, history of science, biographies of scientists and science reference sources.

Ornithology Library 5.52
Address: Zoological Museum, Akeman Street, Tring, Hertfordshire HP23 6AP, UK
Telephone: (044 282) 4181
Enquiries to: Librarian, D.A. Vale
Library facilities: There are two libraries in the museum: Rothschild Library - a comprehensive collection of pre-1937 books and serials on ornithology, travel books, general natural history, history of ornithology and biographies of ornithologists and bird artists; sub-department of ornithology - comprehensive collection of post-1937 books and serials on ornithology.

Zoology Library UK 5.53
Telephone: (01) 589 6323, extension 379
Enquiries to: Keeper of Zoology
Subject coverage: All branches of the animal kingdom (except insects), especially: taxonomic literature; animal anatomy; animal behaviour; animal ecology; animal parasitology; animal physiology; zoogeography; zoos; history of zoology; biographies of zoologists and zoological reference sources.
Information services: Bibliographic services (of a specialized nature relating to taxonomy).
No charge is made when there are reciprocal benefits to the museum.
Consultancy: Telephone answers to technical queries; writing of technical reports; no compilation of statistical trade data; no market surveys in technical areas.
A selective identification service is available, for which enquiries should be made in advance.
A charge for all consultancy services is made to commercial organizations and projects, but no charge is made when there are reciprocal benefits to the museum.

CAB International Mycological Institute 5.54
– CMI
Address: Ferry Lane, Kew, Richmond, Surrey TW9 3AF, UK
Telephone: (01) 940 4086
Telex: 265871 MONREF G
Affiliation: CAB International
Enquiries to: Director, Professor D.L. Hawksworth
Founded: 1920
Subject coverage: Information service and book publishing in fungal taxonomy, plant pathology, medical and veterinary mycology, industrial mycology, fungal biotechnology, and biodeterioration. Worldwide service for identifying microfungi and plant pathogenic bacteria. Sale of fungus cultures (CMI incorporates UK National Collection of Fungus Cultures). Contract studies in mycology, biodeterioration, and biotechnology.
Library facilities: Loan services (via British Library interlending system only - c); reprographic services (photocopy service - c).
The library is open to those with a professional educational interest; users should contact the librarian prior to their arrival.
Library holdings: 6000 bound volumes; 650 current periodicals; 200 000 reprints.
Information services: Bibliographic services (nc); no translation services; literature searches (c); access to on-line information retrieval systems (c).
Consultancy: Telephone answers to technical queries (nc - but more detailed answers and full studies are available on a contract charged basis).
The institute offers an identification service for microfungi. It also offers contract research, consultancy and testing services.
Courses: A variety of technical training courses are held each year on the following topics: mycology, plant pathology, bacteriology, biodeterioration, and culture collection management. Details are available on request.
Publications: *Services and Publications*, annual.

Freshwater Biological Association 5.55
Address: The Ferry House, Far Sawrey, Ambleside, Cumbria LA22 0LP, UK
Telephone: (096 62) 2468
Affiliation: Natural Environment Research Council
Enquiries to: Senior librarian, Ian Pettman
Founded: 1929
Subject coverage: Freshwater biology.
Library facilities: Not open to all users; loan services (c); no reprographic services.
The association houses the Fritsch Collection of Algae.
Information services: Bibliographic services (c); no translation services; literature searches (c); access to on-line information retrieval systems (c).
Dialog and Euronet services.
Consultancy: Yes (c).
Publications: Annual report; *Scientific Publications* series; *Occasional Publications* series; *Annual List of British Papers on Freshwater Biology; Current Awareness* series, monthly.

Institute of Terrestrial Ecology * 5.56
– ITE
Address: Merlewood Research Station, Grange-over-Sands, Cumbria LA11 6JU, UK
Telephone: (044 84) 2264
Telex: 65102 MERITE
Affiliation: Natural Environment Research Council

Enquiries to: Chief Librarian, J. Beckett
Founded: 1973
Subject coverage: Biology applied to the study of natural ecosystems and environmental management.
Library facilities: Open to all users; loan services; reprographic services.
Library facilities are open to genuine researchers only; loan and reprographic services are made through the interlibrary network.
There are libraries at all research stations within the ITE: Monks Wood Experimental Station (Huntingdon); Merlewood Research Station (Grange-over-Sands); Furzebrook Research Station (Wareham); Edinburgh; Banchory; Bangor Research Station; and Culture Centre of Algae and Protozoa (Cambridge).
Information services: No information services.
Consultancy: No consultancy.

Institute of Virology * 5.57

Address: Mansfield Road, Oxford OX1 3SR, UK
Telephone: (0865) 512362
Telex: 83147 (Virox- G)
Affiliation: Natural Environment Research Council
Enquiries to: Librarian, E.F. Hemmings
Kenneth M. Smith Library
Founded: 1981
Subject coverage: Viruses pathogenic to insects, wild plants, and vertebrate animals.
Library facilities: Open to all users; no loan services; reprographic services (c).
Information services: No translation services; no literature searches; access to on-line information retrieval systems (c).
Dialog.
Consultancy: Yes (c).

Marine Biological Association of the United Kingdom 5.58

– MBA
Address: Citadel Hill, Plymouth PL1 2PB, UK
Telephone: (0752) 221761
Affiliation: Natural Environment Research Council
Enquiries to: Head, Library and Information Services
Founded: 1888
Subject coverage: Marine sciences: marine and estuarine biology, ecology, pollution; oceanography; fisheries and related subjects.
Library facilities: Not open to all users; loan services; reprographic services.
The library is open only to marine scientists and loans only material not available from the British Library; reprographic services are limited also to this material.
Library holdings: 60 000 bound volumes; 1400 current periodicals; 65 000 pamphlets and reprints.
There is an extensive collection of expedition reports,

together with a rare books section, archives, colour slides, films, charts, microfilm and microfiche material. The library includes the Marine Pollution Information Centre, with a collection of 32 000 documents on the scientific aspects of marine pollution.
Information services: Bibliographic services (c); no translation services; literature searches (c); access to on-line information retrieval systems (c).
The library is the UK focal point and main input centre for the UN/FAO Aquatic Sciences and Fisheries Information System (ASFIS), contributing abstracts to the computer-searchable data base and monthly publication *Aquatic Sciences and Fisheries Abstracts CASFAD*. It also provides data and tests formats for the Environmental Chemicals Data and Information Network of the EEC (ECDIN), and collaborates along similar lines with the UNEP International Register of Potentially Toxic Chemicals (IRPTC).
Data analysis is undertaken of the effects of chemicals in marine and estuarine environments.
The library makes a charge for bibliographic services only if appreciable work is required.
Consultancy: Telephone answers to technical queries (no charge if queries are brief); no writing of technical reports; no compilation of statistical trade data; no market surveys in technical areas.
Courses: Work experience and practical training are provided for overseas marine librarians and information specialists.
Publications: *Marine Pollution Research Titles*, monthly reference bulletin; bibliographies of marine and estuarine pollution and of estuary and coastal waters biology; library catalogue; publications list on request; booklet, *A Guide to the Library*, available on request.

Microtest Research Limited 5.59

Address: University Road, Heslington, York YO1 5DU, UK
Telephone: (0904) 41805
Enquiries to: Managing director, Dr R.C. Garner
Founded: 1976
Subject coverage: Biological contract services.
Information services: Literature searches (c).
Consultancy: Telephone answers to technical queries (c); writing of technical reports (c).

National Museum of Wales Library 5.60

Address: Cathays Park, Cardiff CF1 3NP, UK
Telephone: (0222) 397951
Enquiries to: Librarian
Founded: 1907
Subject coverage: Includes botany, geology, archaeology, industrial and maritime studies, and zoology.

Library facilities: Open to all users (nc - by arrangement); loan services (nc - via interlibrary loans); reprographic services (c).
The library is open from Tuesday to Friday.
Library holdings: 126 000 bound volumes, including reports; 1100 current periodicals.
Information services: No information services apart from very limited bibliographic services.
Consultancy: No consultancy.
Publications: Publications list on request.

Royal Botanic Garden, Edinburgh   5.61

Address: Inverleith Row, Edinburgh EH3 5LR, UK
Telephone: (031) 552 7171
Affiliation: Scottish Office, Edinburgh
Enquiries to: Librarian
Founded: 1670
Subject coverage: Taxonomic botany; amenity horticulture.
Library facilities: Open to all users (by appointment - c); loan services (limited, interlibrary); reprographic services (limited).
The library is primarily for the staff and students at the establishment. Loan and reprographic services are provided through BLLD forms.
Library holdings: Over 100 000 bound volumes; 1750 current periodicals.
Information services: Bibliographic services (limited - c); no translation services; literature searches (limited - nc); no access to on-line information retrieval systems.
Consultancy: Telephone answers to technical queries (limited - nc); no writing of technical reports; no compilation of statistical trade data; no market surveys in technical areas.
Publications: *Notes from the Royal Botanic Garden, Edinburgh*; *Royal Botanic Garden, Edinburgh* departmental publication series; *British Fungus Flora*; books; guides.

Royal Botanic Gardens, Kew   5.62
– RBG Kew
Address: Kew, Richmond, Surrey TW9 3AB, UK
Telephone: (01) 940 1171
Telex: 296694
Enquiries to: Director
Founded: 1841
Subject coverage: Taxonomic and economic botany; horticulture; plant physiology; cytology; conservation.
Library facilities: Not open to all users (bona fide researchers only by permission from the chief librarian - nc); loan services (via British Library only); reprographic services.
Library holdings: About 750 000 bound volumes; 2000 current periodicals.

Information services: Bibliographic services (limited); no translation services; no literature searches; no access to on-line information retrieval systems.
Consultancy: Identification service.
Publications: Annual report; full list on request.

Sheffield University Biomedical Information Service   5.63
– SUBIS
Address: Sheffield S10 2TN, UK
Telephone: (0742) 78555
Telex: 547216 UGSHEF G (Quote 'SUBIS')
Enquiries to: Head, J.K. Barkla
Founded: 1966
Subject coverage: Cell biology; physiology; immunobiology; neurobiology; biotechnology.
Library facilities: The service forms a part of the university library.
Information services: Bibliographic services (monthly current awareness service - c); literature searches (c); access to on-line information retrieval systems.
Own data base; Medline.
Consultancy: Yes (c).

Ulster Museum   5.64
Address: Botanic Gardens, Belfast BT9 5AB, UK
Telephone: (0232) 668251
Affiliation: Department of Education, Northern Ireland
Enquiries to: Director, Keeper of Botany and Zoology, or Keeper of Geology
Founded: 1831
Subject coverage: Includes botany and zoology, geology, and technology.
Library facilities: Open to all users (nc); loan services (nc); reprographic services (c).
Library holdings: 30 000 bound volumes; 200 current periodicals.
Information services: No bibliographic services; no translation services; literature searches (nc); no access to on-line information retrieval systems.
Consultancy: Telephone answers to technical queries (nc); writing of technical reports (c); no compilation of statistical trade data; no market surveys in technical areas.

YUGOSLAVIA

Muzejski Dokumentacioni Centar *   5.65
[Museum Documentation Centre]
Address: Zagreb, Yugoslavia
Subject coverage: Museology; archaeology; natural history.
Publications: Informatica Museologica.

Prirodnjački Muzej u Beogradu *   5.66
[Belgrade Natural History Museum]
Address: PO Box 401, Njegoševa 51, Beograd, Yugoslavia
Telephone: (011) 4444 2263
Enquiries to: Director, Vojislav Vasić
Library holdings: 54 706 volumes.

6 CHEMICAL AND PHARMACEUTICAL SCIENCES AND INDUSTRIES

AUSTRIA

Fachverband der Chemischen Industrie Österreichs * 6.1
[Austrian Association of Chemical Industries]
Address: Wiedner Hauptstrasse 63, A-1040 Wien, Austria
Telephone: (06505) 3340
Telex: 1 14235 bks
Affiliation: Bundeswirtschaftskammer (Austrian Federal Economic Chamber)
Founded: 1946
Subject coverage: Statistics and reports on the Austrian chemicals industry.
Library facilities: Not open to all users; no loan services; no reprographic services.

BELGIUM

Centre de Recherche, d'Analyse et de Contrôle Chimiques * 6.2
– CERACHIM
[Chemical Research, Analysis and Control Centre]
Address: 55 Boulevard Sainctelette, B-7000 Mons, Belgium
Telephone: (065) 31 22 22
Affiliation: Institut Supérieur des Ingénieurs Industriels du Hainaut
Enquiries to: Director, Professor Dr G. Lembourg
Founded: 1964
Subject coverage: Foodstuffs analysis - quality control; analysis and control of mineral oil products; quality testing of industrial products; technical collaboration with industry to improve quality of industrial production; study of environmental contamination by human and industrial activities; research on applications of pyrolysis treatments of industrial wastes.
Library facilities: Not open to all users; loan services; reprographic services (c).
Information services: Bibliographic services.
Consultancy: Yes.

Fédération des Industries Chimiques de Belgique * 6.3
– FIC
[Federation of Belgian Chemical Industries]
Address: Square Marie-Louise 49, B-1040 Bruxelles, Belgium
Telephone: (02) 230 40 90
Telex: 23167 FECHIM B
Enquiries to: Head, Documentation Department. Bibliothèque François Boudart
Founded: 1919
Subject coverage: The employers association represents the economic, social, scientific and professional interests of the chemicals industry.
Library facilities: Available to member companies.
Library holdings: 10 000 bound volumes; 250 current periodicals.
Information services: Bibliographic services; no translation services; no literature searches; access to on-line information retrieval systems.
On-line searches are available to member companies using Dialog, SDC, Blaise, ESA, STN, and Belindis.

Instituut voor Scheikundig Onderzoek/Institut de Recherches Chimiques 6.4
– ISO
[Chemical Research Institute]
Address: 5 Museumlaan, B-1980 Tervuren, Belgium

Telephone: (02) 767 53 01; (02) 767 93 39

Affiliation: Ministry of Agriculture
Enquiries to: Director, J.R. Istas
Founded: 1928
Subject coverage: Chemical aspects of agricultural research: protection of the rural environment (pollution problems); soil fertility; protection of crops; purity of food products; biotechnological studies (valorization of vegetables and agricultural products) and wastes; some problems related to tropical agricultural products.
Library facilities: Open to all users (c); no loan services; reprographic services (c).
Library holdings: 4140 bound volumes; 115 current periodicals.
Information services: No information services.
Consultancy: Telephone answers to technical queries; writing of technical reports (on certain subjects - c); no compilation of statistical trade data; no market surveys in technical areas.

BULGARIA

Chemical Pharmaceutical Research Institute * 6.5

Address: Kliment Ochridsky 1-a, 1156 Sofia, Bulgaria
Telephone: 72 39 52
Telex: 22653 NIHFI BG
Enquiries to: Director, I. Again; Head, Mila Draganova, Department of Scientific and Technical Information
Founded: 1959
Library facilities: Open to all users; loan services; reprographic services.
Information services: Bibliographic services; literature searches; access to on-line information retrieval systems.
The institute has access to the following on-line information retrieval systems: Biosis; Compendex; Inspec; AGRIS; Chemical abstracts.

CZECHOSLOVAKIA

Ceskoslovenská Společnost Chemická 6.6
[Czechoslovak Chemical Society]
Address: Hradčznské námesti 12, 118 29 Praha 1, Czechoslovakia
Telephone: 53 90 74
Enquiries to: Secretary, D V. Chvalovský

Founded: 1866

DENMARK

Århus Universitet, Kemisk Institut 6.7
[Århus University, Department of Chemistry]
Address: Langelandsgade 140, DK-8000 Århus C, Denmark
Telephone: (06) 124633
Enquiries to: Librarian, Helle Lawridsen
Founded: 1933
Subject coverage: Chemistry.
Library facilities: Open to all users (nc); loan services (nc); reprographic services (nc - only if original not on loan).
The library has microfiche reader-printer facilities.
Information services: Bibliographic services (nc); no translation services; literature searches (nc); access to on-line information retrieval systems (nc).
Consultancy: Telephone answers to technical queries (nc); no writing of technical reports; no compilation of statistical trade data; no market surveys in technical areas.

Danmarks Farmaceutiske Højskoles Bibliotek 6.8
[Royal Danish School of Pharmacy Library]
Address: Universitetsparken 2, DK-2100 København Ø, Denmark
Telephone: (01) 370850
Enquiries to: Librarian, Helga Thomsen
Founded: 1892
Subject coverage: Pharmacy, pharmacognosy, chemistry.
Library facilities: Open to all users (nc); loan services (nc); reprographic services (c).
Library holdings: 48 000 bound volumes; 366 current periodicals.
Information services: Bibliographic services (nc); no translation services; literature searches (nc); no access to on-line information retrieval systems.
Consultancy: No consultancy.

Kemikaliekontrollen * 6.9
[State Chemical Supervision Service]
Address: 26 Moerkhoej Bygade, DK-2860 Soeborg, Denmark
Telephone: (01) 697088
Affiliation: National Agency of Environmental Protection
Enquiries to: Director, H.H. Povlsen
Founded: 1933

Subject coverage: Control and supervision of pesticides, poisons and hazardous materials with regard to labelling, packaging and trade; quality control of pesticides by chemical and physical testing; chemical analysis of organic trace contaminants in drinking water; environmental pollution analysis; testing and development of methods of analysis for pesticides.
Library facilities: Not open to all users; no loan services; reprographic services (c).
Consultancy: Consultancy (c).

FINLAND

Suomen Farmaseuttinen Yhdistys ry/Farmaceutiska Föreningen i Finland rf *   6.10
[Society of Pharmaceutical Sciences in Finland]
Address: Farmasian laitos, Fabianinkatu 35, SF-00170 Helsinki, Finland
Telephone: (90) 1911
Enquiries to: Editor-in-chief, Anino Perälä-Suominen
Founded: 1889
Subject coverage: Promotion of the study of pharmaceutical sciences by the organization of lectures, by the publication of scientific journals, and by scholarships.
Library facilities: Open to all users; loan services (overnight); reprographic services (c).
Library holdings: *Acta Pharmaceutica Fennica* and all its exchange journals are kept in the library of Farmasian laitos (School of Pharmacy), University of Helsinki, which is open to all users.
Publications: *Acta Pharmaceutica Fennica* , in English, quarterly.

Teknillinen Korkeakoulu   6.11
[Helsinki University of Technology, Department of Chemistry]
Address: Kemistintie 1A, SF-02150 Espoo, Finland
Telephone: (90) 451 2743
Telex: 125161
Enquiries to: Librarian, Marjukka Patrakka
Subject coverage: Organic chemistry; biochemistry and food technology; physical chemistry; inorganic and analytical chemistry; chemical technology; chemical engineering.
Library facilities: Open to all users. Loan and reprographic services are available through the main library of Helsinki University of Technology.
Publications: List on request.

FRANCE

Institut des Corps Gras *   6.12
– ITERG
[Fats and Oils Research Institute]
Address: 10a rue de la Paix, 75002 Paris, France
Telephone: (1) 42 96 50 29
Telex: 230905 STABILI
Enquiries to: Director, Service de Documentation
Founded: 1943
Subject coverage: Fats and oils and their derivatives: production methods, properties, applications in foodstuffs (oils, proteins, margarine) and in chemicals industry (soaps, detergents, paints, cosmetics). Chemistry, biochemistry, catalysis and analysis of oil-based products.
Library facilities: Open to all users; no loan services; reprographic services (for exchange only).
Information services: For members only.
Publications: *Revue Française des Corps Gras,* monthly; list on request.

Institut National de Recherche Chimique Appliquée   6.13
– IRCHA
[National Institute of Applied Chemical Research]
Address: BP 1, 91710 Vert le Petit, France
Telephone: (1) 64 93 24 75
Telex: 600820F
Affiliation: Ministry of Industry
Enquiries to: External relations officer, Alain Prats
Founded: 1959
Subject coverage: Fine chemistry, biotechnology, new materials and environment. The institute undertakes research and development, and provides advice and technical assistance.
Library facilities: Open to all users (normally - nc); loan services (normally - nc); reprographic services (nc).
Library holdings: About 10 000 bound volumes; about 500 current periodicals.
Information services: Bibliographic services (c); no translation services; literature searches (c); access to on-line information retrieval systems.
Consultancy: Telephone answers to technical queries; no writing of technical reports; no compilation of statistical trade data; no market surveys in technical areas.
Courses: Training courses, held two or three times per annum, on the following topics: exotoxicology; good laboratory practice; French and foreign chemical regulations.

Société Française de Chimie * 6.14
[Chemical Society of France]
Address: 250 rue Saint-Jacques, 75005 Paris, France
Telephone: (1) 43 25 20 78
Affiliation: Fédération des Sociétés Chimiques Europénnes
Enquiries to: Executive secretary, G. Perreau
Founded: 1984
Subject coverage: Information and documentation on all aspects of chemical research.
Library facilities: No library facilities.

GERMAN DEMOCRATIC REPUBLIC

Chemische Gesellschaft der DDR 6.15
[Chemical Society of the GDR]
Address: Clara-Zetkin-Strasse 105, DDR-1080 Berlin, German Democratic Republic
Enquiries to: Secretary, Dr U. Klein

GERMAN FEDERAL REPUBLIC

Deutsche Gesellschaft für Chemisches Apparatewesen, Chemische Technik und Biotechnologic eV 6.16
– DECHEMA
Address: Postfach 970 146, D-6000 Frankfurt am Main 97, German Federal Republic
Telephone: (069) 7564 0
Telex: 412 490 dchad
Enquiries to: Head, Dr Reiner Eckermann. Information Systems and Databanks Department
Founded: 1926
Subject coverage: Information systems and services in all areas of chemical engineering and biotechnology.
Library facilities: Open to all users (nc); no loan services; no reprographic services.
Library holdings: 15 000 bound volumes; 250 current periodicals.
Information services: Bibliographic services (c); no translation services; literature searches (c); access to on-line information retrieval systems (c).
The centre maintains seven major on-line systems (accessible through DATEX-P, IPSS, Telenet, Tymnet, STN): DECHEMA (Dechema Chemical Engineering and Biotechnology Abstracts Data Bank); DETHERM-SDR (Dechema Thermophysical Property Data Bank - Data Retrieval System);

DETHERM-SDC (Dechema Thermophysical Property Data Bank - Data Calculation System); DETHERM-SDS (Dechema Thermophysical Property Data Bank - Data Synthesis System); DEQUIP (Dechema Equipment Suppliers Data Bank); DETEQ (Dechema Environmental Technology Equipment Data Bank); COALDATA (European Community Data Bank of Coal Data).
Dechema provides four information services in the field of chemical technology, chemical engineering, and biotechnology: Dechema Data Service; Dechema Chemical Engineering and Biotechnology Information Service; Dechema Corrosion Information Service; Dechema Chemical Equipment Suppliers Information Service.
Consultancy: No consultancy.
Publications: *DECHEMA Chemistry Data Series* , 2-3 volumes a year; *DECHEMA Erkstofftabelle* (Corrosion Data Sheets); *DECHEMA Monographien*, 1-2 volumes a year; *Chemical Technology and Biotechnology Abstracts*, monthly; list on request.

Fachinformationszentrum Chemie GmbH 6.17
– FIZ CHEMIE
[Chemical Information Centre Limited]
Address: Postfach 12 60 50, Steinplatz 2, D-1000 Berlin 12, German Federal Republic
Telephone: (030) 3190030
Telex: 181 255
Facsimile: (030) 3132037
Affiliation: The centre was set up jointly by the federal government, the city of Berlin, Gesellschaft Deutscher Chemiker (GDCh), DECHEMA, and Forschungsgesellschaft Kunststoffe.
Enquiries to: Scientific director, Dr Christian Weiske
Founded: 1981
Subject coverage: Chemistry and chemical engineering: theoretical and physical chemistry; analytical chemistry; inorganic chemistry; organic chemistry; macromolecular chemistry; biochemistry; biotechnology, food, agricultural and environmental chemistry; chemical technology; biotechnology.
Library facilities: Not open to all users; no loan services; reprographic services (c).
Library holdings: 5800 bound volumes; 220 current periodicals.
Information services: Bibliographic services (c); no translation services; literature searches (c); access to on-line information retrieval systems (c).
FIZ CHEMIE maintains the following data banks: CAS ONLINE (registry and abstracts file); CAS PROFIL; Chemical Industry Notes (CIN); Dechema Chemical Engineering Abstracts Data Bank (DECHEMA); Kunststoffe Kautschuk Fasern (KKF); Verfahrenstechnische Berichte (VtB); Dechema Thermophysical Property Data Bank (DETHERM);

DEQUIP; DETEQ. Profiles extracted from DECHEMA, KKF and DETHERM are produced in machine-readable, as well as printed form.

FIZ CHEMIE offers services such as SDI, retrospective literature searches, and taped information services, using information extracted from these data bases.

Consultancy: Telephone answers to technical queries (nc); no writing of technical reports; no compilation of statistical trade data; no market surveys in technical areas.

Courses: About sixty workshops and three seminars per annum on the following topics: CAS ONLINE; FIZ CHEMIE Data Bases; DETHERM.

Publications: Literature reference services: *Chemischer Informationsdienst*, weekly; *Literaturkurzberichte Chemische Technik und Biotechnologie* (Chemical Engineering and Biotechnology), monthly; *Literaturschnelldienst Kunststoffe Kautschuk Fasern*, monthly; card index of common and trade names of drugs, dyes and chemical substances; *Biotechnologie - Verfehren, Anlagen, Apparate*; primary journals in chemistry; catalogue of services.

Gmelin-Institut für anorganische Chemie und Grenzgebiete der Max-Planck-Gesellschaft 6.18
[Gmelin Institute of Inorganic Chemistry and Related Fields of the Max Planck Society]
Address: Varrentrappstrasse 40-42, Carl-Bosch-Haus, D-6000 Frankfurt am Main 90, German Federal Republic
Telephone: (0611) 79171
Telex: 4 12 526 gmeli
Enquiries to: Managing director, Professor Ekkehard Fluck
Subject coverage: Inorganic chemistry, physics, mineralogy etc. The Gmelin Handbook (full title given below), edited by the institute, is a documentation of the whole field of inorganic chemistry and its neighbouring fields unique in the world. The work, which at present comprises 502 volumes totalling more than 153 000 pages, has been continued in English alone since 1982.

Inorganic chemists, physicists, mineralogists and other specialized scientists screen the results of international research extracted by modern documentation techniques, evaluate them on the basis of the original documents in the light of the present state of the art and make them available in a broader context. By including direct references to the relevant literature, the handbook constitutes a reliable basis for literature searches.

The Gmelin Institute has a staff of 118 persons, of whom 73 are scientists (including the Chief Editors), plus some 140 external scientific staff members.

Publications: *Gmelin Handbuch der Anorganischen Chemie*.

Institut für Spektrochemie und angewandte Spektroskopie * 6.19
[Spectrochemistry and Applied Spectroscopy Institute]
Address: Bunsen-Kirchhoff-Strasse 11, D-4600 Dortmund 1, German Federal Republic
Telephone: (0231) 129001
Affiliation: Gesellschaft zur Förderung der Spektrochemie und angewandten Spektroskopie eV
Enquiries to: Director, Professor Günther Tölg
Founded: 1952
Subject coverage: Modern analytical chemistry with physical, especially spectroscopic, methods.
Library facilities: Open to all users; no loan services; no reprographic services.
Information services: No information services.
Consultancy: Yes (c).

Pharma-Dokumentations-Service 6.20
– PDS
[Pharma-Documentation Service]
Address: Karlstrasse 21, D-6000 Frankfurt am Main 1, German Federal Republic
Telephone: (069) 2556 268
Telex: 04 12718bpi d
Affiliation: Chemie Wirtschaftsfoerderungs-GmbH
Enquiries to: Information searcher, G. Stanzel
Founded: 1977
Subject coverage: PDS was established to provide customers with on-line search facilities of international data bases, library brokerage service and information management consultancy.
Library facilities: Open to all users (c); no loan services; reprographic services (c).

PDS provides customers with copies of literature from its own and external libraries.

Information services: Bibliographic services (c); no translation services; literature searches (c); access to on-line information retrieval systems (c).

PDS has access to the following: Dialog; DIMDI; European Space Agency (ESA); Fachinformationszentrum Technik (FIZ 16); INKA; SDC.

Consultancy: Telephone answers to technical queries (nc); no writing of technical reports; no compilation of statistical trade data; no market surveys in technical areas.

HUNGARY

Magyar Ásványolaj és Földgáz Kisérleti Intézet *   6.21
– MÁFKI

[Hungarian Oil and Gas Research Institute]

Address: PO Box 167, József Attila útca 34, H-8201 Veszprém, Hungary

Telephone: (080) 12-440

Telex: 32288 mafki-h

Enquiries to: Chief librarian, Dr Csiszár Miklós

Founded: 1948

Subject coverage: Crude oil and natural gas chemistry; petrochemicals.

Library facilities: Not open to all users; loan services; reprographic services.

Information services: No information services.

Consultancy: No consultancy.

Magyar Kémikusok Egyesülete *   6.22
– MKE

[Hungarian Chemical Society]

Address: Anker köz 1, H-1061 Budapest, Hungary

Telephone: (01) 427 343

Telex: 22-5369 mtesz h

Affiliation: Federation of Technical and Scientific Societies (MTESZ)

Enquiries to: Deputy secretary-general, Dr Áron Jakabos

Founded: 1907

Subject coverage: Chemical science and chemical teaching; development of the chemical industry and other industries, applying chemical processes; raising the professional level of experts working in these fields.

Library facilities: No library facilities.

Information services: No information services.

Publications: Journal of the Hungarian Chemical Society; Hungarian Journal of Chemistry; Journal of Secondary School Chemistry.

Magyar Tudományos Akadémia, Bányászati Kémiai Kutatólaboratóriuma   6.23
– MTA BKKL

[Hungarian Academy of Sciences, Chemical Research Laboratory for Mining]

Address: PO Box 2, H-3515 Miskolc-Egyetemváros, Hungary

Telephone: (46) 67 211

Telex: 62-421 mtaok h

Enquiries to: Director, Dr József Tóth

Founded: 1957

Subject coverage: Enhanced oil recovery; reservoir engineering; mineral processing; physical and analytical chemistry of dispersed systems; development and production of relevant instruments.

Library facilities: Open to all users (nc); loan services (c); reprographic services (c).

Library holdings: 5000 bound volumes; 30 current periodicals.

Information services: No information services.

Consultancy: No consultancy.

IRELAND

Pharmaceutical Society of Ireland *   6.24
Address: 37 Northumberland Road, Dublin 4, Ireland

Telephone: (01) 600699

Enquiries to: Assistant registrar

Founded: 1875

Subject coverage: Information on: pharmacy law and practice; medicines.

Library facilities: Not open to all users (reading room only by arrangement); no loan services; reprographic services (c).

Information services: No information services.

Consultancy: No consultancy.

Publications: Irish Pharmacy Journal; Annual Calendar and Register.

ITALY

Associazione Nazionale dell'Industria Chimica *   6.25
– ASCHIMICI

[National Association of the Chemical Industry]

Address: Via Fatebenefratelli 10, 20121 Milano, Italy

Telephone: (02) 63621

Telex: 332488 Aschim I

Founded: 1946

Subject coverage: ASCHIMI, representing the Italian chemicals industry, considers economic problems, industrial relations problems and chemistry research policy.

Publications: Compendio Statistico, statistical analysis, with international comparisons, of the industry; Industria Chimica, bimonthly economic review; Quarterly Chemical Trends; various monographs.

NETHERLANDS

Instituut voor Toegepaste Chemie TNO   6.26

– ITC-TNO

[Institute of Applied Chemistry TNO]

Address: Postbus 5009, 3502 JA Utrecht, Netherlands

Telephone: (030) 882721

Telex: 40022 CIVO-Zeist

Enquiries to: Librarian, Tiny van Dam

Founded: 1948

Subject coverage: Biotechnology; microbiology; biochemistry; organometallic chemistry; polymer chemistry; process technology; crystallization; solar energy; organic syntheses; coordination chemistry; analytical chemistry; chemotherapy; electrochemistry; physical chemistry.

Library facilities: Open to all users (normally no charge); loan services (nc); reprographic services (sometimes no charge).

Library holdings: 7000 bound volumes; 110 current periodicals; 50 reports.

Information services: Bibliographic services (occasional charge); no translation services; literature searches (c); access to on-line information retrieval systems (c).

Access to on-line information retrieval systems is provided not by this institute but by CID-TNO, Postbus 36, 2600 AA Delft.

Consultancy: Telephone answers to technical queries (nc); no compilation of statistical trade data; no market surveys in technical areas.

The institute sometimes undertakes the writing of technical reports; a charge is made for this service.

Koninklijke Nederlandse Maatschappij ter Bevordering der Pharmacie *   6.27

– KNMP

[Royal Dutch Society for the Advancement of Pharmacy]

Address: Alexanderstraat 11, Postbus 30460, 2500 GL 's-Gravenhage, Netherlands

Telephone: (070) 655922

Enquiries to: Head, P. de Smet.

Documentation and Information Service

Founded: 1842

Subject coverage: Pharmaceutical sciences and matters of concern to professional pharmacists.

Library facilities: Not open to all users; no loan services; reprographic services (c).

Library facilities are open to members only.

Information services: Bibliographic services (nc); no translation services; literature searches (nc); access to on-line information retrieval systems (nc).

Questions from members only are answered in detail; occasional questions from non-members are answered in less detail, and a charge is made. On-line access is available to: Medline; Excerpta Medica; Toxline (via DIMDI).

Consultancy: Yes (nc).

Publications: *Informatorium Medicamentorum* , annual textbook on drug therapy.

Technische Hogeschool Delft   6.28

[Delft University of Technology, Department of Chemical Technology]

Address: Julianalaan 136, 2628 BL Delft, Netherlands

Library facilities: Library.

NORWAY

Norsk Kjemisk Selskap *   6.29

– NKS

[Norwegian Chemical Society]

Address: Postboks 1107, Blinden, N-0317 Oslo 3, Norway

Telephone: (02) 45 56 94

Enquiries to: Secretary general, Berit F. Pedersen

Founded: 1893

Library facilities: No library facilities.

Information services: No information services.

POLAND

Instytut Chemii Nieorganicznej *   6.30

– IChN

[Institute of Inorganic Chemistry]

Address: Ulica Sowinński 11, 44-101 Gliwice, Poland

Telephone: 31 30 51/59

Telex: 036132; 036133

Enquiries to: Director, Dr Jan Dubik.

Scientific and Economic Information Centre

Founded: 1948

Subject coverage: Inorganic chemistry; industrial applications and fundamental research.

Library facilities: Open to all users; loan services; reprographic services.

National and international interlibrary loan service; reprographic services of material held by the library and from other Polish sources.

Information services: Bibliographic services; literature searches.

Literature services are available in the following areas: chemical technology of inorganic products, especially ash soda, sulphuric acid, phosphoric acid and phosphate fertilizer industries, titanium dioxide, chromium, silicium, fluorine compounds and other inorganic salts.

Publications: *Biuletyn Informacyjny Instytutu Chemii Nieorganicznej* (Information Bulletin of the Institute of Inorganic Chemistry), bimonthly; *Biezaca Informacja Chemiczna - Seria Nieorganika* (Current Chemical Information - Inorganic Chemistry Series), monthly.

Instytut Chemii Przemysłowej   6.31
– IChP
[Industrial Chemistry Research Institute]
Address: Rydygiera 8, 01-793 Warszawa, Poland
Telephone: 38 80 21; 38 90 81
Telex: IChP 813586 Pl
Affiliation: Ministry of Chemical and Light Industries
Enquiries to: Manager, Scientific-Technical Library. Scientific, Technical and Economic Information Department
Founded: 1922
Subject coverage: Carbon- and petrochemistry; heavy organic synthesis; light organic synthesis; catalysts; polymer synthesis (also adhesives and resins for paints and lacquers); processing of plastics; household chemistry products; chemical engineering and design; corrosion prevention.
Library facilities: Open to all users (nc); loan services (nc); reprographic services (c).
Reprographic and microreading facilities; central information service.
Library holdings: 46 350 bound volumes; 777 current periodicals.
Information services: Bibliographic services (c); no translation services; literature searches (c); no access to on-line information retrieval systems.
Topical information base for plastics, comprising the abstracts of publications from all over the world in the years 1961-1985.
Consultancy: Telephone answers to technical queries (nc); writing of technical reports (nc); compilation of statistical trade data (nc); market surveys in technical areas (nc).
Publications: *Biezaca Informacja Chemiczna, serie: Tworzywa Sztuczne i Kauczuki Syntetyczne; Przetwórstwo Tworzyw Sztucznych* (Current Chemical Information, series: Plastics and Synthetic Rubber; Plastics Processing), monthly journals.

Instytut Przemysłlu Organicznego *   6.32
[Industrial Organic Chemistry Institute]
Address: 6 Ulica Annopol, 03-236 Warszawa, Poland
Telephone: 11 12 31
Telex: 814844 IPO PL
Affiliation: Ministerstwo Przemystu Chemicznego; Lekkiego.
Biblioteka Naukowo - Techniczna
Enquiries to: Director, Dr Wiesłław Moszczyński
Founded: 1952
Subject coverage: Chemistry, biology, physics; main subject: pesticides, auxiliaries, chemical security.
Library facilities: Open to all users (nc); loan services (nc); reprographic services (nc).
Interlibrary loans available within Poland; reprographic services with regard to material from all Polish sources, at request of foreign institutions.
Library holdings: 27 864 bound volumes; 342 current periodicals.
Information services: Bibliographic services (c); no translation services; literature searches (c); no access to on-line information retrieval systems.
Consultancy: Consultancy (nc).
Publications: *Biezaca Informacja Chemiczna, Series Pestycydy* (Current Chemical Information, Pesticides series), monthly.

Instytut Włókien Chemicznych *   6.33
[Chemical Fibres Institute]
Address: Ulica Skłodowska-Curie 19-27, 90-570 Łódź, Poland
Telephone: (Łódź) 339 510
Telex: Pl IWCh 886549
Enquiries to: Director, Dr Henryk Pstrocki
Founded: 1952
Publications: *Biezaca Informacja Chemiczna, seria: Włókna Chemiczne* (Current Chemical Information, series: Chemical Fibres), monthly.

PORTUGAL

Quimigal-Quimica de Portugal EP Centro de Documentacão *   6.34
Address: Avenida 24 Julho 170, 1300 Lisboa, Portugal
Telephone: 609061
Enquiries to: Head, Maria Oclete Barros Heurigues. Documentation Centre
Subject coverage: Chemistry of the following products: nitric acid, fertilizers, pesticides, copper metallurgy, sulphuric acid, precious metals, pyrites, urea, textiles, glass fibre, polymers, soap, edible products.

Library facilities: Open to all users; loan services; reprographic services (c).
Information services: Bibliographic services (nc); translation services; literature searches (c); access to on-line information retrieval systems (c).
The centre provides SDI services and reference service, and has access to the following: Dialog; ESA; Questel; CISI; GCAM; Pergamon InfoLine; Euris; DIMDI; Datastar; Echo.
Consultancy: No consultancy.

Sociedade Farmacêutica Lusitana *   6.35
[National Pharmaceutical Society]
Address: Rua da Sociedade Farmacêutica 18, 1199 Lisboa Codex, Portugal
Telephone: 41424
Enquiries to: Director, Professor Dr Alfredo Albuquerque.
Drug Information Centre
Founded: 1835
Subject coverage: The society represents all Portuguese pharmacists in the following areas: pharmaceutical industries; community pharmacy; hospital pharmacy; clinical chemistry.
Library facilities: Open to all users.
Library holdings: 4800 volumes.
Information services: Advice given to members.
Publications: *Revista Portuguesa de Farmácia?R,* quarterly; *Boletim Informativo* , monthly.

ROMANIA

Institutul de Medicină Si- Farmacie, Biblioteca Centrală   6.36
– IMFBC
[Medical and Pharmaceutical Institute, Central Library]
Address: Bd Dr Petru Groza 8, 76241 Bucureşti, Romania
Telephone: 49 30 30
Affiliation: Ministry of Education
Enquiries to: Director, Petre Silvică
Founded: 1857
Subject coverage: Human medicine, including pharmacy and stomatology.
Library facilities: Not open to all users; loan services; reprographic services.
The library is open only to students, professional staff, and specialist physicians.
Library holdings: 1 029 887 bound volumes; 9600 current periodicals.
Information services: Bibliographic services (c); no translation services.

Consultancy: Telephone answers to technical queries; no writing of technical reports; no compilation of statistical trade data; no market surveys in technical areas.

Oficiul de Informare Documentară pentru Industria Chimică   6.37
– OIDICh
[Chemical Industry Documentary Information Office]
Address: Calea Plevnei 139, 77131 Bucureşti 12, Romania
Telephone: (90) 492108
Telex: 10944 ichim
Affiliation: CCCSAEM (Research Centre for Quality, Standards, Economics and Marketing)
Enquiries to: Manager
Founded: 1956
Library facilities: Open to all users; loan services; reprographic services.
Library holdings: 80 000 volumes.
Information services: Bibliographic services; translation services; literature searches; access to on-line information retrieval systems.
Consultancy: Yes.
Publications: *Revista de chimie; Revista de materiale plastice.*

SPAIN

Asociación Nacional de Químicos de España *   6.38
[National Association of Chemists]
Address: Lagasca 83, 1, Madrid 6, Spain
Enquiries to: Secretary, Joaquín Copado López

SWEDEN

Biomedicinska Centrum   6.39
– BIOMEDICUM
[Biomedical Centre]
Address: Box 570, S-751 23 Uppsala, Sweden
Telephone: 018-17 40 00
Telex: Sweden 76132
Facsimile: 018-15 17 59
Affiliation: Uppsala University
Enquiries to: Librarian, Per Syrén

Founded: 1968
Subject coverage: Biosciences; chemistry; pharmacy.
Library facilities: Open to all users; loan services; reprographic services (c).
Library holdings: over 2500 bound volumes; 650 current periodicals.
Information services: Bibliographic services (nc); no translation services; literature searches (c); access to on-line information retrieval systems (c).
Medlars, ESA-RECON, Dialog, Orbit.
Consultancy: On-the-phone answers to queries - nc.

Ytkemiska Institutet 6.40
[Surface Chemistry Institute]
Address: PO Box 5607, S-114 86 Stockholm, Sweden
Telephone: (08) 22 25 40
Telex: 14375 STURES S
Facsimile: (08) 20 00 63
Affiliation: Swedish Board for Technical Development; Association for Surface Chemistry Research
Enquiries to: Director, Professor Per Stenius; Librarian, Elisabeth Malmberg
Founded: 1969
Subject coverage: Colloid and surface chemistry: pharmaceuticals and cosmetics; surface treatment; chemical technology; polymers; foam research; forest product research, food research, separation technology, minerals, biotechnology.
Library facilities: Available only to members of the Association for Surface Chemistry Research.
Library holdings: 2000 bound volumes; 80 current periodicals; 700 reports.
Information services: Bibliographic services (nc); no translation services; literature searches (c); access to on-line information retrieval systems (c).
Available only to members of the Association for Surface Chemistry Research.
Consultancy: Telephone answers to technical queries (c); writing of technical reports (c).
Courses: Between one and three courses per annum on industrial surface chemistry.

SWITZERLAND

Schweizerische Chemische 6.41
Gesellschaft Société Suisse de
Chimie *
[Swiss Chemistry Society]
Address: Postfach, CH-4002 Basel, Switzerland
Enquiries to: Secretary, Dr E. Sundt

TURKEY

Türkiye Kimya Derneği 6.42
[Chemical Society of Turkey]
Address: PK 829, Harbiye, Halaskârgazi Caddessi 53, Uzay Apt D8, Istanbul, Turkey
Telephone: 1407331
Enquiries to: President, Professor Ali Riza Berkem

UNITED KINGDOM

Biotechnology Centre, Wales 6.43
– BTCW
Address: Singleton Park, Swansea SA2 8PP, UK
Telephone: (0792) 296396
Telex: 48358 ULSWAN-G
Enquiries to: Director, Dr R.N. Greenshields
Founded: 1983
Subject coverage: Biotechnology, with application in the food industry.
Library facilities: No library facilities.
Information services: No information services.
Consultancy: Telephone answers to technical queries (c); writing of technical reports (c); no compilation of statistical trade data; market surveys in technical areas (c).
Biotechnological services (c).
Publications: *International Industrial Biotechnology*, bimonthly bulletin.

British Industrial Biological 6.44
Research Association
– BIBRA
Address: Woodmansterne Road, Carshalton, Surrey SM5 4DS, UK
Telephone: (01) 643 4411
Telex: 25438
Enquiries to: Information Officer.
Information and Advisory Section
Founded: 1961
Subject coverage: Toxicology.
Library facilities: Not open to all users; loan services (nc); reprographic services (normally no charge).
Available to members only.
Library holdings: The library consists of a comprehensive collection of scientific papers on the toxicology of chemicals (well in excess of 100 000 documents) and related worldwide legislation. It holds most of the toxicological journals and there is a small special library of books on toxicology.

Information services: Bibliographic services (available only to members - usually no charge); translation services (translations of German food packaging recommendations and other foreign legislation - nc); literature searches (available only to members - usually no charge); access to on-line information retrieval systems (available only to members: Blaise, Datastar, Dialog, DIMDI, Pergamon InfoLine, ESA-IRS - usually no charge).

Ready prepared authoritative up-to-date summaries of the most relevant data on industrial and other chemicals (updated as new significant information becomes available - c).

Consultancy: Telephone answers to technical queries (nc); writing of technical reports (often nc); no compilation of statistical trade data; no market surveys in technical areas.

Information and advice given to member companies on the toxicology, potential hazards, and safe handling and use of chemicals, including food additives, food contaminants such as packaging migrants, heavy metals, mycotoxins and agricultural chemicals, cosmetics and toiletry ingredients, drug excipients, surgical products, tobacco additives, industrial chemicals and water, air and soil pollutants. International legislation governing these topics is also covered. In addition advice is provided on the design and interpretation of toxicological tests, and on the need for further specific tests on any chemicals.

Written answers are given to technical queries on safety-in-use including assessment of the acceptability of food packaging materials in specific applications.

Publications: *BIBRA Information Bulletin* (10 times a year to members); *Food and Chemical Toxicology.*

Cambridge Crystallographic Data Centre *   6.45

Address: University Chemical Laboratory, Lensfield Road, Cambridge CB2 1EW, UK
Telephone: (0223) 66499
Telex: 81240 CAMSPLG
Affiliation: University of Cambridge
Enquiries to: Director, Dr Olga Kennard
Founded: 1965
Subject coverage: Molecular and crystal structures of organic compounds, organometallic and organo-metal complexes.
Library facilities: Not open to all users (only to users visiting in person); no loan services; no reprographic services (except for very rare reprints).
Information services: Bibliographic services; access to on-line information retrieval systems.
Bibliographic services - for users in the UK only; users in other countries are served by National Data Centres, affiliated to Cambridge. For a list of these centres, see reference *Acta Cryst.* B35, 2331-2339, 1979. Access to on-line information retrieval systems

- for users in countries with National Affiliated Centres only; CIS system (worldwide), CSSR System (UK), SCANDNET (Scandinavia), DARC (France).

The data base comprises evaluated numeric data on 53 000 structures (January 1986) with an annual growth rate of 6000. It is primarily a research tool in chemistry, crystallography, pharmacology and molecular biology.

Programs are available for substructure searches, for geometrical and statistical analysis of the retrieved numeric data and for graphic display of the results, including three-dimensional displays of molecular and crystal structures.

Consultancy: Yes (c).

Institution of Chemical Engineers   6.46
– IChemE

Address: Geo. E. Davis Building, 165-171 Railway Terrace, Rugby CV21 3HQ, UK
Telephone: (0788) 78214
Telex: 311780
Enquiries to: Information Officer, A. Strauch
Founded: 1923
Subject coverage: Chemical engineering: safety and loss prevention; environmental science; energy conservation; physical properties data service.
Library facilities: Open to all users (nc); loan services (charge for postage only); reprographic services (c).
Library holdings: 6000 bound volumes; 100 current periodicals.
Information services: Bibliographic services (charge for non-members); literature searches (charge for non-members); no access to on-line information retrieval systems.

On-the-spot translation assistance is given for Dutch and German.

Literature searches are made using the in-house data base, which contains 32 000 abstracts, input at a rate of about 4000 a year. The library also acts as a clearing house and referral service, and keeps a list of consultant chemical engineers, and a guide to manufacturers and suppliers of chemical plant.

Consultancy: Telephone answers to technical queries (no charge for easily answered queries).

Compilation of a quarterly list of conferences and courses in subjects of interest to chemical engineers.

Publications: Publications list available on request.

Laboratory of the Government Chemist   6.47
– LGC

Address: Cornwall House, Waterloo Road, London SE1 8XY, UK
Telephone: (01) 211 7900
Affiliation: Department of Trade and Industry
Enquiries to: Librarian, P.W. Hammond
Founded: 1842

Subject coverage: Analytical chemistry; research and development in chemistry - related topics and biotechnology.

Library facilities: Open to all users (nc - by appointment only).

Loan services and reprographic services are available only to other government departments; there is no charge for these services.

Library holdings: 10 000 bound volumes; 520 current periodicals.

Information services: No bibliographic services; no translation services; literature searches (c); no access to on-line information retrieval systems.

The laboratory provides the following information services: AMAIS (Agricultural Materials Analysis Information Service); CHAIS (Consumer Hazards Analysis Information Service); CNAS (Chemical Nomenclature Advisory Service); DAGAS (Dangerous Goods Advisory Service); MiCIS (Microbial Culture Information Service).

Consultancy: Telephone answers to technical queries (no charge if little work is involved); no writing of technical reports; no compilation of statistical trade data; no market surveys in technical areas.

Publications: *Report of the Government Chemist* .

Mass Spectrometry Data Centre 6.48
– MSDC

Address: Royal Society of Chemistry, The University, Nottingham NG7 2RD, UK

Telephone: (0602) 507411

Telex: 37488

Affiliation: Royal Society of Chemistry

Enquiries to: Manager, S. Down

Founded: 1966

Subject coverage: Mass spectrometry - bibliography and data.

Library facilities: Reprographic services (photocopy service - c).

Information services: Bibliographic services (c); no translation services; no literature searches; access to on-line information retrieval systems (c).

The MSDC publishes the *Mass Spectrometry Bulletin*, produced from 400 primary scientific journals every month; subscribers are also offered the MSDC photocopy service for copies of original documents, and access to the MSB data base on-line through Pergamon InfoLine. The data base contains about 150 000 references, with 800 new references added every month.

Access is also available to NIH-EPA-CIS (USA), a search system for mass spectrometry, NMR, crystal structure and toxicity.

Consultancy: Telephone answers to technical queries (nc); no writing of technical reports; no compilation of statistical trade data; no market surveys in technical areas.

Publications: *Mass Spectrometry Bulletin*, monthly; *Eight Peak Index of Mass Spectra*, containing 66 000 mass spectra profiles, giving CAS registry numbers, indexed by molecular weight, formula and ion m/z values (spectra are gathered from the following sources: ICI, Thermodynamics Research Centre (TRC), American Society for Testing and Materials (ASTM), Dow Chemical Company, John Wiley and Sons Inc, NIH/EPA); *Mass Spectral Data sheets*, giving 7000 profiles.

30 000 profiles are available on tape as the *Mass Spectra Tape* collection; *Mass Spectrometry Bulletin* and *Eight Peak Index of Mass Spectra* are also available on tape.

Particle Science and Technology 6.49
Information Service *
– PSTIS

Address: Department of Chemical Engineering, University of Technology, Loughborough, Leicestershire LE11 3TU, UK

Telephone: (0509) 263171

Telex: 34319

Affiliation: University of Technology, Loughborough

Enquiries to: Information Officer, R.W. Newbold

Founded: 1967

Subject coverage: Particle science; particle technology; powders; aerosols; colloids; emulsions; bulk solids; air pollution; fluidization; water treatment; filtration; effluent treatment; mineral dressing; chemical engineering; industrial pharmacy.

Library facilities: Open to all users; loan services (c); reprographic services (c).

Library holdings: The library holds over 80 000 documents collected since 1969.

Information services: Bibliographic services (c); no translation services (but informal help given); literature searches (c); access to on-line information retrieval systems (c).

PSTIS maintains a current awareness service, covering over 800 journals; information is stored in the PSTIS data base, which has subject and author indexes, and can be consulted on request.

Consultancy: Yes (c).

Publications: *Current Awareness in Particle Technology*.

Pharmaceutical Society of Great 6.50
Britain
– PSGB

Address: 1 Lambeth High Street, London SE1 7JN, UK

Telephone: (01) 735 9141

Enquiries to: Information pharmacist; Head, P.M. North.

Library and Technical Information Services

Founded: 1841
Subject coverage: Pharmacy and allied subjects.
Library facilities: Not open to all users; no loan services; reprographic services.
The library is open to bona fide members only.
Information services: Information services are available to members only.
The society maintains Martindale On-line, a data bank produced by PSGB on commercial services.

Sheffield Assay Office 6.51
Address: PO Box 187, 137 Portobello Street, Sheffield S1 4DS, UK
Telephone: (0742) 755111
Telex: 54659
Enquiries to: The Assay Master
Founded: 1773
Subject coverage: Hallmarking of platinum, gold and silver; assaying of all precious metal alloys.
Library facilities: Not open to all users (approved users only - nc); no loan services; reprographic services (c).
Library holdings: 1600 bound volumes; 16 current periodicals.
Information services: No bibliographic services; no translation services; literature searches (in certain cases); no access to on-line information retrieval systems.
Consultancy: Telephone answers to technical queries (straightforward enquiries only - nc); writing of technical reports (c); no compilation of statistical trade data; no market surveys in technical areas.
Courses: Specialist courses arranged individually by agreement.

YUGOSLAVIA

Serbian Chemical Society * 6.52
Address: PO Box 462, Karnegijeva 4, 11000 Beograd, Yugoslavia
Enquiries to: Secretary, Professor Branislav Nikolić

7 EARTH SCIENCES

AUSTRIA

Bundesamt für Eich- und Vermessungswesen *

7.1

– BEV

[Federal Office of Metrology and Surveying]

Address: Schiffamtsgasse 1 - 3, A-1025 Wien, Austria
Telephone: (0222) 35 76 11
Telex: 115468
Affiliation: Federal Ministry for Construction and Technology
Enquiries to: President, Dr Friedrich Rotter
Founded: 1923
Subject coverage: Mechanics; frequency; heat; acoustics; radiation; electricity; magnetism; basic surveying; boundary cadastre; topographic surveying; surveying of federal boundaries.
Library facilities: Not open to all users.

Bundesversuchs- und Forschungsanstalt Arsenal/ Geotechnisches Institut *

7.2

– BVFA-Arsenal

[Federal Testing and Research Establishment Arsenal, Geotechnical Institute]

Address: Franz Grill-Strasse 9, POB 8, A-1031 Wien, Austria
Telephone: (0222) 78 25 31
Telex: 136677
Affiliation: Ministry for Construction and Technology
Enquiries to: Head, Professor Dr E. Schroll. Dokumentationsstelle für Strassenwesen und Verkehrstechnik
Founded: 1950
Subject coverage: Geochemistry; applied mineralogy and petrology; geophysics; hydrogeology and applied geology; soil mechanics; road construction techniques; traffic engineering; transport sciences; documentation and information services.

Library facilities: Not open to all users; no loan services; reprographic services (c).
Reading room copies only.
Information services: Bibliographic services (c); translation services (c); literature searches (c); access to on-line information retrieval systems (c).
IRRD - International Road Research Documentation of OECD, Paris; ICTED - International Cooperation in the Field of Transport Economics of ECMT, Paris; IRB - Informationsverbundzentrum Raum und Bau, Stuttgart, Federal Republic of Germany.
Consultancy: Yes (c).
Publications: Bibliographic list.

Geologische Bundesanstalt

7.3

– GBA

[Geological Survey of Austria]

Address: POB 154, Rasumofskygasse 23, A-1031 Wien, Austria
Telephone: (0222) 72 56 74
Telex: 13 29 27
Affiliation: Bundesministerium für Wissenschaft und Forschung
Enquiries to: Director, Dr Traugott E. Gattinger. Bibliothek
Founded: 1849
Subject coverage: Earth sciences; geological mapping; mineral research; applied geosciences; applied geophysics; geochemistry; palaeontology; environmental geology; geotechnics.
Library facilities: Open to all users (nc); loan services (nc); reprographic services (c).
Library holdings: 212 396 volumes; including journals; 938 current periodicals.
Information services: Bibliographic services (c); no translation services; literature searches; access to on-line information retrieval systems (c).
Consultancy: No consultancy.

Naturhistorisches Museum Wien *   7.4
[Natural History Museum of Vienna]
Address: Burgring 7, Postfach 417, A-1014 Wien, Austria
Telephone: (0222) 93 45 41
Affiliation: Ministry of Science and Research
Enquiries to: Head librarian, A.-Ch. Hilgers
Founded: 1748
Subject coverage: Anthropology; botany; geology; mineralogy; palaeontology; prehistory; zoology.
Library facilities: Open to all users; loan services (nc); reprographic services (c).
Information services: Bibliographic services (nc); translation services (nc); literature searches (nc); no access to on-line information retrieval systems. Zoological record.
Consultancy: Yes (nc - scientific staff of the museum may be consulted).

Österreichische Gesellschaft für   7.5
Erdölwissenschaften
– ÖGEW
[Austrian Association for Petroleum Sciences]
Address: PO Box 309, Erdbergstrasse 72, A-1031 Wien, Austria
Telephone: (0222) 73 23 48
Telex: 132138
Enquiries to: Secretary, Dr Herbert Lang
Founded: 1960
Subject coverage: Petroleum sciences.
Library facilities: No library facilities.
Information services: No information services.

Zentralanstalt für Meteorologie   7.6
und Geodynamik *
[Central Meteorology and Geodynamics Establishment]
Address: Hohe Warte 38, A-1190 Wien, Austria
Telephone: (0222) 36 56 70
Telex: 131837
Affiliation: Bundesministerium für Wissenschaft und Forschung
Enquiries to: Publications chief, Dr Alois Machalek
Founded: 1851
Subject coverage: Meteorology; climatology; environmental meteorology; meteorological observations; network and instrumentation; radiosonde; geophysics and seismology.
Library facilities: Open to all users (limited); loan services (nc - within Austria only); reprographic services (office only).
Information services: Bibliographic services (limited); no translation services; literature searches (limited); no access to on-line information retrieval systems.
Consultancy: Yes (nc - limited).
Publications: Yearbook; *Erdmagnetische Berichte*, annually; *Ergebnisse von Strahlungsmessungen in Österreich*, annually; *Monatsübersicht der Witterung in Österreich, Tägliche Beobchtungen, Messungen der*

Radioaktivität der Luft in Bodennähe, all monthly; *Tägliche synoptische Wetterkarte*, daily; *Aerologische Berichte*, quarterly; bibliographic list.

BELGIUM

Centre Belge d'Étude et de   7.7
Documentation des Eaux
– CEBEDEAU
[Belgian Study and Documentation Centre for Water]
Address: 2 Rue Armand Stevart, B-4000 Liège, Belgium

Institut National des Industries   7.8
Extractives
– INIEX
[National Institute Extractive Industries]
Address: Rue du Chéra 200, B-4000 Liège, Belgium
Telephone: (41) 52 71 50
Telex: 41128 B
Affiliation Ministry of Economic Affairs
Enquiries to: Public relations, Loes van Mechelen
Founded: 1967
Subject coverage: Mines; quarries; energy; safety; radiocommunications; pollution; polymers; concrete.
Library facilities: Open to all users (nc); loan services (nc); reprographic services (c).
Library holdings: 8000 bound volumes; 300 current periodicals; 6000 reports.
Information services: Bibliographic services (nc); no translation services; literature searches (nc); no access to on-line information retrieval systems.
Consultancy: Telephone answers to technical queries; no writing of technical reports; no compilation of statistical trade data; no market surveys in technical areas.
Publications: Annual report.

Institut Royal Météorologique de   7.9
Belgique
– IRM
[Royal Meteorological Institute of Belgium]
Address: Avenue Circulaire 3, B-1180 Bruxelles, Belgium
Telephone: (02) 375 24 78
Telex: 21315
Affiliation: Ministère de l'Éducation Nationale
Enquiries to: Librarian, Mr Dale. Bibliothèque
Founded: 1913
Subject coverage: Meteorology; internal and external geophysics; climatology; hydrology; physical oceanography; atmospheric physics.

Library facilities: Not open to all users; loan services (nc); reprographic services (c).

The loan of books and periodicals is authorized to any researcher after a written request mentioning the aim of the research has been accepted by the director of the Royal Observatory of Belgium and after payment of a deposit.

The catalogue of the library periodicals is available (c).

Library holdings: 125 000 bibliographical units; 4226 periodicals.

Information services: Bibliographic services (nc); no translation services; no literature searches; no access to on-line information retrieval systems.

Services are obtainable by written request to the director of the institute.

Consultancy: Consultancy services are available by written request to the director of the institute (nc).

International Centre for Earth Tides   7.10

Address: c/o Observatoire Royal de Belgique, Avenue Circulaire 3, B-1180 Bruxelles, Belgium
Telephone: (02) 375 24 84
Telex: 21565
Affiliation: Federation of Astronomical and Geophysical Services
Enquiries to: Director, Professor Paul Melchior
Founded: 1957
Subject coverage: Problems related to tides - fundamental astronomy; geophysics; geodynamics.
Library facilities: Reprographic services (c).
Information services: Bibliographic services (c); translation services (c).

The centre has a data bank covering earth tidal measurements throughout the world (on magnetic tape or listings - occasional small charge).

Consultancy: Writing of technical reports (c).

Service Géologique de Belgique   7.11
[Belgian Geological Service]
Address: 13 Rue Jenner, Parc Léopold, Bruxelles 4, Belgium

BULGARIA

Bulgarian Geological Society   7.12
Address: Ruski 15, 1000 Sofia, Bulgaria
Telephone: 85 81 295
Enquiries to: Secretary, Z. Iliev
Library holdings: 16 000 volumes.

CYPRUS

Cyprus Geographical Association   7.13
Address: P O Box 3656, Nicosia, Cyprus
Enquiries to: General secretary, Y. Koymides
Library holdings: 500 volumes.

Geological Survey Department   7.14
Address: Nicosia Cyprus
Telephone: (021) 402338
Telex: 4660 MINAGRI CY
Affiliation: Ministry of Agriculture and Natural Resources
Enquiries to: Director
Founded: 1955
Subject coverage: Geological mapping; mineral and water exploration; geotechnical investigations; geophysical and geochemical surveys.
Library facilities: Open to all users (for reference by researchers in earth sciences).
Publications: List on request.

CZECHOSLOVAKIA

Knihovna Ústředního Ústavu Geologického   7.15
[Geological Institute Library]
Address: 19 Malostranské námesti, 118 21 Praha 1, Czechoslovakia
Enquiries to: Librarian, Marie Mezerová
Library holdings: 125 000 volumes.

Ústředního Ústav Geologický   7.16
[Geological Survey of Prague]
Address: Malostranské námesti 19, 118 21 Praha 1, Czechoslovakia
Telephone: (02) 533641-9
Telex: 122540 uug c
Enquiries to: Director, Dr Jaroslav Vacek
Subject coverage: Geological mapping; mineral resource prospecting; ore, non-metallic, and oil deposits; geochemistry; hydrogeology.
Publications: *Journal of Geological Sciences;* bulletin; geological and mineral deposit maps; monographs; bibliographies.

DENMARK

Danmarks Geologiske Undersøgelse   7.17
– DGU
[Geological Survey of Denmark]
Address: Thoravej 31, DK-2400 København NV, Denmark
Telephone: (01) 10 66 00
Telex: 19999 DANGEO DK
Affiliation: Miljøministeriet
Enquiries to: Librarian
Founded: 1888
Library facilities: Open to all users (nc); loan services (limited - nc); reprographic services (c).
No charge is made for photocopies.
Information services: Bibliographic services (limited - nc); translation services (limited - nc); literature searches (nc); access to on-line information retrieval systems (c).
Consultancy: Telephone answers to technical queries (nc); writing of technical reports (nc); compilation of statistical trade data; no market surveys in technical areas.
Publications: *Fortegnelse over skrifter* (list of publications); report series; reprint series.

Dansk Meteorologiske Institut   7.18
[Danish Meteorological Institute]
Address: Lyngbyvej 100, DK-2100 København Ø, Denmark

Farvandsdirektoratet, Nautisk Afeling *   7.19
[Royal Danish Administration of Navigation and Hydrography]
Address: 19 Esplanaden, DK-1263 København K, Denmark
Telephone: (01) 13 51 75
Affiliation: Ministry of Defence
Subject coverage: Hydrographic surveys and physical oceanography in Danish, Faroe, and Greenland waters; navigational safety including aids to navigation, piloting, and lifeboat service. Preparation and production of nautical charts and publications.

Geodaetisk Institut   7.20
[Geodetic Institute]
Address: Gamlehave Alle 22, DK-2920 Charlottenlund, Denmark
Telephone: (01) 63 18 33
Enquiries to: Librarian, Viggo F.M. Lunddahl
Founded: 1816
Subject coverage: Geodesy; mapping and surveying; seismology.

Library facilities: Open to all users (nc); loan services (nc); reprographic services (nc).
Library holdings: About 300 000 bound volumes; 200 current periodicals.
Consultancy: No consultancy.
Publications: List available on request.

FINLAND

Geodeettinen Laitos   7.21
[Finnish Geodetic Institute]
Address: Ilmalankatu 1A, SF-00240 Helsinki, Finland
Telephone: (90) 410433
Affiliation: Ministry of Agriculture and Forestry
Enquiries to: Professor Juhani Kakkuri
Founded: 1918
Subject coverage: Geodetic and photogrammetric research; I order geodetic measurements; legal length metrology; gravimetric survey.
Library facilities: Open to all users (for scientists and students only); loan services (nc - limited); no reprographic services.
Library holdings: 32 000 volumes, articles, etc, covering geodesy, astrometry, gravimetry, photogrammetry, geophysics.
Information services: Bibliographic services (nc - only references for geodetic bibliography); no translation services; no literature searches.
Geodetic observational data is for distribution to scientific research collaboration participants.
Consultancy: Yes (c - limited).

Geologian Tutkimuskeskus   7.22
– GTK
[Geological Survey of Finland]
Address: Kivimiehentie 1, SF-02150 Espoo, Finland
Telephone: (90) 46931
Telex: 12 3185 geolo SF
Affiliation: Ministry of Trade and Industry
Enquiries to: Head, Caj Kortman.
Information Bureau
Founded: 1885
Subject coverage: Geological mapping and investigation; mineral exploration.
Library facilities: Open to all users (nc); loan services (c); reprographic services.
Library holdings: 103 000 bound volumes; 885 current periodicals; 2500 reports.
Information services: Bibliographic services; literature searches; access to on-line information retrieval systems.
SDC; Dialog.

Publications: Survey bulletin; geological maps; maps; annual report; list on request.

Ilmatieteen Laitos *   7.23
– IL
[Finnish Meteorological Institute]
Address: PO Box 503, SF-00101 Helsinki 10, Finland
Telephone: (90) 171 922
Telex: 124436 efkl sf
Enquiries to: Librarian, Ritva Hänninen
Founded: 1841
Subject coverage: Meteorology; climatology; geomagnetism and aeronomy; air pollution and air protection.
Library facilities: Open to all users; loan services; reprographic services (c).
The library holds microfilms of the institute's weather maps from 1946 and of *Kuuhausikatsaus Suomen ilmastoon* (monthly weather bulletin) from 1880.
Information services: Bibliographic services; literature searches (restricted).
Telephone and written queries answered on meteorological data, etc.
Consultancy: Yes.
Publications: Bibliographic list.

Maanmittaushallitus *   7.24
– MMH
[National Board of Survey]
Address: Box 84, Opastinsilta 12, SF-00521 Helsinki 52, Finland
Telephone: (90) 1541
Affiliation: Ministry of Agriculture and Forestry
Enquiries to: Librarian, Birgitta Storgårds
Founded: 1633
Subject coverage: The National Survey Administration, which maintains and develops the real estate system and produces, develops, and updates topographic and thematic maps, is managed by the board.
Library facilities: No library facilities.
Information services: No information services.
Consultancy: Yes (c).
Publications: Bibliographic list.

Merentutkimuslaitos   7.25
[Finnish Institute of Marine Research]
Address: Box 33, SF-00931 Helsinki 93, Finland
Telephone: (90) 331 044
Telex: 125731 imr sf
Enquiries to: Librarian, Marjatta Heinänen
Founded: 1918
Subject coverage: Biological, chemical, and physical oceanography; environmental marine research.
Library facilities: Open to all users (nc); loan services (nc); reprographic services (c).
Library holdings: 39 500 bound volumes; 535 current periodicals.
Information services: Bibliographic services (nc); no translation services; no literature searches; no access to on-line information retrieval systems.
Consultancy: Telephone answers to technical queries (nc); no writing of technical reports; no compilation of statistical trade data; no market surveys in technical areas.

Teknillinen Korkeakoulu   7.26
[Helsinki University of Technology, Department of Mining and Metallurgy]
Address: Vuorimiehentie 2, SF-02150 Espoo, Finland
Telephone: (90) 455 4122
Telex: 125161
Enquiries to: Librarian, Marja Lampi-Dmitriev
Subject coverage: Metal physics; metal forming and heat treatment; theoretical process metallurgy; corrosion prevention; applied electrochemistry; applied process metallurgy; economic geology; mining engineering; mineral engineering.
Library facilities: Open to all users.
Loan and reprographic services are available through the main library of the university.
Publications: List on request.

Vesihallitus, Kirjasto   7.27
[National Water Board, Library]
Address: PO Box 250, SF-00101 Helsinki, Finland
Telephone: (90) 69511
Enquiries to: Information officer, Marja-Liisa Poikolainen
Founded: 1970
Subject coverage: Hydrology; limnology; water chemistry; hydraulic engineering; water resources management; water supply and waste water treatment.
Library facilities: Open to all users; loan services (nc); reprographic services (c).
Library holdings: 40 000 bound volumes; 460 current periodicals; 350 reports.
Information services: No information services available to non-staff users.
Consultancy: No consultancy.

Publications: Bibliographic list; National Water Board report.

FRANCE

Association Française pour l'Étude des Eaux   7.28
– AFEE
[French Water Study Association]
Address: 21 rue de Madrid, 75008 Paris, France
Telephone: (1) 45 22 14 67 (Paris); 93 74 22 23 (Valbonne)
Enquiries to: Secretary.
Alternative address: Sophia-Antipolis,06560 Valbonne, France
Founded: 1949
Subject coverage: Documentation on water resources; water treatment; water supply; sewage treatment; water analysis; hydrobiology; toxicology.
Library facilities: Open to all users; no loan services; reprographic services (c).
Information services: Bibliographic services (bulletin and paperfiles); translation services (c); literature searches (c); access to on-line information retrieval systems. Spidel, Paris.
Consultancy: Yes.
Publications: *Information Eaux*, bulletin, eleven issues per year.

Bureau de Recherches Géologiques et Minières   7.29
– BRGM
[Geological and Mining Research Office]
Address: BP 6009, 45060 Orléans Cedex, France
Telephone: 38 64 34 34
Telex: BRGM 780 258F
Enquiries to: Head of Geological Information and Documentation Department, R. Medioni
Founded: 1960
Subject coverage: Earth sciences; economic geology and mineral exploration; ground water resources; engineering geology.
Library facilities: Not open to all users (open only to users especially authorized); loan services (only in exceptional cases); reprographic services (c).
Information services: Bibliographic services (c); translation services (c); literature searches (c); access to on-line information retrieval systems (c). Pascal-Geode, ECOMINE and GEOBANQUE data bases on host computers ESA-IRS and Télésystèmes-Questel. Retrospective searches; SDI services; ECOMINE - press review on mineral economics.
Consultancy: Yes (c).

Publications: *Géologie de la France; Hydrogéologie, Géologie de l'Ingénieur; Chronique de la Recherche Minière; Géochronique; Géothermie/Actualités* (all quarterly); geological maps of France.

Centre Séismologique Euro-Méditerranéen *   7.30
– CSEM/EMSC
[European Mediterranean Seismological Centre]
Address: 5 rue René Descartes, 67084 Strasbourg Cedex, France
Telephone: 88 61 48 20
Telex: 890 826 F
Enquiries to: Secretary general
Founded: 1976
Subject coverage: Seismicity of Europe and the Mediterranean area.
Library facilities: No library facilities.
Information services: A European-Mediterranean seismicity data bank is under creation.
Consultancy: No consultancy.

Direction de la Météorologie Nationale, Service Central d'exploitation Météorologique, Division Documentation   7.31
[National Meteorology Directorate, Central Meteorology Service, Documentation Division]
Address: 2 avenue Rapp, 75340 Paris Cedex 07, France
Telephone: 45 55 95 02
Telex: 2000 61 F
Affiliation: Secrétariat d'État chargé des Transports
Enquiries to: Chef de Division
Founded: 1878
Subject coverage: Meteorology; climatology; connected sciences.
Library facilities: Open to all users (nc); loan services (nc); reprographic services (c).
Library holdings: 20 000 bound volumes; 1300 current periodicals.
Information services: Bibliographic services (c); no translation services; no literature searches; access to on-line information retrieval systems (c).
Consultancy: Yes.
Publications: Publications catalogue.

Institut National d'Astronomie et de Géophysique   7.32
– INAG
[National Institute of Astronomy and Geophysics]
Address: 77 avenue Denfert Rochereau, 75014 Paris, France
Telephone: (1) 43 20 13 30
Telex: 270070

Affiliation: Centre National de la Recherche Scientifique
Subject coverage: Coordinates research programmes in astronomy and geophysics.
Publications: Annual report.

Laboratoire d'Information et de Documentation en Géographie 'Intergeo' 7.33

['Intergeo' Geographical Documentation and Information Laboratory]
Address: 191 rue Saint-Jacques, 75005 Paris, France
Telephone: (1) 46 33 74 31
Affiliation: Centre National de la Recherche Scientifique
Enquiries to: Director, Paolo Pirazzoli; Head of documentation, Anne-Marie Briend.
'Intergeo' map library and photo library
Founded: 1947
Subject coverage: Geography.
Library facilities: Open to all users (nc); no loan services; reprographic services.
Library holdings: About 900 bound volumes; about 50 current periodicals.
Information services: Bibliographic services (c); translation services (c - from Russian to French); access to on-line information retrieval systems.
Consultancy: Telephone answers to technical queries; no writing of technical reports; no compilation of statistical trade data; no market surveys in technical areas.
Publications: List available on request.

Muséum National d'Histoire Naturelle 7.34

– MNHN
[National Natural History Museum]
Address: 38 rue Geoffroy Saint-Hilaire, 75005 Paris, France
Telephone: (1) 43 31 71 24
Affiliation: Ministère de l'Éducation Nationale
Enquiries to: Central library director, Yves Laissus
Founded: 1635
Subject coverage: Natural history, earth sciences.
Library facilities: Open to all users; loan services (nc); reprographic services (c).
Information services: Bibliographic services (c); no translation services; no literature searches; access to on-line information retrieval systems (c).
Consultancy: Yes (nc).

Service de la Documentation et des Publications 7.35

– SDP
[Documentation and Publications Service]
Address: Boite Postal 337, 29273 Brest Cedex, France
Telephone: 98 22 40 13
Telex: OCEANEX 940627 F
Affiliation: Institut de Recherche pour l'Exploitation de la Mer
Enquiries to: Head, Raoul Piboubes
Founded: 1984
Subject coverage: Oceanology: marine geology, biology, ecology, geochemistry, geophysics, physics and chemistry; marine technology; marine pollution; fisheries; aquaculture; law of the sea; economics.
Library facilities: Open to all users (nc); loan services (serials only); reprographic services (c).
Information services: Bibliographic services (c); no translation services; literature searches (c); access to on-line information retrieval systems (c).

GERMAN DEMOCRATIC REPUBLIC

Bergakademie Freiberg, Wissenschaftliche Informationszentrum 7.36

[Freiberg Mining Academy, Scientific Information Centre]
Address: PSF47 Agricola-Strasse 10, DDR-9200 Freiberg, German Democratic Republic
Telephone: 51 28 16
Telex: 078 535
Enquiries to: Library director, Dr Dieter Schmidmaier
Founded: 1765
Library facilities: Open to all users; loan services; reprographic services.

GERMAN FEDERAL REPUBLIC

Bundesanstalt für Geowissenschaften und Rohstoffe * 7.37

– BGR
[Federal Institute for Geosciences and Natural Resources]
Address: Postfach 51, D-3000 Hannover 51, German Federal Republic

Telephone: (0511) 464-0
Telex: 923730
Founded: 1959
Subject coverage: Geosciences; earth sciences.
Library facilities: Open to all users; loan services; reprographic services (nc).
Information services: Bibliographic services (c); no translation services; literature searches (c); access to on-line information retrieval systems (c).
GEOLINE data base is available through INKA.
Consultancy: Yes (c).
Publications: Bibliographic list; year books; geological maps of German Federal Republic.

Bundesanstalt für Gewässerkunde 7.38
[Federal Institute of Hydrology]
Address: Kaiserin-Augusta-Anlagen 15-17, D-5400 Koblenz, German Federal Republic
Telephone: (0261) 12431
Telex: 8-62 499
Affiliation: Bundesminister für Verkehr
Founded: 1948
Subject coverage: Hydrology, water resources, water pollution control and related fields.
Library facilities: Not open to all users; reprographic services (c - if more than 20 pages).
Library holdings: 40 000 bound volumes.
Information services: No information services.

Deutscher Wetterdienst, 7.39
Bibliothek
[German Weather Service, Library]
Address: Frankfurter Strasse 135, D-6050 Offenbach am Main, German Federal Republic
Telephone: (069) 8062 354
Enquiries to: Hans-Detlef Kirch
Founded: 1952
Subject coverage: Meteorology; climatology.
Library facilities: Open to all users (nc); loan services (but not for private individuals - nc); reprographic services (c).
Library holdings: The library houses 140 000 publications in meteorology and climatology and their applications in science and technology; fundamental publications in meteorology since the 16th century; about 1100 current periodicals.
Information services: Bibliographic services (c); no translation services; literature searches (c); no access to on-line information retrieval systems.
Consultancy: Yes (c).

Deutsches Hydrographisches 7.40
Institut
[German Hydrographic Institute]
Address: Postfach 220, Bernhard-Nocht-Strasse 78, D-2000 Hamburg 4, German Federal Republic
Telephone: (040) 3190-1

Telex: 211 138 bmvhh d
Facsimile: (040) 3190 5150
Affiliation: Federal Ministry for Transport
Enquiries to: Information officer, G. Heise.
Bibliothek im Deutschen Hydrographischen Institut
Founded: 1945
Subject coverage: Sea surveying; oceanography; pollution monitoring.
Library facilities: Open to all users (nc); loan services (nc); reprographic services (nc).
Library holdings: 113 000 bound volumes, including reports; 1500 current periodicals.
Information services: Bibliographic services (c - Hydrographische Dokumentation); no translation services; literature searches (nc); access to on-line information retrieval systems (c).
Consultancy: Telephone answers to technical queries (c); writing of technical reports (nc); compilation of statistical trade data (nc); no market surveys in technical areas.
Publications: Navigational charts; *Seehandbücher* (Pilots); notices to mariners; (list of lights); *Deutsche Hydrographische Zeitschrift*, a scientific journal; *Meereskundliche Beobachtungen* (Oceanographic Observations); annual report; *Ozeanographie* (collected reprints on oceanography); *Überwachung des Meeres* (Monitoring the Seas).

Dokumentationszentrale Wasser 7.41
– DZW
[Water Documentation Centre]
Address: Rochusstrasse 36, D-4000 Düsseldorf N, German Federal Republic
Telephone: (0211) 482041
Telex: 08-584281
Affiliation: Fraunhofer-Gesellschaftzur Förderung der angewandten Forschung eV, München
Enquiries to: E. Weinstock
Founded: 1960
Subject coverage: Documentation of literature and information on hydrology and related sciences.
Library facilities: Open to all users; no loan services; reprographic services (c - paper copies, microfilms).
Information services: Translation services; literature searches.
Abstracting service; full-text supply.
Publications: *Dokumentation Wasser (DW)*, twelve issues per year, each with 320 index cards; *DZW-Information*, quarterly.

Gesellschaft Deutscher 7.42
Metallhütten- und Bergleute
– GDMB
[German Society of Metallurgical and Mining Engineers]
Address: POB 210, D-3392 Clausthal-Zellerfeld, German Federal Republic
Telephone: (05323) 3438

Telex: 953828 tu clz d (GDMB)
Enquiries to: Secretary
Founded: 1912
Subject coverage: Exploration; mining; ore dressing; metallurgy (non-ferrous); analysis of metals; industrial minerals.
Library facilities: Open to all users; loan services (postage charged); reprographic services (c).
Information services: Bibliographic services (c); literature searches (c).
Consultancy: Yes (c).
Publications: Erzmetall, monthly.

Gmelin-Institut für anorganische Chemie und Grenzgebiete der Max-Planck-Gesellschaft 7.43

[Gmelin Institute of Inorganic Chemistry and Related Fields of the Max Planck Society]
Address: Varrentrappstrasse 40-42, Carl-Bosch-Haus, D-6000 Frankfurt am Main 90, German Federal Republic
Telephone: (0611) 79171
Telex: 4 12 526 gmeli
Enquiries to: Managing director, Professor Ekkehard Fluck
Subject coverage: Inorganic chemistry, physics, mineralogy etc. The Gmelin Handbook (full title given below), edited by the institute, is a documentation of the whole field of inorganic chemistry and its neighbouring fields unique in the world. The work, which at present comprises 502 volumes totalling more than 153 000 pages, has been continued in English alone since 1982.
Inorganic chemists, physicists, mineralogists and other specialized scientists screen the results of international research extracted by modern documentation techniques, evaluate them on the basis of the original documents in the light of the present state of the art and make them available in a broader context. By including direct references to the relevant literature, the handbook constitutes a reliable basis for literature searches.
The Gmelin Institute has a staff of 118 persons, of whom 73 are scientists (including the Chief Editors), plus some 140 external scientific staff members.
Publications: Gmelin Handbuch der Anorganischen Chemie.

Institut für Angewandte Geodäsie 7.44

– IfaG
[Applied Geodesy Institute]
Address: Richard-Strauss-Allee 11, D-6000 Frankfurt am Main 70, German Federal Republic
Telephone: (069) 6333-1
Telex: ifag-d 0413592
Affiliation: Bundesminister des Innern
Founded: 1952

Subject coverage: Cartography; cartographic research; geodetic research; photogrammetric research.
Library facilities: Open to all users; loan services; no reprographic services.
Information services: Bibliographic services.
Publications: Gerzeichnis der Veroffentlichungen (publications list).

GREECE

Institute of Geology and Mineral Exploration 7.45

Address: 70 Messoghion Street, Athinai 115 27, Greece
Telephone: (01) 7798412
Telex: 216357
Affiliation: Ministry of Industry and Energy
Founded: 1950
Publications: Geological and Geophysical Research; bulletin; annual report.

HUNGARY

Földmérési és Távérzékelési Intézet 7.46

– FÖMI
[Geodesy, Cartography and Remote Sensing Institute]
Address: Postafiók 546, H-1373 Budapest, Hungary
Telephone: (01) 113 431
Telex: 22-4964
Enquiries to: Deputy scientific director, Dr Tibor Lukács
Founded: 1967
Subject coverage: Geodesy; surveying; photogrammetry; engineering surveying; satellite geodesy; remote sensing.
Library facilities: Open to all users (nc); loan services (nc); reprographic services (c).
Library holdings: 20 500 bound volumes; 94 current periodicals; 1550 reports.
Information services: Bibliographic services (nc); no translation services; literature searches (nc).
Literature searches are made by means of the reference journal Geodinform which has an annual input of about 1500 abstracts.
Consultancy: No consultancy.

Magyar Állami Földtani Intézet 7.47
[Hungarian Geological Institute]
Address: Népstadion Útca 14, Postafiók 106, H-1442 Budapest, Hungary
Telephone: (01) 836 912
Telex: 225 200 mafi h
Subject coverage: Geological survey of Hungary; geological documentation; reconnaissance survey, estimation, and calculation of mineral resources.
Publications: Annual report.

Magyar Orszagos Meteorologiai 7.48
Intézet Konyvtara
[Hungarian Meteorological Institute]
Address: Kitaibel Pal U 1, Budapest 2, Hungary

Magyar Tudományos Akadémia, 7.49
Bányászati Kémiai
Kutatólaboratóriuma
– MTA BKKL
[Hungarian Academy of Sciences, Chemical Research Laboratory for Mining]
Address: PO Box 2, H-3515 Miskolc-Egyetemváros, Hungary
Telephone: (46) 67 211
Telex: 62-421 mtaok h
Enquiries to: Director, Dr József Tóth
Founded: 1957
Subject coverage: Enhanced oil recovery; reservoir engineering; mineral processing; physical and analytical chemistry of dispersed systems; development and production of relevant instruments.
Library facilities: Open to all users (nc); loan services (c); reprographic services (c).
Library holdings: 5000 bound volumes; 30 current periodicals.
Information services: No information services.
Consultancy: No consultancy.

Vízgazdalkodasi Tudományos 7.50
Kutató Központ
– VITUKI
[Water Resources Development Research Centre]
Address: Postafiók 27, H-1453 Budapest 92, Hungary
Telephone: (01) 338-160
Telex: 224959-h
Enquiries to: J. Sztrókay
Founded: 1952
Subject coverage: Water resources development; data on surface and ground water resources for systematic water management; hydrological events; surface and subsurface waters; commercial, industrial, and agricultural use of water; self-purification of watercourses; purification of wastewater before emission into streams; hydrotechnology.
Library facilities: Open to all users; no loan services; no reprographic services.

Information services: Bibliographic services; translation services (c); literature searches (c); access to on-line information retrieval systems (through International Relations Department).
Consultancy: Yes (transactions on an exchange basis).
Publications: Proceedings; hydrographic year book; daily hydrographic map; groundwater information map, monthly; map of karstic levels in the trans-Danubian Central Range, annually; reports in Hungarian, English, French, German, and Russian; complete list on request.

ICELAND

Hafrannsóknastofnunin 7.51
[Marine Research Institute]
Address: Skúlagata 4, PO Box 1390, 121 Reykjavík, Iceland
Telephone: (91) 20240
Telex: ISRADO IS 2066
Enquiries to: Librarian, Eiríkur Th. Einarsson
Founded: 1937; name changed in 1965
Subject coverage: Marine biology; physical and chemical oceanography; marine geology; fish-farming.
Library facilities: Open to all users (nc); no loan services; reprographic services (c).
Library holdings: About 7500 bound volumes; 560 current periodicals.
Information services: Bibliographic services (nc); no translation services; literature searches (c); access to on-line information retrieval systems (Dialog, DIMDI and Pergamon InfoLine - c).
Consultancy: Telephone answers to technical queries (nc); no writing of technical reports; no compilation of statistical trade data; no market surveys in technical areas.
Publications: Rit Fiskideildar (occasional, in English or German with English summary); Hafrannsóknir (occasional, in Icelandic, with English summary; includes annual report of the institute and fishing prospects reports); Fjölrit Hafrannsóknastofnunar (mimeographed papers published occasionally, in Icelandic).

Vedurstofa Íslands * 7.52
[Icelandic Meteorological Office]
Address: Bústaðavegi 9, 105 Reykjavík, Iceland
Telephone: (91) 86000
Enquiries to: Librarian
Founded: 1920
Subject coverage: Meteorology; seismology.
Library facilities: Open to all users; loan services (nc); no reprographic services.

Information services: No information services.
Consultancy: Yes (nc).
Publications: *Veðráttan* (monthly meteorological bulletin with a yearly summary); *Hafís við strendur Íslands* (Sea Ice off the Icelandic Coasts); *Seismological Bulletin - the Icelandic Stations.*

IRELAND

Geological Survey of Ireland   7.53
Address: Beggar's Bush, Haddington Road, Dublin 2, Ireland
Affiliation: Department of Energy
Enquiries to: Librarian
Founded: 1845
Subject coverage: Geology.
Library facilities: Open to all users; loan services; reprographic services.
Publications: *Geological Survey of Ireland Bulletin,* annual; *Geological Survey of Ireland Special Paper,* occasional; *Geological Survey of Ireland Guide Series,* occasional; *Geological Survey of Ireland Report Series,* occasional; *Geological Survey of Ireland Information Circular,* occasional.

Irish Meteorological Service   7.54
Address: Glasnevin Hill, Dublin 9, Ireland
Telephone: (01) 424411
Telex: 25239
Affiliation: Department of Communications
Enquiries to: Librarian
Founded: 1936
Subject coverage: Meteorology; climatology; atmospheric physics.
Library facilities: Open to all users (by arrangement - nc); loan services (to other libraries - BLLD forms); reprographic services (c).
Library holdings: 7000 bound volumes; 100 current periodicals; 10 000 reports.
Information services: No bibliographic services; no translation services.
Searches on-line of the service's computer catalogue which contains catalogue indexes, books, and reports, but excludes articles in journals - nc.
Consultancy: Telephone answers to technical queries (nc); writing of technical reports (c); no compilation of statistical trade data; no market surveys in technical areas.
Specialized forecasts, tailor-made for special interests, can be obtained. Technical queries and enquiries about the writing of technical reports should be addressed to the officer-in-charge of the appropriate section (climate division; forecasting office; agrometeorological section; marine section).

Publications: List available on request.

National Museum of Ireland   7.55
Address: Kildare Street, Dublin 2, Ireland
Telephone: (01) 765521
Parent: Department of the Taoiseach
Enquiries to: Librarian, F. Devlin
Founded: 1877
Subject coverage: Archaeology; ethnography; fine arts; zoology; geology.
Library facilities: Open to all users (nc); no loan services; reprographic services (c).
The library provides facilities for consultation of material not available elsewhere, and has a collection of aerial photographs of archaeological and geographical features available for consultation.
Library holdings: 20 000 bound volumes; 100 current periodicals.
Information services: No bibliographic services; no translation services; literature searches (based on museum holdings only - nc); no access to on-line information retrieval systems.
Consultancy: Telephone answers to technical queries (nc); no writing of technical reports; no compilation of statistical trade data; no market surveys in technical areas.

ITALY

Istituto di Ricerca sulle Acque *   7.56
– IRSA
[Water Research Institute]
Address: Via Reno 1, 00198 Roma, Italy
Telephone: (06) 841451
Telex: 614588
Affiliation: Consiglio Nazionale delle Ricerche
Enquiries to: Director, Professor Roberto Passino
Founded: 1968
Subject coverage: Water supply and water resources management; water pollution and water quality; wastewater treatment; sludge disposal.
Library facilities: No library facilities.
Information services: No information services.
Consultancy: Yes (nc).
Publications: Bibliographic list on request. The institute's publications can be acquired at CNR-Ufficio Vendita Pubblicazioni, Piazzale Aldo Moro 7, 00185 Roma, Italy.

Servizio Geologico d'Italia   7.57
[Geological Survey of Italy]
Address: Largo S Susanna 13, 00187 Roma, Italy
Telephone: (06) 4744645
Telex: 611637

Affiliation: Ministero dell'Industria, del Commercio e dell'Artigianato
Enquiries to: Director, Professor A. Jacobacci
Founded: 1873
Library facilities: Open to staff only.
Information services: For internal use only.

Società di Studi Geografici 7.58
[Geographical Society]
Address: Via Laura 48, 50121 Firenze, Italy
Telephone: (055) 21 26 15
Library facilities: Library.

LUXEMBOURG

Service Géologique 7.59
[Geological Survey]
Address: 43 boulevard G.-D. Charlotte, Luxembourg
Telephone: 44 41 26
Affiliation: Direction des Ponts et Chaussées
Enquiries to: Ingénieur-géologue, chef de division, Jacques Bintz
Founded: 1947

MONACO

Centre Scientifique de Monaco * 7.60
– CSM
[Monaco Scientific Centre]
Address: 16 Boulevard de Suisse, MC Monte-Carlo, Monaco
Telephone: (93) 30 33 71
Founded: 1960
Subject coverage: Oceanography, environment; meteorology; seismology; marine microbiology; marine pollution.
Library facilities: No library facilities.

Musée Océanographique * 7.61
[Oceanographic Museum]
Address: Avenue Saint-Martin, Monaco-Ville, MC - 98000, Monaco
Telephone: (93) 30 15 14
Telex: 469037 F
Enquiries to: Head librarian, J. Carpine-Lancre
Founded: 1906
Subject coverage: Oceanography; marine biology; marine environment; aquarium sciences; history of oceanography.
Library facilities: Open to all users; loan services (c); reprographic services (c).

Information services: Bibliographic services; no translation services; no literature searches; access to on-line information retrieval systems (c).
Consultancy: No consultancy.

NETHERLANDS

Instituut voor Cultuurtechniek en Waterhuishouding 7.62
– ICW
[Land and Water Management Research Institute]
Address: Postbus 35, Marijkeweg 11, 6700 AA Wageningen, Netherlands
Telephone: (08370) 19100
Telex: 75230 (start message with ICW)
Affiliation: Ministry of Agriculture and Fisheries
Enquiries to: Librarian, G. Naber.
Library of the Staring Building, Postbus 45, 6700 AA Wageningen
Founded: 1955
Subject coverage: Hydrology, water quality, water management; soil technology; land use planning.
Library facilities: Open to all users (nc); loan services (occasional charge); reprographic services (c).
Library holdings: 100 000 bound volumes; 1200 current periodicals.
Information services: Bibliographic services (occasional charge); no translation services; literature searches (occasional charge); access to on-line information retrieval systems (occasional charge).
Consultancy: Telephone answers to technical queries (nc); no writing of technical reports; no compilation of statistical trade data; no market surveys in technical areas.
Publications: Current awareness list; bibliographies.

International Institute for Land Reclamation and Improvement 7.63
– ILRI
Address: Postbus 45, 6700 AA Wageningen, Netherlands
Telephone: (08370) 19100
Telex: VISI-NL 75230
Enquiries to: Librarian, G. Naber.
Library of the Staring Building
Founded: 1956
Subject coverage: Land reclamation and water management.
Library facilities: Open to all users; loan services (c); reprographic services (c).
Information services: Bibliographic services (nc); no translation services; literature searches (c); access to on-line information retrieval systems (c).

On-line literature search. Bibliographic data base of the Directorate of Agricultural Research of the Ministry of Agriculture and Fisheries and the Agricultural University of Wageningen.
Consultancy: Yes (c).
Publications: Current awareness list, six times a year; bibliographies, occasionally.

Koninklijk Nederlands Meteorologisch Instituut   7.64
– KNMI
[Royal Netherlands Meteorological Institute]
Address: Postbus 201, De Bilt, near Utrecht, Netherlands
Telephone: (030) 766911
Telex: 47096
Enquiries to: Librarian, J. van der Lingen
Founded: 1854
Subject coverage: Meteorology; climatology; physical oceanography; geophysics; seismology; earth-magnetism.
Library facilities: Open to all users (nc); loan services (nc); reprographic services (c).
Library holdings: 55 000 bound volumes, including reports; 600 current periodicals.
Information services: Bibliographic services (c); no translation services; literature searches (c); access to on-line information retrieval systems (c).
Consultancy: Telephone answers to technical queries (c); writing of technical reports (c).

Rijks Geologische Dienst   7.65
– RGD
[Geological Survey of the Netherlands]
Address: Postbus 157, 2000 AD Haarlem, Netherlands
Telephone: (023) 319362
Telex: 71105 GEOLD
Affiliation: Ministry of Economic Affairs
Enquiries to: Librarian, Annelies Duivenvoorden
Founded: 1903
Subject coverage: Geology of the Netherlands, in particular oil and gas; quaternary geology.
Library facilities: Open to all users (nc); loan services (nc); reprographic services (nc).
Library holdings: 10 000 bound volumes; 500 current periodicals; 2000 reports.
Information services: No information services.
Consultancy: No consultancy.
Publications: List on request.

Stichting voor Bodemkartering   7.66
[Dutch Soil Survey Institute]
Address: Postbus 98, 6700 AB Wageningen, Netherlands
Telephone: (08370) 19100
Telex: VISI-NL 75230

Enquiries to: Librarian, G. Naber.
Library of the Staring Building
Founded: 1945
Subject coverage: Soil science; landscape research; cartography; earth sciences.
Library facilities: Open to all users; loan services (c); reprographic services (c).
Information services: Bibliographic services (nc); no translation services; literature searches (c); access to on-line information retrieval systems (c).
On-line literature search. Bibliographic data base of the Directorate of Agricultural Research of the Ministry of Agriculture and Fisheries and the Agricultural University of Wageningen.
Consultancy: Yes (c).
Publications: Current awareness list, six times a year; annotated bibliographies.

Technische Hogeschool Delft   7.67
[Delft University of Technology, Department of Mining Engineering]
Address: Mijnbouwstraat 120, 2628 RX Delft, Netherlands
Library facilities: Library.

Technische Hogeschool Delft   7.68
[Delft University of Technology, Department of Geodesy]
Address: Thysseweg 11, 2629 JA Delft, Netherlands
Library facilities: Library.

Topografische Dienst   7.69
– TDN
[Topographical Service]
Address: Bendienplein 5, 7815 SM Emmen, Drente, Netherlands
Telephone: (05910) 96911
Affiliation: Ministry of Defence
Enquiries to: Director, P.W. Geudeke
Founded: 1815
Subject coverage: Aerial photographs; topographical maps.
Library facilities: Open to all users (nc); loan services (books only, not maps - nc); reprographic services (c).
Library holdings: The service has a book collection on cartography and related subjects, a map collection of topographical maps of the Netherlands and a large part of the world, and a collection of aerial photographs. It houses a total of 7000 bound volumes, about 250 000 maps and about 150 000 aerial photographs. In addition, there are 40 current periodicals.
Information services: No information services.
Consultancy: Telephone answers to technical queries (nc); no writing of technical reports; no compilation of statistical trade data; no market surveys in technical areas.

NORWAY

International Union of Geological Sciences/Union Internationale des Sciences Géologiques
7.70

– IUGS

Address: Geological Survey of Norway, Postboks 3006, N-7001 Trondheim, Norway
Telephone: (07) 92 15 00
Affiliation: International Council of Scientific Unions
Enquiries to: Secretary-general, Professor R. Sinding-Larsen
Subject coverage: Marine geology; meteorites; stratigraphy; systematics in petrology; experimental petrology; tectonics; geology teaching; geological documentation; storage, automatic processing and retrieval of data; history of geological sciences; geoscience and man.

Norges Geografiske Oppmåaling *
7.71

– NGO

[Geographical Survey of Norway]
Address: Monserudveien N-3500 Hønefoss, Norway

Telephone: (067) 24100
Affiliation: Ministry of the Environment
Founded: 1773
Subject coverage: Geodesy, photogrammetry, cartography.
Library facilities: Open to all users; loan services (nc); reprographic services (c).
Information services: No information services.
Consultancy: Yes.

Norges Geologiske Undersøkelse *
7.72

– NGU

[Geological Survey of Norway]
Address: Postboks 3006, N-7001 Trondheim, Norway
Telephone: (07) 921611
Enquiries to: Librarian
Founded: 1858
Subject coverage: Geology; geophysics; geochemistry.
Library facilities: Open to all users; loan services (nc); reprographic services (c).
Information services: Bibliographic services (nc); no translation services; literature searches (nc - manual); no access to on-line information retrieval systems.
Consultancy: Yes (nc).
Publications: *Publikasjoner og kart 1891-1983* (publications and maps).

Norges Geotekniske Institutt
7.73

– NGI

[Norwegian Geotechnical Institute]
Address: PO Box 40, Taasen, N-0801 Oslo 8, Norway
Telephone: (02) 230388
Telex: 19787 ngi n
Affiliation: Royal Norwegian Council for Scientific and Industrial Research
Enquiries to: Librarian
Founded: 1953
Subject coverage: Soil mechanics; foundation engineering; rock mechanics; engineering geology; snow mechanics; avalanche investigations; site investigations; earth- and rockfill dams; foundation of offshore structure.
Library facilities: Open to all users; loan services (nc); reprographic services (c).
Information services: Bibliographic services; literature searches; access to on-line information retrieval systems.
Access to Polydoc, Dialog, and InfoLine.
Consultancy: Yes.

Norsk Oseanografisk Datasenter *
7.74

– NOD

[Norwegian Oceanographic Data Centre]
Address: Postboks 1870/72, N-5011 Bergen-Nordnes, Norway
Telephone: (05) 230300
Telex: 42297 OCEAN N
Affiliation: Fiskeridirektoratets Havforsknings-instituttet
Enquiries to: Consultant, Inger J. Svendsen
Founded: 1972
Subject coverage: Storing and processing of hydrographical data.
Library facilities: Open to all users; no loan services; no reprographic services.
Library holdings: The centre has a catalogue of on-going projects in oceanography.
Information services: No bibliographic services; no translation services; literature searches (c); access to on-line information retrieval systems (c).
NOD is a Scandinavian ASFA input centre.
Consultancy: Yes (c).

POLAND

Biblioteka Główna Akademii Górniczo-Hutniczej im Stanisława Staszica w Krákowie *

7.75

– AGH

[Central Library of the Stanislaus Staszic University of Mining and Metallurgy]
Address: Al Mickiewicza 30, 39-059 Kraków, Poland
Telephone: 34 14 04
Telex: 0322311
Enquiries to: Director, Maria Świerczyńska
Founded: 1919
Subject coverage: Foundry, geology, metallurgy, mineral industry, mining, materials technology.
Library facilities: Open to all users; loan services; reprographic services.
National and international interlibrary loan service and reprographic service of scientific materials held by the library.
Information services: Bibliographic services.
Publications: List of Foreign Acquisitions, quarterly; List of Current Periodicals in the Central Library and the Institution Libraries, annual.

Główny Instytut Górnictwa *

7.76

– GIG

[Central Mining Institute]
Address: PO Box 3672, Plac Gwarków 1, 40-951 Katowice, Poland
Telephone: 581 631
Telex: 0312359 GIG PL
Enquiries to: Director, Władysław Chładzyński.
Scientific-Technical and Economic Information Centre
Founded: 1945
Subject coverage: Hard coal mining and related sciences.
Library facilities: Open to all users; loan services; reprographic services.
Loan services are available to staff members, scientists and employees of the Ministry of Mines and Power; interlibrary loan services.
Information services: Bibliographic services; translation services; literature searches.
Publications: Biuletyn Informacyjny (Bulletin of Information); Przeglad Informacyjny (Mining Abstracts); Prace GIG-Komunikaty (GIG Transaction-Bulletin); Informacja dla Kadry Kierowniczej (Information for Leading Executives); Bibliography of Polish Mining Literature; publications lists on request.

Instytut Geologiczny

7.77

– IG

[Geological Institute]
Address: Ulica Rakowiecka 4, 00-975 Warszawa, Poland
Telephone: 022-49 53 51
Telex: 815541 igol
Affiliation: Ministry of Environmental Protection and Mineral Resources
Enquiries to: Grazyna Burchart
Founded: 1919
Subject coverage: Palaeogeography; palaeontology; stratigraphy; tectonics; petrography; geochemistry; mineralogy; mineral resources; economic geology; geophysics; hydrogeology; engineering geology.
Library facilities: Open to all users (nc); loan services (nc); reprographic services (c).
Library holdings: 228 250 bound volumes; 700 current periodicals; 255 800 reports.
Information services: Bibliographic services (c); translation services (c); literature searches (c); no access to on-line information retrieval systems.
Consultancy: No consultancy.
Publications: Bibliografia Geologiczna Polski; Biuletyn Instytutu Geologicznego; Prace Instytutu Geologicznego; Geology of Poland; cartographic publications.

Instytut Górnictwa Naftowego i Gazownictwa

7.78

– IGNiG

[Oil and Gas Institute]
Address: Ul. Lubicz 25a, 30-960 Kraków, Poland
Telephone: 094-21 00 33
Telex: 0325276
Enquiries to: Librarian, Dr Wincenty Pawłowski.
Główna Biblioteka Branzowa, Branzowy Ośrodek Informacji Naukowej, Technicznej i Ekonomicznej
Founded: 1944
Subject coverage: Geology, geophysics, geochemistry, and geomicrobiology in relation to oil and gas exploration and engineering; processing and treatment of fuel gases; transport, distribution, and storage of gaseous fuels; drilling equipment and technology; exploitation of oil and gas deposits; applications of gas fuels.
Library facilities: Open to all users (nc); loan services (nc); reprographic services (c).
Microfilming.
Library holdings: 57 000 bound volumes, including brochures and pamphlets; 421 current periodicals.
Information services: Bibliographic services (c); translation services (c); literature searches (c); no access to on-line information retrieval systems.
Consultancy: Telephone answers to technical queries (nc); writing of technical reports (c); compilation of statistical trade data (c); no market surveys in technical areas.

Publications: *Prace Instytutu Górnictwa Naftowego i Gazownictwa; Biuletyn Informacyjny Górnictwa Naftowego i Gazownictwa,* quarterly periodical.

Instytut Meteoroliogii i Gospodarki Wodnej   7.79
– IMGW
[Meteorology and Water Management Institute]
Address: 61 Podlésna Street, 01-673 Warszawa, Poland
Telephone: 35 28 13
Telex: 81 43 31
Affiliation: Ministry for Environmental Protection and Nature Resources
Enquiries to: Head, Dr Hanna Mycielska.
Information Centre
Founded: 1919
Subject coverage: Meteorology; hydrology; oceanology; water management; water engineering.
Library facilities: Open to all users (nc); loan services (nc); reprographic services (c).
Information services: Bibliographic services (c); no translation services; literature searches (c).
Consultancy: No consultancy.

PORTUGAL

Instituto Geográfico e Cadastral   7.80
– IGC
[Geographical and Cadastral Institute]
Address: Praça da Estrela, 1200 Lisboa, Portugal
Telephone: (1) 609925/6/7
Affiliation: Ministério das Finanças
Enquiries to: Director-general
Founded: 1926
Subject coverage: Geodesy; cartography; photogrammetry; cadastral survey.
Library facilities: Open to all users; loan services (c); reprographic services (c).
Loan and reprographic services are available only when authorized by the director-general.
Information services: Bibliographic services; translation services (c - if authorized by the director-general).
Consultancy: Yes.
Publications: *Cartas e Publicações* (maps and publications).

Instituto Hidrográfico   7.81
– IH
[Hydrographic Institute]
Address: Rua das Trinas 49, 1296 Lisboa Codex, Portugal
Telephone: (019) 601191

Telex: INSTHIDROMAR CENCOM 12587P
Affiliation: Portuguese navy
Enquiries to: Director, Captain António Manuel Ribeiro Rosa.
Information and Documentation Centre
Founded: 1970
Subject coverage: Hydrography; navigation; oceanography.
Library facilities: Not open to all users (students only); no loan services; reprographic services (c).
Library holdings: 8000 bound volumes; 150 current periodicals; 130 reports.
Information services: Bibliographic services (nc); no translation services; literature searches (nc); no access to on-line information retrieval systems.
Consultancy: Telephone answers to technical queries (nc); no writing of technical reports; no compilation of statistical trade data; no market surveys in technical areas.
Publications: Bibliographic lists.

Instituto Nacional de Meteorologia e Geofísica   7.82
– INMG
[National Institute of Meteorology and Geophysics]
Address: Rua C do Aeroporto, 1700 Lisboa Codex, Portugal
Telephone: (019) 802221
Enquiries to: Director general, Professor Mendes Victor
Founded: 1947
Subject coverage: Meteorology; geophysics.
Library facilities: Open to all users; loan services; reprographic services.
Information services: Bibliographic services; translation services; literature searches.
Consultancy: Yes.
Publications: *Anuário Climatológico de Portugal; Anuário Sismológico de Portugal; Boletim Actinométrico de Portugal,* monthly; *Resumos Meteorológicos para a Aeronáutica,* monthly; *Boletim Geomagnético Preliminar,* monthly; *Boletim Meteorológico para a Agricultura,* three times a month; *Boletim Meteorológico,* daily; *Revista do Instituto Nacional de Meteorologia e Geofísica,* quarterly; *Boletim do Centro de Física de Lisboa,* monthly; *Boletim Informatico,* monthly; *Observacos Mágnéticos de S Miguel-Açores,* annually.

Instituto de Investigação Científica Tropical, Centro de Documentação e Informação   7.83
[Institute for Tropical Scientific Research, Documentation and Information Centre]
Address: Rua Jau 47, 1300 Lisboa, Portugal
Telephone: 645321

Affiliation: Ministério da Educação
Enquiries to: Director, F. Almeida Ribeiro
Founded: 1957
Subject coverage: Earth sciences; geographic engineering science; biology; agronomy; ethnological sciences.
Library facilities: Open to all users; loan services (only to staff); reprographic services (c).
Library holdings: 33 200 bound volumes; 980 current periodicals.
Information services: Bibliographic services (nc); no translation services; literature searches (nc); no access to on-line information retrieval systems.
Consultancy: Telephone answers to technical queries; no writing of technical reports; no compilation of statistical trade data; no market surveys in technical areas.
Publications: List of publications available on request.

Serviços Geológicos de Portugal 7.84
– SGP
[Geological Survey of Portugal]
Address: Rua Academia das Ciências 19-20, 1200 Lisboa, Portugal
Telephone: (019) 363915
Affiliation: Ministério da Indústria e Commércio
Enquiries to: Dr Maria de Fatima Gomes da Silva Beato.
Library
Founded: 1857
Subject coverage: Geology and related subjects.
Library facilities: Open to all users (researchers); no loan services; reprographic services (photocopying).
Library holdings: 80 500 bound volumes; 2 300 periodicals.
Information services: Bibliographic services; no translation services; no access to on-line information retrieval systems.
Consultancy: Yes.
Publications: Various geological publications; *Comunicações, Memórias* and geological maps, with explanatory notes.

ROMANIA

Institutul de Geologie și 7.85
Geofizica
[Geology and Geophysics Institute]
Address: Strada Caransebeş 1, Bucureşti, Romania
Telephone: 65 66 25
Enquiries to: Director, I. Bercia
Library holdings: 180 000 volumes.

Institutul de Meteorologie și 7.86
Hidrologie
– IMH
[Meteorology and Hydrology Institute]
Address: Soseauna Bucureşti-Ploieşti 97, Bucureşti, Romania
Telephone: (90) 79 32 40; 33 35 40
Telex: 06500-10460 IMH-R
Parent: Consiliul Naţional al Apelor - CNA
Enquiries to: Director
Founded: 1884.
Subject coverage: Mathematical models for numerical weather forecasts; design of automatic systems of meteorological and hydrological information; studies of solar radiation, atmospheric electricity, radar and satellite meteorology, microclimatology, and agrometeorology; studies into forecasts and warning of specified phenomena dangerous to aviation; research in air pollution; hail suppression; modelling studies for stream flow simulation and ground water resources.
Library facilities: Open to all users (for scientists only); loan services; reprographic services. Exchange of publications.
Information services: Bibliographic services; translation services; literature searches.
Consultancy: Yes.

SPAIN

Instituto Español de Oceanografía 7.87
– IEO
[Spanish Institute of Oceanography]
Address: Calle Alcalá 27, Madrid 14, Spain
Telephone: 91-222 74 93
Telex: 44460
Affiliation: Ministerio de Agricultura y Pesca
Enquiries to: Head, oceanographic data centre, Dr F. Fernandez
Subject coverage: Marine biology; fisheries biology and technology; marine pollution; physical oceanography; marine geology.

Instituto Geológico y Minero de 7.88
España *
– IGME
[Geological and Mining Institute of Spain]
Address: Rios Rosas 23, Madrid 3, Spain
Telephone: 90-44 16 500
Telex: 48054 igme e
Enquiries to: Head of documentation and information, Mercedes Barreno Ruiz
Founded: 1849
Subject coverage: Geological and mineral research.

Library facilities: Open to all users; no loan services; reprographic services.
Information services: Bibliographic services (nc); no translation services; no access to on-line information retrieval systems.
Will soon be on-line to Pascal-Geode.
Consultancy: Yes (nc).
Publications: Bibliographic list.

Laboratorio de Geotecnia 7.89
– LTMS
[Geotechnical Laboratory]
Address: Alfonso XII 3, 28014 Madrid, Spain
Telephone: 4682400
Enquiries to: J. Enrique Dapena Garcia
Founded: 1944
Library facilities: Open to all users (nc); no loan services; reprographic services (c).
Library holdings: 13 026 bound volumes; 96 current periodicals.
Information services: Bibliographic services (nc); no translation services; literature searches (nc); no access to on-line information retrieval systems.
Consultancy: No telephone answers to technical queries; writing of technical reports (c).
Courses: Foundations with problems, biannual; geotextiles use, biannual.

Servicio Meteorológico Nacional 7.90
[National Meteorological Service]
Address: Cuidad Universitaria, Apartado 285, Madrid, Spain

SWEDEN

Statens Geotekniska Institut * 7.91
– SGI
[Swedish Geotechnical Institute]
Address: S-581 01 Linköping, Sweden
Telephone: 013-11 51 00
Telex: 50125 VTISGI S
Enquiries to: Librarian, Ingrid Carlsson
Founded: 1944
Subject coverage: Research, information, and consulting in geotechnics.
Library facilities: Open to all users; loan services (c); reprographic services (c).
Information services: Bibliographic services (c); literature searches (c).
On-line information retrieval system for internal use only.
Publications: Reports; information documents; accessions list, bi-monthly.

Sveriges Geologiska Undersökning * 7.92
– SGU
[Geological Survey of Sweden]
Address: Library, Box 670, S-751 28 Uppsala, Sweden
Telephone: 018-17 93 96
Telex: 76154 Geoswed
Affiliation: Ministry of Industry
Enquiries to: Librarian, Martin Ostling
Founded: 1858
Subject coverage: Geology.
Library facilities: Open to all users; loan services; reprographic services.
Information services: Bibliographic services; literature searches.
Consultancy: Yes.
Publications: Bibliography and index of geology.

Sveriges Meteorologiska och Hydrologiska Institut * 7.93
– SMHI
[Swedish Meteorological and Hydrological Institute]
Address: Box 923, S-601 19 Norrköping, Sweden
Telephone: 011-15 80 00
Telex: 64 400 SMHI S
Enquiries to: Librarian, Centa Forsman
Founded: 1873
Subject coverage: Meteorology; climatology; hydrology; oceanography.
Library facilities: Open to all users; loan services (nc); reprographic services (c).
Information services: No information services.
Consultancy: No consultancy.

SWITZERLAND

Eidgenössisches Institut für Schnee- und Lawinen Forschung/ Institut Fédéral pour l'Étude de la Neige et des Avalanches/ Istituto Federale per lo Studio della Neve e delle Valanghe 7.94
[Federal Institute for Snow and Avalanche Research]
Address: Weissfluhjoch, CH-7260 Davos-Dorf, Switzerland
Telephone: (083) 5 32 64
Telex: 74 309
Affiliation: Bundesamt für Forstwesen
Enquiries to: Director, Dr C. Jaccard
Founded: 1936

Subject coverage: Research, development and consultancy on snow, snowpack, avalanches, avalanche protection, mountain afforestation.
Library facilities: Library facilities for staff only.
Library holdings: 5000 bound volumes; 100 current periodicals; 600 reports.
Information services: No information services.
Consultancy: Telephone answers to technical queries (nc); writing of technical reports (c); no compilation of statistical trade data; no market surveys in technical areas.
Courses: Biennial course on snow and avalanches.

Geologische Landesaufnahme/ Cartographie Géologique   7.95
[Geological Mapping]
Address: Birmannsgasse 8, CH-4055 Basel, Switzerland
Telephone: (061) 25 53 30
Affiliation: Landeshydrologie und -geologie/Service Hydrologique et Géologique National
Founded: 1860; 1986 transferred to the Confederation
Subject coverage: Coordinating office for geological mapping in Switzerland.
Library facilities: Small library for internal use only.
Information services: No information services.
Consultancy: No consultancy.
Publications: Geological maps; explanatory notes; *Beiträge zur Geologische Karte der Schweiz.*

Muséum d'Histoire Naturelle, Bibliothèque   7.96
[Natural History Museum, Library]
Address: Case Postale 434, Route de Malagnou, CH-1211 Genève 6, Switzerland
Telephone: (022) 35 91 30
Enquiries to: Head librarian, C. Favarger
Founded: 1820
Subject coverage: Zoology; entomology; geology; mineralogy; palaeontology.
Library facilities: Open to all users (nc); loan services (nc); reprographic services (c).
Library holdings: 50 000 bound volumes; 1100 current periodicals.
Information services: Bibliographic services; no translation services; literature searches; no access to on-line information retrieval systems.

Schweizerisches Meteorologisches Anstalt   7.97
[Swiss Meteorological Institute]
Address: Karhbulstrasse 58, Postfach, CH-8044 Zürich, Switzerland
Enquiries to: Director, A. Junod
Founded: 1880
Library holdings: 35 000 volumes.

Publications: *Annalen.*

TURKEY

Harita Genel Komutanliği *   7.98
[General Command of Mapping]
Address: Cebeci, Ankara, Turkey
Telephone: 197740
Affiliation: Ministry of National Defence
Enquiries to: Librarian, Cananözdenkos
Founded: 1925
Subject coverage: Map production.
Library facilities: Not open to all users; loan services; no reprographic services.
As this establishment is a government agency, the library is open to employees only. It is also a military headquarters and therefore restricted.
Information services: Bibliographic services; no translation services; no literature searches; no access to on-line information retrieval systems.
Consultancy: No consultancy.

Seyir, Hidrografi ve Oşinografi Dairesi Başkanliği *   7.99
[Navigation, Hydrology and Oceanography Department]
Address: Başkanglği, Čubuklu-Istanbul, Turkey
Telephone: 331 02 29; 331 02 30
Affiliation: Turkish navy
Enquiries to: Rear Admiral Ševket Güçlüer
Founded: 1909
Subject coverage: Oceanography; navigation; geophysics; submarine geology.
Library facilities: Open to all users; no loan services; no reprographic services.
Consultancy: Yes.

TC Maden Tetkik ve Arama General Müdürlüğü   7.100
– MTA
[Mineral Research and Exploration Institute of Turkey]
Address: Ismet Inönü Bulvari, Ankara,, Turkey
Telephone: 234255
Telex: 42741 MTA TR
Enquiries to: Chief librarian, Sevim Özertan
Founded: 1935
Library facilities: Open to all users; loan services (to staff only); no reprographic services.
Information services: No information services.
Consultancy: No consultancy.

UNITED KINGDOM

British Geological Survey   7.101
– BGS
Address: Nicker Hill, Keyworth, Nottingham NG12 5GG, UK
Telephone: (06077) 6111; (01) 589 3444 (Information Point)
Telex: 378173 BGSKEY G
Affiliation: Natural Environment Research Council
Enquiries to: Information Services group manager, J. Bain; Chief librarian, National Reference Library of Geology, G. McKenna
Founded: 1835
Subject coverage: Geological sciences.
The London section of the British Geological Survey Library, at present housed at the Geological Museum, will in due course be transferred to Keyworth. However, the existing Information Point will be retained in London for public access to geological maps, reports and other data. The ordering of larger scale maps will also continue to be conducted through this Information Point, which is at the following address: Geological Museum, Exhibition Road, South Kensington, London SW7 2DE.
Library facilities: Open to all users (nc); no loan services; reprographic services (c).
Reprographic services are also available at the Information Point and BGS Information Services (c).
Library holdings: 500 000 bound volumes; 3000 current periodicals; 5000 reports.
Information services: Bibliographic services (no charge if work takes less than four hours); no translation services; literature searches.
No charge is made by the library for literature searches that take less than four hours; the Information Point and BGS Information Services occasionally charge for literature searches.
Access to on-line information retrieval systems will become available through the Information Point and BGS Information Services, with no charge for in-house retrieval systems.
Consultancy: Telephone answers to technical queries (nc); writing of technical reports (c); no market surveys in technical areas.
Copies can be made of published and unpublished BGS literature.
All these services are available through the Information Point and BGS Information Services.
Publications: Map catalogue; publications list; leaflets, including a leaflet on the BGS National Geosciences Data Centre.

British Museum (Natural History), Department of Library Services   7.102
Address: Cromwell Road, London SW7 5BD, UK
Telephone: (01) 589 6323
Enquiries to: Head, A.P. Harvey; deputy head, R.E.R. Banks
Founded: 1881
Subject coverage: The department is responsible for six subject libraries, including one in earth sciences. The five holding material in the biological sciences are listed below.
Library facilities: Open to all users (if bona fide research workers); no loan services; reprographic services (c).
Library holdings: The museum libraries contain approximately 750 000 volumes, 18 000 serial titles (of which some 10 500 are currently received), 70 000 maps, tens of thousands of prints, drawings andmanuscripts. The Museum Library is the repository for the archivesof the museum under the Public Records Act 1958 and 1967.
There are six subject libraries, including five in the biological sciences.
Information services: Bibliographic services (c); literature searches (c); access to on-line information retrieval systems (c).
Dialog.
Consultancy: Yes (c).
Publications: Catalogue of the books, manuscripts, maps and drawings in the British Museum (Natural History) 8 volumes; Serial Catalogue of the Books, Manuscripts, Maps and Drawings in publications in the British Museum (Natural History) Library the British Museum (Natural History) 8 volumes; Serial 3 volumes (a microfiche edition updated annually is available); Publications in the British Museum (Natural History) Library three Bulletin of the British Museum (Natural History); extensive volumes and a microfiche edition updated publishing programme; list of publications available. annually; Bulletin of the British Museum (Natural History); extensive publishing programme; list of publications available.

General Library   7.103
Telephone: (01) 589 6323, extension 382
Enquiries to: Librarian, S.L. Goodman
Subject coverage: The literature of natural history apart from that limited to one particular aspect.

Palaeontology and Mineralogy Library   **7.104**
Telephone: (01) 589 6323, extension 207, 679
Enquiries to: Librarian, A. Lum
Subject coverage: Palaeontology; regional geology; stratigraphy; history of geology; biographies of geologists and geological reference sources; physical anthropology and palaeoanthropology.
Mineral chemistry, mineral analysis, crystallography, mineral deposits, clay deposits, geochemistry, geophysics, light and colour, mineralogy, petrology, volcanology, tectonics, gemmology, meteoritics, lunar sciences, historical biographical and reference material relating to these areas.
Consultancy: Consultancy and advice are available from the Keeper of Palaeontology, Department of Palaeontology, and the Deputy Keeper of Mineralogy, Dr D.R.C. Kempe, Department of Mineralogy. On-the-phone answers to technical queries are available from the Department of Mineralogy; normally there is no charge. The writing of technical reports, for which there is a charge, is undertaken by the Department of Palaeontology, by arrangement only. The Department of Mineralogy will help with the writing of scientific papers only.
The Department of Palaeontology will identify fossil specimens; for this service a charge is made to institutions, but not normally to individuals.
The Department of Mineralogy will identify rocks, minerals and meteorites; this service, together with answers to related questions, forms part of the department's normally free enquiry service.

Institute of Oceanographic   7.105
Sciences *
– IOS
Address: Brook Road, Wormley, Godalming, Surrey GU8 5UB, UK
Telephone: (042 879) 4141
Telex: 858833 OCEANS G
Affiliation: Natural Environment Research Council
Enquiries to: Librarian, D.W. Privett
Founded: 1949
Subject coverage: Scientific research on physical, chemical, geological, and biological oceanography, both deep sea and coastal.
Library facilities: Open to all users (to bona fide research workers by appointment); loan services (provided items are not available through British Library Lending Division - nc); reprographic services (10 per print).
Library holdings: Comprehensive author, subject, and region indexes to all aspects of marine sciences are maintained. Extensive chart and atlas collection.
Information services: Bibliographic services (weekly listing of documents received and of items selected for indexing in library catalogues available on application); no translation services; access to on-line information retrieval systems.

Comprehensive library indexes are supplemented by access to computer-on-line data bases, such as ASFA, Biosis, Georef, etc.
Consultancy: Yes, provided by the institute's Marine Information and Advisory Service, MIAS - c. MIAS provides numerical oceanographic data and advice with emphasis on waves, currents, tides, sea levels.
Publications: *IOS Report; MIAS New Bulletin*; annual report; collected reprints, annually.

Institution of Mining and   7.106
Metallurgy
– IMM
Address: 44 Portland Place, London W1N 4BR, UK
Telephone: (01) 580 3802
Telex: 261410
Enquiries to: Librarian
Founded: 1892
Subject coverage: Economic geology, mining, mineral processing, and extractive metallurgy of non-ferrous and industrial minerals excluding coal.
Library facilities: Open to all users (non-members may be charged); loan services (nc); reprographic services (c).
Information services: Bibliographic services (c); no translation services; literature searches (c); access to on-line information retrieval systems (c).
Extensive and detail indexes to the international minerals industry literature available. On-line minerals industry bibliographic database, Immage, available.
Consultancy: No consultancy.
Publications: *IMM Abstracts*, 6 issues per year; microfiche version of 1892 - 1949 indexes available for sale in whole or part.

National Meteorological Library   7.107
Address: London Road, Bracknell, Berkshire RG12 2SZ, UK
Telephone: (0344) 420242
Telex: 849801
Affiliation: Meteorological Office
Enquiries to: Librarian, M.E. Crewe
Founded: 1870
Subject coverage: Meteorological and climatological data and literature; more limited coverage in related disciplines: fluid dynamics, hydrology, oceanography, planetary atmospheres.
Library facilities: Open to all users (prior notice to visit preferred - nc); loan services (within the UK only - nc); reprographic services (limited - c).
Visual aids library contains a large collection of slides and photographs for loan; a search fee is likely to be charged.
Library holdings: 150 000 bound volumes; 350 current periodicals; numerous reports.
Information services: Bibliographic services (c); no translation services (but many translations in

meteorology held); literature searches (c); access to on-line information retrieval systems (limited facility - c).

There may, if appropriate, be no charge for bibliographic services and literature searches.

Consultancy: Telephone answers to technical queries (may be referred to specialist staff of the Meteorological Office - c); writing of technical reports (on referral and under contract - c); no compilation of statistical trade data; n o market surveys in technical areas.

Introduction arranged to comprehensive range of weather services - c).

National Museum of Wales Library   7.108

Address: Cathays Park, Cardiff CF1 3NP, UK
Telephone: (0222) 397951
Enquiries to: Librarian
Founded: 1907
Subject coverage: Includes botany, geology, archaeology, industrial and maritime studies, and zoology.
Library facilities: Open to all users (nc - by arrangement); loan services (nc - via interlibrary loans); reprographic services (c).
The library is open from Tuesday to Friday.
Library holdings: 126 000 bound volumes, including reports; 1100 current periodicals.
Information services: No information services apart from very limited bibliographic services.
Consultancy: No consultancy.
Publications: Publications list on request.

Ordnance Survey   7.109
– OS

Address: Romsey Road, Maybush, Southampton SO9 4DH, UK
Telephone: (0703) 792687
Telex: 477843
Enquiries to: Manager, information and public relations
Founded: 1791
Subject coverage: Geodetic and topographic surveying.
Library facilities: Not open to all users; loan services (available to other libraries on written application or via BLLD); reprographic services (c).
The library is open to research workers - prior written application required.
Information services: Permanent exhibition centre - tours by prior arrangement. Ordnance Survey videos available for free loan.
Consultancy: Yes (c).

Soil Survey of England and Wales   7.110

Address: Rothamsted Experimental Station, Harpenden, Hertfordshire, UK
Telephone: (058 27) 63133
Affiliation: Agriculture and Food Research Council
Enquiries to: Head
Founded: 1939
Subject coverage: Soils and agriculture, forestry and other uses.
Library facilities: No library facilities.
Information services: No information services.
Consultancy: Telephone answers to technical queries; writing of technical reports; no compilation of statistical trade data; no market surveys in technical areas.
Soil surveys and interpretations; environmental studies.

Ulster Museum   7.111

Address: Botanic Gardens, Belfast BT9 5AB, UK
Telephone: (0232) 668251
Affiliation: Department of Education, Northern Ireland
Enquiries to: Director, Keeper of Botany and Zoology, or Keeper of Geology
Founded: 1831
Subject coverage: Includes botany and zoology, geology, and technology.
Library facilities: Open to all users (nc); loan services (nc); reprographic services (c).
Library holdings: 30 000 bound volumes; 200 current periodicals.
Information services: No bibliographic services; no translation services; literature searches (nc); no access to on-line information retrieval systems.
Consultancy: Telephone answers to technical queries (nc); writing of technical reports (c); no compilation of statistical trade data; no market surveys in technical areas.

YUGOSLAVIA

Hidrografski Institut Ratne Mornarice *   7.112
[Hydrographic Institute of the Yugoslav Navy]

Address: Zrinsko Frankopanska 66, 5800 Split, Yugoslavia
Telephone: (058) 444 33
Telex: 26270
Affiliation: Ministry of National Defence
Enquiries to: Assistant director, M. Sc. Benković
Founded: 1922

Subject coverage: Hydrographic surveys; cartography; oceanographic surveys; navigation; meteorology.
Library facilities: Open to all users; no loan services; no reprographic services.
Information services: Bibliographic services (nc); no translation services; no literature searches; no access to on-line information retrieval systems.
Consultancy: Yes (nc).

Muzejski Dokumentacioni Centar *   7.113
[Museum Documentation Centre]
Address: Zagreb Yugoslavia
Subject coverage: Museology; archaeology; natural history.
Publications: *Informatica Museologica.*

Srpsko Geološko Društvo   7.114
[Serbian Geology Society]
Address: POB 227, Kamenička 6/VI, Beograd, Yugoslavia
Library facilities: Library.

8 ENERGY SCIENCES

ALBANIA

Instituti i Fizikës Bërthamore 8.1
[Nuclear Physics Institute]Tiranë, Albania
Enquiries to: Director, S. Koja

AUSTRIA

Institut für Umweltforschung 8.2
– ifu
[Environmental Research Institute]
Address: Elisabethstrasse 11, Graz, Styria, Austria
Telephone: (316) 36030
Affiliation: Forschungsgesellschaft Joanneum - Graz
Enquiries to: Director, Dr Mert König
Founded: 1970
Subject coverage: Environmental research; regional planning; energy systems; energy technologies; biotechnology; housing research.
Consultancy: Telephone answers to technical queries (nc); writing of technical reports (c).

International Atomic Energy Agency 8.3
– IAEA
Address: PO Box 100, Vienna International Centre, Wagramerstrasse 5, A-1400 Wien, Austria
Telephone: (0222) 23600
Telex: 1-12 645
Enquiries to: Director general, Dr Hans Blick
Publications: Quarterly bulletin.

Österreichische Gesellschaft für Sonnenenergie und Weltraumfragen Gesellschaft mbH * 8.4
[Austrian Solar and Space Agency]
Address: Garnisongasse 7, A-1090 Wien, Austria
Telephone: (0222) 4381770
Telex: 116560
Enquiries to: Helmut Hummer
Founded: 1972
Subject coverage: Solar energy, space research and technology.
Information services: No bibliographic services; no translation services; literature searches; access to on-line information retrieval systems.
Quest (IRS of ESA); Dialog (Lockheed).
Consultancy: Yes.

Österreichische Gesellschaft für Erdölwissenschaften 8.5
– ÖGEW
[Austrian Association for Petroleum Sciences]
Address: PO Box 309, Erdbergstrasse 72, A-1031 Wien, Austria
Telephone: (0222) 73 23 48
Telex: 132138
Enquiries to: Secretary, Dr Herbert Lang
Founded: 1960
Subject coverage: Petroleum sciences.
Library facilities: No library facilities.
Information services: No information services.

Österreichisches Forschungszentrum Seibersdorf GmbH 8.6
– FZS
[Seibersdorf Research Centre of Austria]
Address: Lenaugasse 10, A-1082 Wien, Austria
Telephone: (0222) 427511
Telex: 07-5400
Enquiries to: INIS Liaison officer, Dr Maria Tisljar

Founded: 1956
Subject coverage: Peaceful applications of nuclear energy and radiation; basic research in nuclear theory, reactor components, nuclear fuels, measuring, control, and information techniques; environmental research; radiation protection; social diseases and pharmaceutics; isotope application.
Library facilities: No library facilities.
Information services: Literature searches; access to on-line information retrieval systems;.
Inis Inspec; ESA-RECON; Lockheed; SDC; INKA data bases. Inis SDI services.
Consultancy: Yes.

Vienna International Centre Library * 8.7
– VICL
Address: PO Box 100, Wagramerstrasse, A-1400 Wien, Austria
Telephone: (0222) 2360
Telex: 12645
Affiliation: International Atomic Energy Agency; United Nations Industrial Development Organization; and United Nations agencies in Vienna.
Enquiries to: Head, Library, Wilson H. Neale
Founded: 1979
Subject coverage: Nuclear and allied sciences; applied economics; social sciences; drugs and narcotics.
Library facilities: Open to all users; loan services (nc - inter-library loans) reprographic services.
Current awareness bulletins; selected dissemination of information service; periodicals circulation service.
Information services: Bibliographic services (nc); no translation services; literature searches (nc); access to on-line information retrieval systems (c).
Library on-line (LION) - VIC items catalogued; library acquisitions (LIAC) - VIC Library purchases; United Nations documents (UNDOC).
INIS (world nuclear literature), *AGRIS* (world agricultural literature), and *IDA* (UNIDO documents), are maintained by other VIC units but used by VIC library.
Consultancy: No consultancy.
Publications: Film catalogue 1984; films on the peaceful uses of atomic energy. Serials titles 1984. VIC library on-line bibliographic data bases directory 1982. Library services brochure in English/1982 and German/1983.

BELGIUM

Association Royale des Gaziers Belges 8.8
– ARGB
[Royal Belgian Gas Association]
Address: PO Box 11, B-1640 Rhode-Saint-Gènese, Belgium
Telephone: (02) 358 37 66
Enquiries to: Head, Ir M. Desprets
Founded: 1877
Subject coverage: Gas technology.
Library facilities: Not open to all users; no loan services; reprographic services (c).
The library is not available to non-staff users, but information is available on request. Current bibliographic file of the Association Technique de l'Industrie du Gaz en France.
Information services: No bibliographic services; no translation services; no literature searches; no access to on-line information retrieval systems.
Consultancy: Yes.

Fédération Européenne des Producteurs Autonomes et des Consommateurs Industriels d'Énergie/Europäischer Verband für industrielle Energiewirtschaft * 8.9
– FEPACE
[European Federation of Autonomous Producers and Industrial Consumers of Energy]
Address: 57 Boulevard Louis Schmidt, B-1040 Bruxelles, Belgium
Telephone: (02) 734 56 65
Telex: 63 511
Enquiries to: Secretary-general, Henri van der Haert
Founded: 1955
Library facilities: Not open to all users.
Information services: Bibliographic services; translation services; literature searches.
Consultancy: Yes.

Institut National des Industries Extractives Institut National des Industries Extractives 8.10

– INIEX – INIEX
[National Institute Extractive Industries National Institute of Extractive Industries]
Address: Rue du Chéra 200, B-4000 Rue du Chéra 200, Liège, Liége, Belgium Belgium
Telephone: (41) 52 71 50 *Telephone:* (41) 52 71 50

Telex: 41128 B *Telex:* 41128 B
Affiliation Ministry of Economic Affairs *Affiliation:*
Ministry of Economic Affairs
Enquiries to: Public relations, Loes van Mechelen
Public relations, Loes van Mechelen
Founded: 1967 *Founded:* 1967
Subject coverage: Mines; quarries; energy; safety;
radiocommunications; pollution; polymers; concrete.
Mines; quarries; energy; safety; radiocommunications;
pollution; polymers; concrete.
Library facilities: Open to all users (nc); loan services
(nc); reprographic services (c) Open to all users (nc);
loan services (nc); reprographic services (c).
Library holdings: 8000 bound volumes; 300 current
periodicals; 6000 reports. 8 000 bound volumes; 300
current periodicals; 6 000 reports.
Information services: Bibliographic services (nc); no
translation services; literature searches (nc); no access
to on-line information retrieval systems Bibliographic
services (nc); no translation services; literature searches
(nc); no access to on-line information retrieval systems.
Consultancy: Telephone answers to technical
queries; no writing of technical reports; no compilation
of statistical trade data; no market surveys in technical
areas Telephone answers to technical queries; no
writing of technical reports; no compilation of
statistical trade data; no market surveys in technical
areas.
Publications: Annual report. Annual report.

Studiecentrum voor Kernenergie/ 8.11
Centre d'Étude de l'Énergie
Nucléaire
– SCKG
[Nuclear Energy Study Centre]
Address: Boeretang 200, B-2400 Mol, Belgium
Telephone: (014) 31 18 01
Telex: Atomol 31922
Affiliation: Ministry of Economic Affairs
Enquiries to: M.-Cl. Bouten
Founded: 1952
Subject coverage: Nuclear energy; energy sources;
pollution.
Library facilities: Open to all users; loan services;
reprographic services (c).
Information services: Bibliographic services (c);
literature searches (c); access to on-line information
retrieval systems (c).
Lockheed; ESA/IRS; Systems; Inis.
Consultancy: Yes (c).
Publications: *BLG-reports*; annual report.

BULGARIA

Scientific and Technical Union of 8.12
Energetics
Address: Rakovski 108, 1000 Sofia, Bulgaria

CZECHOSLOVAKIA

Ústav pro Výzkum a Využití 8.13
Paliv
– ÚVP
[Fuel Research Institute]
Address: 250 97 Praha 9 Běchovice, Czechoslovakia
Telephone: 73 53 51
Telex: 121333 ins b
Facsimile: USTAVPALIV PRAHA
Enquiries to: Head of Information Centre, Miroslav
Starý
Subject coverage: Gas industry; coal upgrading; fuel
utilization.
Library facilities: Open to all users; loan services
(c); reprographic services (c).
Services can be provided free of charge on an exchange
basis.
Information services: Bibliographic services (c);
translation services (c); literature searches (c); no
access to on-line information retrieval systems.
Consultancy: Yes (nc).

Ústřední Informační Středisko pro 8.14
Jaderný Program
– ÚISJP
**[Czechoslovak Atomic Energy Commission, Centre for
Scientific and Technical Information]**
Address: Zbraslav nad Vitavou Czechoslovakia
Telephone: 59 15 83

DENMARK

DANATOM * 8.15
**[Danish Association for the Industrial Development of
Atomic Energy]**
Address: Baltorpvej 154, 2750 Ballerup, Denmark
Telephone: (02) 36 00 22
Telex: 35177
Affiliation: Danish Academy of Technical Sciences
Enquiries to: General secretary, Arne Jensen

Founded: 1956
Library facilities: No library facilities.
Consultancy: Yes.

Danmarks Tekniske Højskole, 8.16
Laboratoriet for Varmeisolering *
[Technical University of Denmark, Thermal Insulation Laboratory]
Address: Building 118, DK-2800 Lyngby, Denmark
Telephone: (02) 883511
Telex: 37529 DTHDIA DK
Enquiries to: Librarian and manager, Bjarne Saxhof

Founded: 1959
Subject coverage: Research and higher education in the field of thermal insulation and energy saving (low-energy houses, solar energy, thermal storage, thermal comfort).
Library facilities: Open to all users (by appointment only); loan services (nc, but occasionally a deposit is asked for); no reprographic services (except for out-of-stock laboratory publications; users may copy material at the laboratory - charge made).
Consultancy: Yes (c).
Publications: List on request.

Risø Bibliotek * 8.17
[Risø Library]
Address: PO Box 49, DK-4000 Roskilde, Denmark
Telephone: (02) 37 12 12
Telex: 43116 risoe dk
Affiliation: Forsøgsanlag (Risø National Laboratory)
Enquiries to: Chief librarian, Eva Pedersen
Founded: 1957
Subject coverage: Nuclear energy and alternative energy sources.
Library facilities: Open to all users; loan services; reprographic services.
Library holdings: 500 000 technical reports.
Information services: Bibliographic services; literature searches; access to on-line information retrieval systems.
Inis, IRS, Dialog, SDC, DIANE. The library is the technical processing centre for NEI (Nordic Energy Index).
Consultancy: Yes.

FINLAND

Säteilyturvakeskus 8.18
– STUK
[Finnish Centre for Radiation and Nuclear Safety]
Address: PO Box 268, SF-00101 Helsinki 10, Finland
Telephone: (90) 61671

Telex: 122691 STUK-SF
Enquiries to: Librarian, Armi Länkelin
Founded: 1958
Subject coverage: Radiation protection; nuclear safety.
Library facilities: Open to all users; loan services; reprographic services.
Library holdings: 20 000 bound volumes; 290 current periodicals; 15 000 reports.
Information services: Bibliographic services (nc); no translation services; literature searches (nc); no access to on-line information retrieval systems.
Consultancy: Telephone answers to technical queries (nc); no writing of technical reports; no compilation of statistical trade data; no market surveys in technical areas.
Publications: List on request.

Teknillinen Korkeakoulu 8.19
[Helsinki University of Technology, Department of Technical Physics]
Address: Rakentajanaukio 2 C, SF-02150 Espoo, Finland
Telephone: (90) 4512474
Telex: 125161
Enquiries to: Librarian, Silja Rummukainen
Subject coverage: Material physics; nuclear engineering; information technology; energy.
Library facilities: Loan and reprographic services are available through the main university library.
Library holdings: 500 000 reports in the field of nuclear science and energy.
Publications: List on request.

FRANCE

Association Technique de 8.20
l'Industrie du Gaz en France *
– ATG
[French Gas Association]
Address: 62 rue de Courcelles, 75008 Paris, France
Telephone: (1) 47 66 03 51
Telex: 642621F
Affiliation: Union Internationale de l'Industrie du Gaz en France
Enquiries to: Chef du département documentation-éditon, M.C. Dedeystere
Founded: 1874
Subject coverage: Production, treatment, storage, transport, distribution, and utilization of natural gas, liquid petroleum gas, manufactured gas, biogas, and hydrogen.
Library facilities: Open to all users; no loan services; reprographic services (c).

Information services: Bibliographic services (c); translation services (c); literature searches (c); access to on-line information retrieval systems (c). Minitel.

Commissariat a'AAC l'Énergie Atomique, Service de Documentation   8.21
– CEA
[Atomic Energy Commission, Documentation Service]
Address: CEA/CEN Saclay, F-91191 Gif-sur-Yvette Cedex, France
Telephone: (69) 08 22 08
Telex: 691379 F CEN Saclay
Enquiries to: Head
Founded: 1952
Subject coverage: Nuclear science and engineering; energy.
Library facilities: Open to all users (nc); no loan services; reprographic services.
Library holdings: 70 000 bound volumes; 1800 current periodicals; 450 000 reports.
Information services: Bibliographic services (c) (SDI - Inis); no translation services; literature searches (c); access to on-line information retrieval systems (c).
International Nuclear Information System (Inis) of the AIEA, Vienna, Energy Data Base (EDB) of the US DOE, and other systems. The service produces the data base *MEETING AGENDA* which is available on-line through Questel, and contains announcements of scientific and technical meetings and exhibitions.
Consultancy: No consultancy.
Publications: Catalogue des collections de périodiques de la bibliothèque centrale de CEA; CEA reports and notes.

Électricité de France, Direction des Études et Recherches   8.22
– EdF
[French Electricity, Study and Research Section]
Address: 1 avenue du Général de Gaulle, 92141 Clamart, France
Telephone: (1) 47 65 41 37
Telex: 20434 FEDFNORM
Facsimile: (1) 47 65 31 24
Enquiries to: Head, J.P. Allard
Founded: 1946
Subject coverage: Generation transmission, and distribution of electric power.
Library facilities: Open to all users (by special arrangement); no loan services; reprographic services (only original documents referenced in EDF-DOC data base - c).
Library holdings: About 68 000 bound volumes; 1500 current periodicals; 100 000 reports.

Information services: No bibliographic services; no translation services; no literature searches; access to on-line information retrieval systems.
The EDF data base is available through two hosts: ESA/IRS; Questel.
Consultancy: Yes (occasionally).
Publications: Documentation Technique, monthly; Thesaurus EDF.

Gaz de France   8.23
Address: 23 rue Philibert-Delorme, Paris 17e, France

Institut Français du Pétrole   8.24
– IFP
[French Petroleum Institute]
Address: 1 et 4 avenue de Bois-Preau, BP 311, 92506 Rueil-Malmaison Cedex, France
Telephone: (1) 47 49 02 14
Telex: IFP A 203050 F
Facsimile: (1) 47 32 30 92
Enquiries to: Head, M. Moureau.
Documentation and Information Analysis Centre
Founded: 1945
Subject coverage: All scientific, technical, and economic subjects related to oil, natural gas, fossil energy, and renewable energy.
Library facilities: Open to all users (nc); loan services (nc) (interlibrary only); reprographic services (subject to 'fair use' and current copyright laws - c).
Library holdings: 230 000 bound volumes; 3300 series (bound except for the current year); millions of patents, standards, reports, theses, laws etc. The library is particularly rich in Eastern European countries' literature, especially the USSR, and literature from China is increasingly being added to its holdings.
Information services: Bibliographic services (charge for major enquiries); translation services (c - only existing translations, indexed in *World Transindex*, none on demand); literature searches (c); access to on-line information retrieval systems (c).
Specialized data banks on tanker and platform accidents - c.
Consultancy: Telephone answers to technical queries (nc); writing of technical reports (c); no compilation of statistical trade data; no market surveys in technical areas.
Statistical Study of patents trends worldwide (c).
Courses: The institute participates in courses run by French universities and specialized schools.
Publications: List and further details on request from Head, Publication Service of IFP; annual report.

Institut Laue-Langevin   8.25
– ILL
[Laue-Langevin Institute]
Address: 156 X, 38042 Grenoble Cedex, France
Telephone: (76) 48 70 20
Telex: ILL A320-621F

Affiliation: Centre National de la Recherche Scientifique (France); Kernforschungszentrum Karlsruhe (German FR); Science and Engineering Research Council (UK)
Enquiries to: Librarian, C. Castets
Founded: 1969
Subject coverage: Solid-state physics; nuclear physics; physical chemistry; materials science; biochemistry. Central facility: thermal neutron research reactor.
Library facilities: Open to all users (nc); no loan services; reprographic services (c - articles less than 20 pages).
Library holdings: 20 000 bound volumes; 200 current periodicals; 9000 reports, including theses.
Information services: Bibliographic services (in-house bibliographic information on thermal neutron scattering experiments); no translation services; no literature searches; no access to on-line information retrieval systems.
Consultancy: Telephone answers to technical queries (nc - restricted to in-house research); writing of technical reports (30 in-house technical reports per annum); no compilation of statistical trade data; no market surveys in technical areas.
Publications: Annual report; *ILL Experimental Reports and Theory College Activities; News for Reactor Users*, newsletter; *Neutron Research Facilities at the ILL High Flux Reactor*; bibliography: publications related to ILL neutron scattering measurements.

OECD Nuclear Energy Agency *   8.26
– OECD/NEA
Address: 38 Boulevard Suchet, 75016 Paris, France
Telephone: (1) 45 24 82 00
Telex: 630 668
Affiliation: Organization for Economic Cooperation and Development
Enquiries to: Librarian, S. Godwin
Founded: 1958
Subject coverage: Applications of nuclear energy; safety, radioactive waste management, economics of nuclear power, nuclear law.
Library facilities: Not open to all users; no reprographic services.
Very restricted loan or reference facilities.
Information services: Very restricted information services.
Consultancy: No consultancy.

Societé Française pour la Gestion   8.27
des Brevêts d'Application
Nucléaire
[French Nuclear Patents Society]
Address: 25 rue de Ponthieu, 75008 Paris, France
Telex: Spibrev 660028 F
Parent: Commissariat à l'Énergie Atomique

Enquiries to: A. Mongredien.
Patent Department
Subject coverage: Patent agent.

GERMAN DEMOCRATIC REPUBLIC

Wissenschaftlich-Technisches   8.28
Gesellschaft für
Energiewirtschaft der Kammer
der Technik
[Energy Industry Scientific and Technical Society of the Chamber of Technology]
Address: Clara-Zetkin-Strasse 115/117, 1080 Berlin, German Democratic Republic
Enquiries to: Secretary, D. Kothe

GERMAN FEDERAL REPUBLIC

Fachinformationszentrum   8.29
Energie, Physik, Mathematik
GmbH
– FIZ Karlsruhe
[Energy, Physics, and Mathematics Information Centre]
Eggenstein-Leopoldshafen 2, D-7514, German Federal Republic
Telephone: (07247) 82 4600/4601
Telex: 17724710
Enquiries to: Online and Marketing Manager, Dr B. Jenschke
Founded: 1977
Subject coverage: Production of data bases and data compilations in energy and technology, aeronautics and astronautics, space research, physics and astronomy, mathematics and computer science.
Library facilities: Open to all users (nc); loan services (c); reprographic services (c).
Library holdings: 1.9m documents of non-conventional literature (reports, conference papers, theses, standards, patents, publications of firms).
Information services: Bibliographic services; no translation services; literature searches (c); access to on-line information retrieval systems (c).
INKA; STN International (on-line service offered jointly by FIZ Karlsruhe and the American Chemical Society). Magnetic tape services, information supply (retrospective searches, SDI profiles, technical information). Editing, distribution of printed

information; document delivery.

Consultancy: Telephone answers to technical queries; no writing of technical reports; no compilation of statistical trade data; no market surveys in technical areas.

Publications: *Facts and Figures*, information leaflet.

Institut für Erdölforschung 8.30
[Petroleum Research Institute]

Address: Walther-Nernst-Strasse 7, D-3392 Clausthal-Zellerfeld, German Federal Republic

Telephone: (05323) 711100

Founded: 1942

Library facilities: Open to all users (nc); no loan services; reprographic services (c).

Library holdings: 8220 bound volumes; 26 current periodicals.

Information services: No information services.

Consultancy: Telephone answers to technical queries (nc); no writing of technical reports; no compilation of statistical trade data; no market surveys in technical areas.

Publications: Annual report.

Institut für Kernenergetik und 8.31
Energiesysteme
– IKE
[Nuclear Power and Energy Systems Institute]

Address: Postfach 80 11 40, D-7000 Stuttgart 80, German Federal Republic

Telephone: (0711) 685 1 2490

Telex: 07255 445 univ

Affiliation: Universität Stuttgart

Enquiries to: Dr Inge Friedrich.

Bibliothek

Kerntechnischer Ausschuss 8.32
– KTA
[Nuclear Safety Standards Commission]

Address: Postfach 10 16 50, Schwertnergasse 1, D-5000 Köln, German Federal Republic

Telephone: (0221) 2068-0

Telex: 8881807 GRSD

Enquiries to: Executive director, W. Schwarzer

Founded: 1972

Subject coverage: Development of nuclear safety standards.

Library facilities: No library facilities.

Information services: No information services.

Consultancy: No consultancy.

Publications: Bibliographic list.

Physikalisch-Technische 8.33
Bundesanstalt Braunschweig und
Berlin
– PTB
[Federal Institute of Metrology in Brunswick and Berlin]

Address: Postfach 3345, Bundesallee 100, D-3300 Braunschweig, German Federal Republic

Telephone: Braunschweig: (0531) 5920; Berlin: (030) 34811

Telex: Braunschweig 9-52822 ptb d

Parent: Bundesminister für Wirtschaft

Enquiries to: Deputy head, library, E. Bretnütz.

Alternative address for PTB: Abbestrasse 2-12, D-1000 Berlin 10, German FR

Founded: 1887

Subject coverage: The PTB is the national institute for science and technology and the highest technical authority for metrology in the Federal Republic of Germany.

Research into: physics and technology, realization and dissemination of SI units, realization and propagation of legal time, realization of the international practical temperature scale, precision determination of physical constants, type testing and approval of measuring devices, slot machines and civilian firearms, type testing in the field of safety technology, radiation protection, medicine and traffic control, technical testing and storage of nuclear fuel, long-term storage and final disposal of radioactive waste. Cooperation in national and international technical committees, elaboration of technical regulations and directives, test work on commission and scientific and technological consultation, aid in metrological training and development overseas.

Library facilities: Open to all users (nc); loan services (nc); reprographic services (nc).

Special library for selected areas in science and technology, mainly for providing staff members with desired literature and information from external information centres. Reference library with restricted loans. The PIB is incorporated in the national library loan service.

Library holdings: 100 800 bound volumes; 860 current periodicals; 5000 reports.

Information services: No bibliographic services; no translation services; no literature searches; access to on-line information retrieval systems (for internal use only).

Consultancy: No consultancy.

Publications: *PTB-Mitteilungen Forschen und Prüfen*, bimonthly; *Jahresbericht*, annual; *PTB-Berichte*, irregular series of reports in the various research areas of the PTB; *PTB-Prüfregeln* etc, various directives and instructions on calibration and technical standards.

GREECE

Greek Atomic Energy Commission 8.34

Address: PO Box 60228, 15310 Aghia Paraskevi Attikis, Athinai, Greece
Telephone: (01) 6513111
Telex: 216199
Affiliation: Ministry of Industry, Research and Technology
Enquiries to: Head librarian, Noria Christophoridou

Founded: 1954
Subject coverage: Use of nuclear energy in various fields of science.
Library facilities: Open to all users (nc); no loan services; reprographic services (c).
Library holdings: 54 528 bound volumes (40 556 journals, 13 972 books); 1187 current periodicals; 107 349 reports.
Information services: Bibliographic services (nc); no translation services; no literature searches; no access to on-line information retrieval systems.
Consultancy: Telephone answers to technical queries (nc); no writing of technical reports; no compilation of statistical trade data; no market surveys in technical areas.
Publications: Research reports.

HUNGARY

Magyar Ásványolaj és Földgáz Kisérleti Intézet * 8.35
– MÁFKI
[Hungarian Oil and Gas Research Institute]
Address: PO Box 167, József Attila útca 34, H-8201 Veszprém, Hungary
Telephone: (080) 12-440
Telex: 32288 mafki-h
Enquiries to: Chief librarian, Dr Csiszár Miklós
Founded: 1948
Subject coverage: Crude oil and natural gas chemistry; petrochemicals.
Library facilities: Not open to all users; loan services; reprographic services.
Information services: No information services.
Consultancy: No consultancy.

Magyar Tudományos Akadémia Atommag Kutató Intézete 8.36
[Nuclear Research Institute of the Hungarian Academy of Sciences]
Address: Bemtér 18 c, 4026 Debrecen, Hungary
Telephone: (52) 17 266
Telex: 72210 atom h
Enquiries to: Director, Professor D. Berényi
Library holdings: 48 000 volumes.

ICELAND

Orkustofnun 8.37
– OS
[National Energy Authority]
Address: Grensásvegur 9, 108 Reykjavík, Iceland
Telephone: 83600
Telex: 2339 orkust is
Affiliation: Ministry of Industry
Enquiries to: Director general, Jakob Björnsson
Subject coverage: Geological, geodetic, hydrological, and hydraulic research for hydro-power developments; geophysical and geological research for geothermal projects; energy statistics.

IRELAND

Bord Fuinnimh Nuicleigh * 8.38
[Nuclear Energy Board]
Address: 20-22 Lower Match Street, Dublin 2, Ireland
Telephone: (01) 766223
Telex: 30610
Enquiries to: Librarian, Eleanor Griffin
Founded: 1973
Subject coverage: Nuclear energy.
Library facilities: Open to all users; loan services (nc); reprographic services (nc).
Information services: No translation services; literature searches (nc).
Consultancy: Yes (nc).
Publications: Annual report; periodical newsletters; *Measurements of Radioactivity of Precipitation-settled Dust and Airborne Particles in Ireland; Report on Radioactivity in the Irish Sea.*

Bord na Mona 8.39
[Peat Development Authority]
Address: Lower Baggot Street, Dublin 2, Ireland
Telephone: (01) 688555
Telex: 30206 monaei

Enquiries to: Technical information officer
Founded: 1951
Subject coverage: Peat production and use for fuel and in horticulture and agriculture; peatland reclamation.
Library facilities: Open to all users; loan services; reprographic services.
Information services: Bibliographic services; translation services (translations made for staff are available to non-staff); literature searches.
Consultancy: Consultancy (in peat fuel utilization only - c).
Publications: *Peat Abstracts*, triannual; annual reports.

Electricity Supply Board 8.40
– ESB
Address: Lower Fitzwilliam Street, Dublin 2, Ireland
Telephone: (01) 765831
Telex: 25313
Enquiries to: Head of Library and Information Services, James C. O'Reilly
Founded: 1926
Subject coverage: Electricity generation and supply.
Library facilities: Open to all users (by appointment only); loan services (to libraries only - nc); reprographic services (nc).
Information services: No bibliographic services; translation services (c); no literature searches; no access to on-line information retrieval systems.
Consultancy: Telephone answers to technical queries (nc); no writing of technical reports; no compilation of statistical trade data; no market surveys in technical areas.

ITALY

Centro Informazioni Studi 8.41
Esperienze SpA *
– CISE
[Information, Studies and Experiments Centre]
Address: PO Box 12081, 39 Via Reggio Emilia, 20100 Milano, Italy
Telephone: (02) 2167 1
Telex: 311643 CISE I
Affiliation: ENEL (National Electricity Agency)
Enquiries to: Head, documentation office, Dr P.A. Comero.
Centre of Bibliographic Information
Founded: 1946
Subject coverage: Energy conservation; solar energy; nuclear energy; thermohydraulics; industrial diagnostics; materials; environmental surveillance; lasers; electronics components and systems; data

acquisition and processing; analytical chemistry.
Library facilities: Not open to all users; no loan services; reprographic services (c).
Access to non-staff users is granted case by case. The centre keeps KWOC indexes of books, periodicals and technical reports.
Information services: Bibliographic services (c); no translation services; literature searches (c); access to on-line information retrieval systems (c).
The centre has access to the following: ESA Quest; Dialog; SDC Orbit; INKA; Data-Star; Télésystèmes-Questel; Pergamon InfoLine; Echo.
Consultancy: Yes (c).

Fiat TTG SpA, Documentazione 8.42
e Biblioteca *
[Fiat TTG SpA, Documentation Office and Library]
Address: PO Box 599, Via Cuneo 20, 10152 Torino, Italy
Telephone: (011) 2600290
Telex: 221050
Enquiries to: Librarian and Information Officer, Dr F. Ferrero
Founded: 1906
Subject coverage: Manufacture of electric energy and cogeneration plants; energy in general.
Library facilities: Not open to all users; loan services; reprographic services (c).
The library is not available to non-staff users (some particular cases excepted). Cooperation with technical and scientific organizations; liaison with national professional associations.
Information services: No bibliographic services; no translation services; literature searches; access to on-line information retrieval systems.
Bibliographic data base organization in progress; bibliographic and subject-based computerized searches.
Consultancy: Yes (nc).

Forum Italiano dell'Energia 8.43
Nucleare *
– FIEN
[Italian Nuclear Energy Forum]
Address: Via Paisiello 26-28, 00198 Roma, Italy
Telephone: (06) 844 2587
Telex: 610183 ENEA (for FIEN)
Affiliation: FORATOM - Forum Atomique Européen

Enquiries to: Secretary General, Pietro Bullio
Founded: 1958
Subject coverage: Nuclear energy and nuclear power.
Library facilities: Not open to all users (to non-FIEN members).
The Forum Library covers all aspects of nuclear energy since 1950, but particularly the industrial, economic, administrative, management, public

information and international aspects.

Publications: *Atoma e Industria* (semi-monthly in Italian and in English); *Energia nucleare: una scelta responsabile* (Nuclear Energy: a Responsible Choice - information book in Italian on nuclear power); *Energia: Energia Nucleare* (information brochure, published in Italian in cooperation with the Swiss Association for Atomic Energy); *Enrico Fermi: the Meaning of a Discovery; Congress Proceedings* (annually, in cooperation with the National Committee for Nuclear Energy and Alternative Energy Sources (ENEA), the proceedings of the Nuclear Congress of Rome); reports and other documentary material, such as the booklet *Nuclear Italy* illustrating Italian research and development and industrial activities in the nuclear field.

Laboratori Nazionali di Frascati   8.44
– LNF

[Frascati National Laboratories]
Address: PO Box 13, 00044 Frascati (Roma), Italy
Telephone: (06) 94031
Telex: 614122 LNFI
Affiliation: Istituto Nazionale di Fisica Nucleare
Enquiries to: Librarian
Founded: 1955
Subject coverage: Nuclear physics.
Library facilities: Open to all users; no loan services; no reprographic services.
Consultancy: Yes.
Publications: *Bollettino d'Informazione*, 3 issues a year.

Stazione Sperimentale per i   8.45
Combustibili *

[Fuels Experimental Station]
Address: Viale Alcide De Gasperi 3, 20097 San Donato Milanese, Italy
Telephone: (02) 510031
Telex: 321622 SSC
Affiliation: Ministero dell'Industria, del Commercio e dell'Artigianato
Enquiries to: Librarian, Matteo Fasano
Founded: 1940
Subject coverage: Research analysis, testing, and consulting in the field of solid, liquid, and gaseous fuels.
Library facilities: Open to all users; no loan services; reprographic services (c).
Information services: Bibliographic services; no translation services; literature searches; no access to on-line information retrieval systems.
Consultancy: Yes.
Publications: *La Rivista dei Combustibili*, monthly; miscellaneous monographs.

NETHERLANDS

Gemeenschappelijk Centrum voor   8.46
Onderzoek GCO/JRC *

[Joint Research Centre]
Address: Postbus 2, 1755 ZG Petten, Netherlands
Telephone: (02246) 5656
Telex: 57211
Affiliation: Commission of the European Communities

Enquiries to: Head, B. Seysener.
Library and Documentation
Founded: 1959
Subject coverage: Nuclear science; materials science.
Library facilities: Open to all users; loan services (c); reprographic services (c).
Information services: Bibliographic services (c); translation services (c); literature searches (c); access to on-line information retrieval systems (c).
The centre has access to the following on-line retrieval systems: LIRS, ESA, INKA, CAS, Echo.
Consultancy: Yes (c).

NV Kema Library *   8.47
Address: Postbus 9035, 6800 ET Arnhem, Netherlands
Telephone: (085) 45 70 57
Telex: 450 16
Enquiries to: A.J. te Grotenhuis
Founded: 1920
Subject coverage: Electrical engineering; nuclear science; environmental pollution; materials testing.
Library facilities: Not open to all users; loan services; reprographic services.
Information services: Literature searches; access to on-line information retrieval systems.
The library has on-line access to the following: SDC; ESA; Inis.

Nationaal Instituut voor   8.48
Kernfysica en Hoge-Energie
Fysica, sectie-Kernfysica
– NIKHEF-K

[National Institute for Nuclear and High-Energy Physics, Nuclear Physics Division]
Address: Postbus 4093, Ooster Ringdijk 18, 1009 AJ Amsterdam, Noord-Holland, Netherlands
Telephone: (020) 5922080
Telex: IKO 11538 NL
Enquiries to: Librarian, N. Kuijl
Founded: 1946
Subject coverage: Nuclear physics; radiochemistry.
Library facilities: Open to all users (nc); loan services (nc); reprographic services (c).

Library holdings: 12 000 bound volumes; 147 current periodicals; 1000 reports.
Information services: No bibliographic services; no translation services; no literature searches; access to on-line information retrieval systems (c).
Consultancy: No consultancy.
Publications: Annual report; list on request.

Nederlandse Gasunie NV * 8.49
[Netherlands Gas Union]
Address: Laan Corpus den Hoorn 102, PO Box 19, 9700 MA Groningen, Netherlands
Telephone: (050) 212035
Telex: 53448
Enquiries to: Librarian, J.G. Bourgonje
Founded: 1963
Subject coverage: Transportation of natural gas; pipeline construction; chemical engineering.
Library facilities: Open to all users (by prior arrangement); loan services (nc); reprographic services (nc).
Information services: No information services.
Consultancy: No consultancy.

Rijks Geologische Dienst 8.50
– RGD
[Geological Survey of the Netherlands]
Address: Postbus 157, 2000 AD Haarlem, Netherlands
Telephone: (023) 319362
Telex: 71105 GEOLD
Affiliation: Ministry of Economic Affairs
Enquiries to: Librarian, Annelies Duivenvoorden
Founded: 1903
Subject coverage: Geology of the Netherlands, in particular oil and gas; quaternary geology.
Library facilities: Open to all users (nc); loan services (nc); reprographic services (nc).
Library holdings: 10 000 bound volumes; 500 current periodicals; 2000 reports.
Information services: No information services.
Consultancy: No consultancy.
Publications: List on request.

NORWAY

Institutt for Energiteknikk 8.51
– IFE
[Energy Technology Institute]
Address: PO Box 40, N-2007 Kjeller, Norway
Telephone: (02) 71 25 60
Telex: 16361 energ n
Enquiries to: Librarian, Solveig Opheim

Founded: 1953 (as Institutt for Atomenergi; name and mandate changed in *Founded:* 1980)
Subject coverage: Energy modelling and system studies; energy conservation; new energy technologies; reactor safety, fuelling and control; isotope production and applications; irradiation technology; process simulation and control; materials technology; petroleum reservoir studies; multi-phase flow studies; condensed matter (physics).
Library facilities: Open to all users; loan services (c); reprographic services (c).
The institute's library is a depositary library for publications issued by the International Atomic Energy Agency; it has a special collection of scientific and technical reports. The main part of the collection is on microfiche.
Information services: No bibliographic services; no translation services; literature searches (c); access to on-line information retrieval systems (c).
The library has on-line access to the following: Nordic Energy Index (NEI); INKA On-line Service; Dialog.
Consultancy: Yes (informal).

Norges Vassdrags- og Energiverk 8.52
– NVE
[Norwegian Water Resources and Energy Office]
Address: Postboks 5091, Majorstua 0301, Oslo 3, Norway
Telephone: (02) 46 98 00
Telex: 71912 nveso n
Enquiries to: Head, Public Relations Office, Øystein Skarheim
Founded: 1920
Subject coverage: Water resources; electricity.
Library facilities: Open to all users; loan services (inter-library); reprographic services.
Information services: Literature searches.
Consultancy: Yes.

Oljedirektoratet * 8.53
[Norwegian Petroleum Directorate]
Address: PO Box 600, N-4001 Stavanger, Norway
Telephone: (04) 533160
Telex: 33 100 NOPED
Enquiries to: Head Librarian, Nina Korbu
Founded: 1974
Subject coverage: Exploration and production of oil and gas; petroleum geology; all aspects of petroleum activities in Norway.
Library facilities: Open to all users; loan services (nc); reprographic services (nc).
Information services: Bibliographic services (nc); no translation services; literature searches (nc); access to on-line information retrieval systems (only in-house users, apart from searches in library's own on-line catalogue which is open to all users - nc).
Consultancy: No consultancy.

POLAND

Centrum Mechanizacji Górnictwa Komag - Glicwe * 8.54
– CMG KOMAG
[KOMAG Mining Mechanization Centre]
Address: Ulica Pszczýnska 37, 44-101 Gliwice, Poland
Telephone: 374-334
Telex: 03629 KOMAG PL
Enquiries to: Manager, Karol Osádnik.
Scientific, Technical and Economic Information Centre
Founded: 1945
Subject coverage: Mining mechanization, mining machines and devices.
Library facilities: Open to all users (c); loan services (c); reprographic services (c).
Library holdings: 35 240 bound volumes; 394 current periodicals; 2329 reports.
Information services: Bibliographic services (c); translation services (c); literature searches (c); no access toon-line information retrieval system
Consultancy: Yes.
Publications: *Informacja Expressowa* (Express Information), current bibliography on mining mechanization based on Polish and foreign mining journals, monthly.

Instytut Chemii i Techniki Jadrowej 8.55
[Nuclear Chemistry and Technology Institute]
Address: Ulica Dorodna 16, 03-195 Warszawa, Poland
Telephone: 11 06 56
Telex: 813 027
Enquiries to: Director, Dr J. Leciejewicz
Subject coverage: Radiobiology; nuclear fuel processing.
Library holdings: 30 000 volumes.

Instytut Energetyki, Branzowy Ósrodek Informacji Naukowej, Technicznej i Ekonomicznej * 8.56
[Power Institute, Branch Centre of Scientific, Technical and Economic Information]
Address: Ulica Mory 8, 01-330 Warszawa, Poland
Telephone: 022-36 75 51
Telex: 813 824
Enquiries to: Head of Branch Centre, Wieslawa Szulc
Subject coverage: Economy of energy, organization and management, demand and utilization; energy sources; energy systems; production of heat and electrical energy; transmission and distribution;

automatic control, measurement, automatic regulation.
Library facilities: Open to all users; loan services; reprographic services.
Information services: Bibliographic services; translation services; literature searches; no access to on-line information retrieval systems.
Consultancy: Yes.

Instytut Górnictwa Naftowego i Gazownictwa 8.57
– IGNiG
[Oil and Gas Institute]
Address: Ul. Lubicz 25a, 30-960 Kraków, Poland
Telephone: 094-21 00 33
Telex: 0325276
Enquiries to: Librarian, Dr Wincenty Pawlłowski.
Głłówna Biblioteka Branzowa, Branzowy Ośrodek Informacji Naukowej, Technicznej i Ekonomicznej
Founded: 1944
Subject coverage: Geology, geophysics, geochemistry, and geomicrobiology in relation to oil and gas exploration and engineering; processing and treatment of fuel gases; transport, distribution, and storage of gaseous fuels; drilling equipment and technology; exploitation of oil and gas deposits; applications of gas fuels.
Library facilities: Open to all users (nc); loan services (nc); reprographic services (c).
Microfilming.
Library holdings: 57 000 bound volumes, including brochures and pamphlets; 421 current periodicals.
Information services: Bibliographic services (c); translation services (c); literature searches (c); no access to on-line information retrieval systems.
Consultancy: Telephone answers to technical queries (nc); writing of technical reports (c); compilation of statistical trade data (c); no market surveys in technical areas.
Publications: *Prace Instytutu Górnictwa Naftowego i Gazownictwa; Biuletyn Informacyjny Górnictwa Naftowego i Gazownictwa*, quarterly periodical.

Instytut Problemóur Jadrowych 8.58
[Nuclear Studies Institute]
Address: 05-400 Otwock-Swierk, Poland
Telephone: 793 481
Telex: 813 244
Enquiries to: Director, Professor Jan Turkiewicz
Library holdings: 26 000 volumes.

Resortowy Ośrodek Informacji Naukowej, Technicznej i Ekonomicznej Górnictwa i Energetyki 8.59
[Mining and Energy Information Centre]
Address: Palace of Culture and Science, 00-901 Warszawa, Poland
Telephone: (022) 269673
Telex: 816416
Affiliation Ministry of Mining and Energy
Enquiries to: Manager, J. Kiełłbasiński
Founded: 1958
Subject coverage: Energy.
Library facilities: Open to all users (nc); loan services (nc); reprographic services (c).
Library holdings: 13 405 bound volumes; 5 450 current periodicals; 395 806 reports.
Information services: Bibliographic services (c); translation services (c); literature searches (c); no access to on-line information retrieval systems.
Consultancy: Telephone answers to technical queries (nc); writing of technical reports (c); compilation of statistical trade data (c); market surveys in technical areas (c).

PORTUGAL

Centro de Informação Técnica para a Indústria * 8.60
– CITI
[Technical Information Centre for Industry]
Address: Azinhaga dos Lameiros à Estrada do Paço do Lumiar, 1699 Lisboa Codex, Portugal
Telephone: (01) 758 61 41
Telex: 42 486
Affiliation: Laboratório Nacional de Engenharia e Tecnologia Industrial
Enquiries to: Director, Dr Ana Maria Ramalho Correia
Founded: 1979
Subject coverage: Industrial technology (chemical, food, metallurgy, electrical and electronics); energy (conventional, renewable, and nuclear).
Library facilities: Open to all users; loan services (nc); reprographic services (c).
Information services: Bibliographic services (c); no translation services; literature searches (c); access to on-line information retrieval systems (c).
CITI has access to Dialog and Télésystèmes - Questel, and expects to have access to other systems in the near future. Patent information is also supplied.
Consultancy: Yes (c).
Publications: Bibliographic list.

Departamento de Energia Nuclear * 8.61
[Nuclear Energy Department]
Address: Avenida da República 45 - 5o, 1000 Lisboa, Portugal
Telephone: (01) 769753
Affiliation: Direcção Geral de Energia
Enquiries to: Deputy Director-General for Energy, H. Carreira Pich
Founded: 1978
Subject coverage: Coordination and planning of activities concerning nuclear power plants and the nuclear fuel cycle; public information and international relationships in the field of nuclear energy.
Library facilities: Not open to all users; loan services (nc); reprographic services (nc).
The library facilities are available mainly to staff users.
Information services: No information services.
Consultancy: No consultancy.

Direccão Geral de Energia * 8.62
– DGE
[General Directorate of Energy]
Address: 241 Rua da Beneficência, 1093 Lisboa Codex, Portugal
Telephone: (01) 771091
Telex: 14755 ENERG P
Affiliation: Ministério da Industria, Energia e Exportação
Enquiries to: Maria da Piedade Roberto
Subject coverage: Energy planning; legislation; alternative sources of energy; energy policy; energy economics; energy conservation; energy statistics.
Library facilities: Open to all users; loan services (nc); reprographic services (nc).
Information services: Bibliographic services (nc); translation services (nc); literature searches (nc); access to on-line information retrieval systems (c).
Consultancy: No consultancy.
Publications: Économie de l'Énergie.

Electricidade de Portugal * 8.63
– EDP
[Electricity of Portugal]
Address: Avenida José Malhoa, A13, 1000 Lisboa, Portugal
Telephone: (01) 723013
Telex: 15563 EDP EC P
Enquiries to: Head of Information and Documentation Department, Olímpio Neves Gonçalves
Founded: 1976
Subject coverage: Electricity.
Library facilities: Not open to all users; loan services; reprographic services.
Exhibition and reproduction of microfilms: extracts of articles; compilation of bibliographies; systematic

search; retrospective search; search through terminal.
Information services: Bibliographic services;
translation services; literature searches; access to on-
line information retrieval systems.
Access through terminal to the following data banks:
NSBS - Economy of Energy; EDP data bank,
Operational Body of Generation and Transmission
Equipment.
Consultancy: Yes.

PETROGAL, Petróleos de Portugal EP   8.64
[Petrogal, Oil of Portugal]
Address: Pátio do Pimenta 25, 1200 Lisboa, Portugal
Telephone: (01) 363131
Telex: 12521 AEGALPP
Enquiries to: Manager, Manuela Azevedo.
Centro de Documentacão e Informaçã o
Founded: 1976
Subject coverage: Oil industry; aromatics.
Library facilities: Open to all users (nc); loan services
(nc); reprographic services (c).
Library holdings: 25 000 bound volumes; 500 current
periodicals.
Information services: Bibliographic services (c); no
translation services; literature searches (c); access to
on-line information retrieval systems (c).
Consultancy: Telephone answers to technical queries
(c); no writing of technical reports; compilation of
statistical trade data (c); no market surveys in technical
areas; E (nc).
Publications: Bibliographic bulletin, monthly; list on
request.

Petroquímica e Gás de Portugal EP *   8.65
– PGP
[Petrochemicals and Gas of Portugal]
Address: Apartado 1933, Avenida António Augusto
de Aguiar 104 - 40, 1004 Lisboa Codex, Portugal
Telephone: (01) 538801
Telex: 12864 PETRO P
Enquiries to: Head of Documentation Service
Founded: 1957
Subject coverage: Production of town gas, ammonia,
and industrial gases; import, transport, and
distribution of piped gas; production and sale of
phthalates and phthalate plasticizers.
Library facilities: Open to all users; loan services
(nc); reprographic services (nc).
Information services: Bibliographic services (nc); no
translation services; literature searches (nc); no access
to on-line information retrieval systems.
Consultancy: No consultancy.

ROMANIA

Institutal Central de Cercetări Energetice, Officiul de Informare Documentară Pentru Energetică   8.66
– ICCE-OIDE
**[Energy Documentary Information Office, Central
Energy ResearchInstitute]**
Address: Bulevardul Energeticienilor 8, sector 3, cd
79619, Bucureşti, Romania
Telephone: 206730/235
Telex: icce r 10783
Enquiries ·to: Head, Vasile Pleşca
Founded: 1974
Library facilities: No library facilities.
Information services: Bibliographic services (c);
translation services (c); literature searches (c); access
to on-line information retrieval systems (c).
Consultancy: No consultancy.

SPAIN

Junta de EnergíaNuclear *   8.67
[Nuclear Energy Commission]
Address: Avenida Complutense 22, Madrid - 3, Spain
Telephone: 91-4496400
Telex: 23555 JUVIG E
Affiliation: Ministerio de Industria y Energía
Enquiries to: Head, Servicio Documentación
Founded: 1954
Subject coverage: Energy and allied sciences and
technology.
Library facilities: Open to all users; loan services
(staff only); reprographic services.
Publications: *Energía Nuclear*, bi-monthly review;
Reports JEN; Guíassobre Seguridad Nuclear.

SWEDEN

AB Ångpanneföreningen *   8.68
– ÅF
[Swedish Steam Users' Association]
Address: PO Box 8133, S-104 20 Stockholm, Sweden

Telephone: 08-234600
Telex: 10361 Energi S
Enquiries to: Librarian, Karin Gartzell
Founded: 1895

Subject coverage: Energy - technology and supply; process technology - pulping; environmental pollution; design of electrical and heating/ventilating installations for industry.
Library facilities: Open to all users; loan services (c); reprographic services (c).
Information services: Bibliographic services (c); translation services (c); literature searches (c); access to on-line information retrieval systems (c).
Access to the following data base systems: Dialog; ESA-Quest; SDC Orbit; Nordic Energy Index; Byggdok/BODIL; Va-Nytt.
Consultancy: Yes (c).

Energiforskningsnämnden 8.69
– EfN
[Energy Research Commission]
Address: Box 43020, S-100 72 Stockholm, Sweden
Telephone: 08-7449725
Telex: 15531 enrecoms
Enquiries to: Librarian, Åke Hügard
Founded: 1982
Library facilities: Open to all users (nc); loan services (nc); reprographic services (c).
Library holdings: 7000 bound volumes; 100 current periodicals.
Information services: Bibliographic services (nc); no translation services; literature searches (nc); no access to on-line information retrieval systems.
Consultancy: No consultancy.
Publications: Newsletter (in Swedish); reports (mainly in Swedish).

Svenska Elverksföreningen, 8.70
Föreningen för Elektricitetens
Rationella Användning
– SEF/FERA
[Swedish Association of Electricity Supply, Society for Rational Use of Electricity]
Address: Box 6405, S-113 82 Stockholm, Sweden
Telephone: 08-225890
Telex: 13593 gaselle s
Enquiries to: Editor, Bo Gustrin
Founded: 1903
Subject coverage: Production, distribution, and rational and safe use of electrical energy.
Library facilities: Open to all users; no loan services; reprographic services (c).
Information services: No information services.
Consultancy: Yes (c).

Svenska Petroleum Institutet * 8.71
– SPI
[Swedish Petroleum Institute]
Address: Sveavägen 21, S-111 34 Stockholm, Sweden
Telephone: 08-235800
Telex: 103 24

Enquiries to: Information manager, Lars Pehrzon
Founded: 1951
Library facilities: No library facilities.
Information services: No information services.
Consultancy: No consultancy (except to member organizations).

SWITZERLAND

Eidgenössisches Institut für 8.72
Reaktorforschung/Institut
Fédéral de Recherches en
Matière de Réacteurs *
– EIR
[Swiss Federal Institute for Reactor Research]
Address: CH-5303 Würenlingen, Switzerland
Telephone: (056) 99 21 11
Telex: 537 14 eir ch
Affiliation: Federal Institute of Technology
Enquiries to: Chief librarian, Elisabeth Dubois; Assistant for information and documentation, Rolf Schmid
Founded: 1955
Subject coverage: Nuclear energy and its applications, and related topics.
Library facilities: Not open to all users; loan services (Swiss inter-library loans and international loans between libraries); reprographic services.
Information services: Bibliographic services; no translation services; literature searches; access to on-line information retrieval systems.
Lockheed (Dialog) and ESA-IRS (and others through Euronet). The institute is in charge of Swiss input activities for Inis.

Organisation Européenne pour la 8.73
Recherche Nucléaire *
– CERN
[European Organization for Nuclear Research]
Address: CH-1211 Genève 23, Switzerland
Telephone: (022) 83 61 11
Telex: 419 000 CER CH
Enquiries to: Head, Dr Alfred Günther.
Scientific Information Service
Founded: 1954
Subject coverage: High-energy physics.
Library facilities: Open to all users (with prior agreement); loan services (interlibrary only); no reprographic services.
Information services: Bibliographic services (by special agreement); literature searches (by special agreement).

Publications: List on request.

Schweizerische Vereinigung für Atomenergie   8.74

– SVA

[Swiss Association for Atomic Energy]

Address: PO Box 2613, CH-3001 Bern, Switzerland

Telephone: (031) 22 58 82

Telex: 912 110 atag

Facsimile: (031) 22 92 03

Affiliation: Secretariat of SVA run by: Atag - General Auditing Company Limited, Department for Public Relations and Information

Enquiries to: Dr P. Bucher; Dr P. Haehlen

Founded: 1958

Subject coverage: Nuclear energy; nuclear engineering; application of ionizing radiation and/or radioactive isotopes.

Library facilities: Open to all users; loan services (for members and organizations); reprographic services (c).

Information services: Bibliographic services (c); translation services (from and into German, French and English - c).

Consultancy: Yes (direct to members of SVA, to non-members through Atag - c.).

Publications: *Schweizer Atomjahrbuch,* annually; *SVA Bulletin/Bulletin ASPEA,* fortnightly; *Kernpunkte/Flash Nucléaire,* every 10 days.

TURKEY

Ankara Nükleer Araştirma ve Eğitim Merkezi   8.75

– ANAEM

[Ankara Nuclear Research and Training Centre]

Address: Beşevler, Ankara, Turkey

Telephone: (41) 234439

Telex: atom tr 42581

Affiliation: Turkish Atomic Energy Council

Enquiries to: Director, Professor Uğur Büget

Founded: 1966

Subject coverage: Research in physics, chemistry, agriculture, electronics, etc, based on peaceful uses of atomic energy.

Library facilities: Open to all users; loan services; no reprographic services.

Information services: Literature searches; access to on-line information retrieval systems. Inis.

Consultancy: Yes.

Publications: Annual progress report; *Turkish Journal of Nuclear Sciences,* biannual.

UNITED KINGDOM

Atomic Energy Research Establishment   8.76

Address: Harwell, Didcot, Oxfordshire OX11 0RB, UK

Telephone: (0235) 24141

Telex: 83135

Affiliation: United Kingdom Atomic Energy Authority

Founded: 1946

Subject coverage: Nuclear science and technology; metallurgy; ceramics and materials in general; engineering; inorganic and analytical chemistry; chemical engineering; physics; health physics; computer science; hazardous materials; energy technology; environmental science.

Library facilities: The Harwell Library facilities are intended primarily for staff members.

Information services: Bibliographic services; no translation services; no literature searches; no access to on-line information retrieval systems.

RECAP - data base of UKAEA publications; BULLETIN - data base of publications relevant to AERE programme of work (also available as a weekly printed information bulletin); NDT - data base of information on non-destructive testing; LIBCAT - on-line library catalogue; Inis SDI - SDI service from Inis magnetic tapes.

Consultancy: No consultancy.

Publications: List of publications available.

British Carbonization Research Association   8.77

– BCRA

Address: Wingerworth, Chesterfield, Derbyshire S42 6JS, UK

Telephone: (0246) 76821

Telex: BCRA BSC Chem Stavly 54211

Enquiries to: Information officer

Founded: 1944

Subject coverage: Coke production and uses; coal tar processing and uses; allied fuels.

Library facilities: Open to all users (non-members by appointment); loan services; reprographic services.

Information services: Bibliographic services; translation services (translations prepared for staff available to non-members; D (subject to demand from staff and members).

Consultancy: Yes (generally provided by non-library staff - c).

Publications: List on request.

British Gas Corporation, Library * 8.78

Address: 59 Bryanston Street, London W1A 2AZ, UK
Telephone: (01) 723 7030
Telex: 261710
Enquiries to: Librarian
Subject coverage: Nationalized gas industry.
Library facilities: Not open to all users; loan services; reprographic services.
Information services: Bibliographic services; translation services; literature searches (very limited); access to on-line information retrieval systems.

British Nuclear Forum * 8.79
– BNF
Address: 1 St Alban's Street, London SW1Y 4SL, UK
Telephone: (01) 930 6888/9
Telex: 264476
Enquiries to: Assistant director, J.H. Green
Founded: 1963
Subject coverage: All aspects of nuclear power.
Library facilities: No library facilities.
Information services: No bibliographic services; no translation services; literature searches (nc); no access to on-line information retrieval systems.
Provision of information on nuclear power to all sectors of the public, including schools.
Consultancy: Telephone answers to technical queries (nc); no writing of technical reports; no compilation of statistical trade data; no market surveys in technical areas.
Courses: Technical symposia at irregular intervals.
Publications: Monthly bulletin.

Coal Research Establishment * 8.80
– CRE
Address: Stoke Orchard, Cheltenham, Gloucestershire GL52 4RZ, UK
Telephone: (0242 67) 3361
Telex: 43568
Affiliation: National Coal Board
Enquiries to: Head of Reports and Intelligence Branch, D.C. Davidson
Founded: 1948
Subject coverage: Coal utilization; development of new and improved systems for combustion and handling of coal; fuel and appliance testing; liquefaction of coal by solvent extraction and hydrocracking; pyrolysis of coal and coal-derived materials; metallurgical coke and smokeless fuel manufacture; development of new products; utilization of colliery wastes; measurement and control of atmospheric pollution.
Library facilities: Open to all users; loan services (nc); reprographic services.

Information services: Bibliographic services (c, occasionally); no translation services; no literature searches; no access to on-line information retrieval systems.
Consultancy: Yes (c).
Publications: Annual report.

Department of Energy 8.81
Address: Thames House South, Millbank, London SW1P 4QJ, UK
Telephone: (01) 211 3394
Telex: 918777 ENERGY G
Enquiries to: Librarian
Subject coverage: Energy resources policy and technology; economics of energy.
Library facilities: Loan services (through national lending scheme only); reprographic services (restricted service - c).
The library is open to bona fide researchers by appointment only.
Information services: No information services.
Consultancy: No consultancy.
Publications: *Publications in Print*, annual listing of all Department of Energy publications in print; *Current Energy Information*, weekly - contains contents pages of energy-related journals, some references to relevant articles, forthcoming conferences; library reading lists.

Electricity Council * 8.82
Address: 30 Millbank, London SW1P 4RD, UK
Telephone: (01) 834 2333
Telex: 23385
Enquiries to: R.G. Hancock
Founded: 1958
Subject coverage: Public electricity supply.
Library facilities: Open to all users (by prior arrangement); no loan services; no reprographic services.
Information services: Bibliographic services (limited to abstracts bulletins and bibliographies); translation services (for the electricity supply industry).
Primarily intended for Council Headquarters, the information services are also available to the Electricity Boards and Consultative Councils; and, so far as appropriate, to external enquirers.
Publications: *Electricity Supply Statistics*, annual; library bulletin; occasional papers.

Institute of Petroleum 8.83
– IP
Address: 61 New Cavendish Street, London W1M 8AR, UK
Telephone: (01) 636 1004
Telex: 264380
Enquiries to: Head of Library/Information Department, Jean Etherton
Founded: 1913

Subject coverage: All aspects of the oil industry, both technical and commercial.
Library facilities: Open to all users; loan services (to members); reprographic services (c).
Information services: Bibliographic services; no translation services (but will refer to external translator); literature searches (c); access to on-line information retrieval systems (c).
Access to SDC, Dialog and Datastar systems, each of which are hosts to numerous data bases.
Consultancy: Yes.
Publications: *Petroleum Review*, monthly; technical papers; data sheets; booklets.

Institution of Chemical Engineers 8.84
– IChemE
Address: Geo. E. Davis Building, 165-171 Railway Terrace, Rugby CV21 3HQ, UK
Telephone: (0788) 78214
Telex: 311780
Enquiries to: Information Officer, A. Strauch
Founded: 1923
Subject coverage: Chemical engineering: safety and loss prevention; environmental science; energy conservation; physical properties data service.
Library facilities: Open to all users (nc); loan services (charge for postage only); reprographic services (c).
Library holdings: 6000 bound volumes; 100 current periodicals.
Information services: Bibliographic services (charge for non-members); literature searches (charge for non-members); no access to on-line information retrieval systems.
On-the-spot translation assistance is given for Dutch and German.
Literature searches are made using the in-house data base, which contains 32 000 abstracts, input at a rate of about 4000 a year. The library also acts as a clearing house and referral service, and keeps a list of consultant chemical engineers, and a guide to manufacturers and suppliers of chemical plant.
Consultancy: Telephone answers to technical queries (no charge for easily answered queries).
Compilation of a quarterly list of conferences and courses in subjects of interest to chemical engineers.
Publications: Publications list available on request.

National Centre for Alternative 8.85
Technology *
– NCAT
Address: Llwyngwern Quarry, Machynlleth, Powys, UK
Telephone: (0654) 2400
Affiliation: Society for Environmental Improvement
Enquiries to: Information officer, Felicity Shooter
Founded: 1974

Subject coverage: Alternative sources of energy; organic gardening; sewage disposal; building energy-conserving structures.
Library facilities: Not open to all users; no loan services; reprographic services (c).
The library is a fairly small part of the information service; much of the information is in files, accessible through enquiries only at present. There is a large alternative technology bookshop with international mail order service.
Information services: An enquiry answering service is available by telephone or post.
Consultancy: Yes (c).
Publications: Bibliographic list.

National Coal Board 8.86
Address: Hobart House, Grosvenor Place, London SW1X 7AE, UK
Telephone: (01) 235 2020

National Engineering Laboratory 8.87
– NEL
Address: East Kilbride, Glasgow G75 0QU, UK
Telephone: (03552) 20222
Telex: 777888
Affiliation: Department of Trade and Industry
Enquiries to: Head, T. Archbold.
Library and Information Services
Founded: 1947
Subject coverage: Mechanical engineering. Topics covered include flow measurement and distribution, turbomachinery, offshore engineering, alternative energy (wave and wind), materials technology, and power systems engineering. NEL is the custodian of the national standards of flow.
Library facilities: Open to all users (at librarian's discretion); loan services (through British Library Document supply Centre).
Information services: Bibliographic services; no translation services; literature searches.
Discretionary charges are made for bibliographic services and literature searches.
The library has on-line access to the following data bases: Orbit (SDC); Dialog (Lockheed); IRS (ESA); INKA GRIPS (Fachinformationszentrum Technik eV); Pergamon InfoLine; Télésystèmes-Questel.
Consultancy: Telephone answers to technical queries; no writing of technical reports; no compilation of statistical trade data; no market surveys in technical areas.
NEL undertakes consultancy for a wide range of United Kingdom manufacturing industry.
Courses: A wide range of conferences, seminars and courses.
Publications: *Index of NEL Publications; A Resource for Industry; Introducing NEL.*

National Radiological Protection Board

8.88

– NRPB
Address: Chilton, Didcot, Oxfordshire OX11 0RQ, UK
Telephone: (0235) 831600
Telex: 837124
Facsimile: (0235) 833891
Enquiries to: Information officer
Founded: 1970
Subject coverage: Radiological protection, including health and safety aspects of nuclear power; biological/ medical effects of ionizing and non-ionizing radiations; radioactivity in consumer products; environmental radioactivity including natural radiation; dosimetry.
Library facilities: Open to all users (by arrangement - nc); loan services (to other libraries - nc); no reprographic services.
The library answers queries on publications excluding those of the board. The information office answers queries of a general nature and on the publications of the board.
Library holdings: 10 000 bound volumes; 100 current periodicals; 5000 reports.
Information services: Bibliographic services (nc); no translation services; no literature searches; no access to on-line information retrieval systems.
Consultancy: No consultancy.
Courses: Regular courses on various aspects of radiological protection.

United Kingdom Atomic Energy Authority

8.89

– UKAEA
Address: 11 Charles II Street, London SW1Y 4QP, UK
Telephone: (01) 930 5454
Telex: 22565 ATOMLO
Facsimile: 01-930 5454 extension 274
Affiliation: Department of Energy
Enquiries to: Press Office
Founded: 1954
Subject coverage: Nuclear power.
Library facilities: Open to all users (by appointment only); no loan services; no reprographic services.
Information services: No information services.
Consultancy: Telephone answers to technical queries (if answer cannot be given immediately, information can usually be obtained - nc); no writing of technical reports; no compilation of statistical trade data; no market surveys in technical areas.
Publications: List on request.

YUGOSLAVIA

Institut za Nuklearne nauke 'Boris Kidric'

8.90

[Boris Kidric Institute of Nuclear Sciences]
Address: POB 522, Vinca, Beograd, Yugoslavia
Telephone: 440 871
Subject coverage: Low energy nuclear physics; solid-state physics; nuclear power problems reactor engineering; systems science and information.
Library holdings: 26 000 volumes.

9 ENGINEERING, CIVIL

AUSTRIA

Bundesversuchs- und Forschungsanstalt Arsenal/ Geotechnisches Institut *

9.1

– BVFA-Arsenal

[Federal Testing and Research Establishment Arsenal, Geotechnical Institute]

Address: Franz Grill-Strasse 9, POB 8, A-1031 Wien, Austria

Telephone: (0222) 78 25 31

Telex: 136677

Affiliation: Ministry for Construction and Technology

Enquiries to: Head, Professor Dr E. Schroll. Dokumentationsstelle für Strassenwesen und Verkehrstechnik

Founded: 1950

Subject coverage: Geochemistry; applied mineralogy and petrology; geophysics; hydrogeology and applied geology; soil mechanics; road construction techniques; traffic engineering; transport sciences; documentation and information services.

Library facilities: Not open to all users; no loan services; reprographic services (c). Reading room copies only.

Information services: Bibliographic services (c); translation services (c); literature searches (c); access to on-line information retrieval systems (c).

IRRD - International Road Research Documentation of OECD, Paris; ICTED - International Cooperation in the Field of Transport Economics of ECMT, Paris; IRB - Informationsverbundzentrum Raum und Bau, Stuttgart, Federal Republic of Germany.

Consultancy: Yes (c).

Publications: Bibliographic list.

Forschungsinstitut des Vereins der Österreichischen Zementfabrikanten

9.2

[Research Institute of the Association of Austrian Cement Manufacturers]

Address: Reisnerstrasse 53, A-1030 Wien, Austria

Telephone: (0222) 75 66 81-0

Affiliation: Verein der Österreichischen Zementfabrikanten

Founded: 1951

Subject coverage: Cement and concrete research and information.

Library facilities: Open to all users (upon request - nc); loan services (nc); reprographic services (c).

Library holdings: 600 bound volumes; about 300 reports.

Information services: No bibliographic services; no translation services; no literature searches; access to on-line information retrieval systems (c). Advisory service for the building industry (nc).

Consultancy: Telephone answers to technical queries (nc); writing of technical reports (normally no charge); no compilation of statistical trade data; no market surveys in technical areas.

Courses: Courses on concrete technology, annual.

Publications: Papers; research reports. List available on request.

Österreichisches Institut für Bauforschung *

9.3

[Austrian Institute for Building Research]

Address: An den langen Lüssen 1/6, A-1190 Wien, Austria

Telephone: (0222) 32 57 88

Enquiries to: Librarian, Dr Franz Rottensteiner

Founded: 1959

Subject coverage: Building research; housing research; standard tenders; EDV application in architecture; solar energy.

Library facilities: Open to all users; no loan services; reprographic services (c).

Information services: No information services.

BELGIUM

Centre Belge d'Étude et de Documentation de l'Eau   9.4
– CEBEDEAU
[Belgian Centre for Water Research and Documentation]
Address: 2 Rue Armand Stévart, B-4000 Liège, Belgium
Telephone: (041) 52 12 33
Enquiries to: D. van den Ackerveken
Founded: 1947
Subject coverage: Water and corrosion technology: analyses; wastewater treatment; industrial and domestic uses of water; corrosion measurement; self purification of rivers; waste treatment; valorization.
Library facilities: Open to all users; no loan services; reprographic services.
Information services: Bibliographic services; literature searches.
Consultancy: Yes.

Centre d'Information du Bitume   9.5
[Bitumen Information Centre]
Address: 351 Boulevard Emile Bockstael, B-1020 Bruxelles, Belgium
Telephone: (02) 478 25 89
Enquiries to: Manager, J.P. Dabin
Founded: 1958
Subject coverage: All technical information relating to bitumen and asphalt.
Library facilities: Open to all users (nc); loan services (nc - within Belgium only); reprographic services (nc).
Information services: Bibliographic services (nc - normally within Belgium only); no translation services; literature searches (normally within Belgium only); no access to on-line information retrieval systems.
Consultancy: Telephone answers to technical queries (nc); writing of technical reports (nc); no compilation of statistical trade data; no market surveys in technical areas.
These services normally apply within Belgium only.

Centre National de Recherches Scientifiques et Techniques pour l'Industrie Cimentière   9.6
– CRIC
[National Centre for Scientific and Technical Research for the Cement Industry]
Address: Rue César Franck 46, B-1050 Bruxelles, Belgium
Telephone: (02) 649 98 50
Enquiries to: Director, Y. Dechamps
Founded: 1960

Subject coverage: Research and quality control in the field of cement fabrication (including quarrying and grinding), cement-based materials production, concretes (precast, reinforced, prestressed) and concrete construction (roads, bridges, housing, etc).
Library facilities: Open to all users (nc); no loan services; reprographic services (c).
Library holdings: 11 000 bound volumes; 100 current periodicals; 700 reports.
Information services: Bibliographic services (c); no translation services; literature searches (c); no access to on-line information retrieval systems.
Consultancy: Telephone answers to technical queries (nc); no writing of technical reports; no compilation of statistical trade data; no market surveys in technical areas.
Publications: Research reports of the CRIC (in French and Dutch); list on request.

Centre Scientifique et Technique de la Construction/ Wetenschappelijk en Technisch Centrum voor het Bouwbedrijf   9.7
– CSTC/WTCB
[Building Scientific and Technical Centre]
Address: Rue du Lombard 41/Lombardstraat 41, B-1000 Bruxelles, Belgium
Telephone: (02) 513 60 50
Telex: 25416 cetex b
Enquiries to: Librarian, D. Forton
Founded: 1959
Subject coverage: Scientific and technical research and studies (engineering, building, main walls, foundations, heating, ventilation, air-conditioning, plumbing, sanitary engineering, joinery, glazing, painting, coating, insulation, acoustics, water- and air- tightness, etc); management of building contractor firms; technical assistance to contractors; documentation.
Library facilities: Open to all users (nc); no loan services; reprographic services (c).
Library holdings: 6600 books; 600 current periodicals; 200 reports.
Information services: Bibliographic services (nc); no translation services; literature searches (nc); access to on-line information retrieval systems (c).
Consultancy: Telephone answers to technical queries (nc); writing of technical reports (c).
Publications: *Activités et Perspectives*; list on request.

CYPRUS

Cyprus Civil Engineers and Architects Association

9.8

Address: PO Box 1825, Zena de Tyras Palace, Nicosia, Cyprus
Telephone: 41 221
Library facilities: Library.

CZECHOSLOVAKIA

Československé Stredisko Vystavoy a Architektury
– ČSVA

9.9

[Czechoslovak Building Centre]
Address: Václavská Náměstí 31, 111 21 Praha 1, Czechoslovakia
Telephone: (02) 265841-3
Telex: 122056 csva c
Enquiries to: Head of Information Department, Lubomír Obr
Founded: 1968
Subject coverage: National information system of products for building construction.
Library facilities: Open to all users (nc); no loan services; no reprographic services.
Library holdings: 6000 bound volumes; 220 current periodicals.
Information services: No bibliographic services; no translation services; literature searches (c); no access to on-line information retrieval systems.
Consultancy: Telephone answers to technical queries (nc); writing of technical reports (nc); no compilation of statistical trade data; market surveys in technical areas (nc).
Material testing (c).
Courses: Seminars or publicity days for local and foreign manufacturers, about nine per annum.

Ústav Stavebních Informací*
– ÚSI

9.10

[Building Information Centre]
Address: Národní Třída 10, 116 87 Praha 1, Czechoslovakia
Enquiries to: Director, Václav Ontl
Founded: 1967
Subject coverage: Central information institution in the building and civil engineering fields (building materials, structures, management, investment construction).

Library facilities: Open to all users; loan services (nc); reprographic services (nc).
Library holdings: The library collects all the available Czechoslovak literature and information sources in the building fields and a selection from foreign sources - books, periodicals, reports, patents, standards.
Information services: Bibliographic services (nc); no translation services; literature searches (nc); access to on-line information retrieval systems (nc).
ÚSI maintains the off-line bibliographic base of the building fields, which is produced through the international cooperation of the MCNTI-members (International Centre of Scientifical and Technical Information in Moscow, USSR). ÚSI also has the on-line access to foreign data bases accessible in ČSSR through an agreement with the following data base centres: Datastar; INKA-FIZ; Questel.
Consultancy: No consultancy.
Publications: Bibliographic list.

DENMARK

Byggeriets Studiearkiv
– BSA

9.11

[Danish National Centre for Building Documentation]
Address: Peder Skramsgade 2D, DK-1054 København, Denmark
Telephone: (01) 126860
Affiliation: Royal Academy of Fine Arts
Enquiries to: Head, Steen Estvad
Founded: 1930
Subject coverage: Architecture; housing; planning; structural and civil engineering.
Library facilities: Open to all users (nc); loan services (nc); reprographic services (c).
VIDEOTEK - regular video shows (nc).
Information services: Bibliographic services (c); no translation services; literature searches (no charge for students); access to on-line information retrieval systems (no charge for students).
Consultancy: Telephone answers to technical queries (no charge for students); no writing of technical reports; no compilation of statistical trade data; no market surveys in technical areas.
Courses: Lectures on architecture; courses in library use; seminars on topical matters.
Publications: *Byggreferat*, abstracts journal, twelve issues per year; list on request.

Danmarks Tekniske Højskole, Laboratoriet for Varmeisolering * 9.12
[Technical University of Denmark, Thermal Insulation Laboratory]
Address: Building 118, DK-2800 Lyngby, Denmark
Telephone: (02) 883511
Telex: 37529 DTHDIA DK
Enquiries to: Librarian and manager, Bjarne Saxhof
Founded: 1959
Subject coverage: Research and higher education in the field of thermal insulation and energy saving (low-energy houses, solar energy, thermal storage, thermal comfort).
Library facilities: Open to all users (by appointment only); loan services (nc, but occasionally a deposit is asked for); no reprographic services (except for out-of-stock laboratory publications; users may copy material at the laboratory - charge made).
Consultancy: Yes (c).
Publications: List on request.

FINLAND

Rakennustieto Saatio * 9.13
[Building Information Centre]
Address: Ruusankatu 3, 00250 Helsinki 25, Finland
Telephone: 90-64 84 01

Teknillinen Korkeakoulu * 9.14
[Helsinki University of Technology, Department of Civil Engineering]
Address: Rakentajanaukio 4, SF-02150 Espoo, Finland
Telephone: 90-4512414
Telex: 125161
Enquiries to: Librarian, Eeva-Liisa Parkkonen
Subject coverage: Highway and railway engineering; bridge engineering; hydraulic engineering; water resources engineering; structural engineering; foundation engineering and soil mechanics; structural mechanics; construction economics and management; traffic and transportation engineering; concrete technology and steel structures.
Library facilities: Loan and reprographic services are available through the main library of Helsinki University of Technology.
Publications: List on request.

Vesihallitus, Kirjasto 9.15
[National Water Board, Library]
Address: PO Box 250, SF-00101 Helsinki, Finland
Telephone: (90) 69511
Enquiries to: Information officer, Marja-Liisa Poikolainen
Founded: 1970

Subject coverage: Hydrology; limnology; water chemistry; hydraulic engineering; water resources management; water supply and waste water treatment.
Library facilities: Open to all users; loan services (nc); reprographic services (c).
Library holdings: 40 000 bound volumes; 460 current periodicals; 350 reports.
Information services: No information services available to non-staff users.
Consultancy: No consultancy.
Publications: Bibliographic list; National Water Board report.

FRANCE

Centre d'Étude et de Recherche de l'Industrie du Béton Manufacturé * 9.16
– CERIB
[Technical Centre of the French Precast Concrete Industry]
Address: BP 59, 28230 Épernon, France
Telephone: (37) 83 52 72
Telex: 782 048
Enquiries to: Chief engineer, Raphaël Setton
Founded: 1967
Subject coverage: Applied research, quality control, technical assistance, education, and information in the field of precast concrete.
Library facilities: Open to all users; no loan services; reprographic services (c).
Scientific and technical articles retrieved in the centre's data base put on microforms.
Information services: Bibliographic services (c); translation services (c); literature searches (c).
Data base on precast concrete, with 12 000 references - abstracts in French.
Consultancy: Yes (c).
Publications: Catalogue on request.

International Road Research Documentation * 9.17
– IRRD
Address: 2 rue André Pascal, 75775 Paris Cedex 16, France
Telephone: (1) 45 24 92 44
Telex: 620160 OCDE PARIS
Affiliation: Organization for Economic Cooperation and Development
Enquiries to: Acting head, B. Horn.
OECD Road Transport Research Programme Division
Founded: 1965

Subject coverage: Worldwide accessible information and documentation system for road transport research.
Library facilities: No library facilities.
Information services: Bibliographic services (c); literature searches (c); access to on-line information retrieval systems (c).
Services are supplied via IRRD members and IRS of ESA, ESRIN, Frascati, Italy.
Consultancy: No consultancy.
Publications: *Code of Practice*; list on request.

ITBTP-CATED * 9.18

[Technical Institute for Building and Public Works, Technical Assistance and Documentation Centre]
Address: 9 rue La Pérouse, 75784 Paris Cedex 16, France
Telephone: (1) 47 20 88 00
Telex: FEDEBAT 611975 F
Affiliation: Fédération Nationale du Bâtiment
Enquiries to: J. Devoge
Subject coverage: Technical assistance, documentation, and information management concerning building techniques (conception, practice and technology, improvements), regulation and standards, and building environment.
Library facilities: No library facilities.
Information services: No bibliographic services; no translation services; no literature searches; access to on-line information retrieval systems (c).
Interactive factual data bank - Ariane contains information (in French) required by the building industry in the following areas: building technology, building know-how; regulation and standards governing construction in France; building products (14 000 manufacturers and societies, 140 000 trademarks, 3500 families of products); extensions about social and economic problems (intended users: building practitioners, contractors, architects in France and abroad).
Consultancy: Consultancy (generally given by telephone to subscribers).
Publications: CATED catalogue.

Laboratoire Central des Ponts et 9.19
Chaussées *

– LCPC
[Road and Bridge Central Laboratory]
Address: 58 boulevard Lefèbvre, 75732 Paris Cedex 15, France
Telephone: (1) 532 31 79
Telex: LCPARI 200361 F
Affiliation: Ministère de l'Urbanisme et du Logement
Enquiries to: Head of Documentation Section, Charlotte Nadel
Founded: 1949
Subject coverage: Civil engineering; earthworks; pavements; engineering structures; bridges; materials

(aggregates, hydraulic and bituminous binders, concretes, steels, paints, byproducts and wastes); environment and quality of life; town planning.
Library facilities: Not open to all users (open to some occasional users by appointment and exchanges made in some cases with similar organizations); loan services (nc); reprographic services.
Information services: Bibliographic services (nc); translation services (nc); literature searches (c); access to on-line information retrieval systems.
Data base: International Road Research Documentation of Road Research Programme - OECD; French Coordinating Centre of IRRD cooperation and French member of IRRD.
Consultancy: No consultancy.
Publications: Catalogue on request; *Bulletin de Liaison des Laboratoires des Ponts et Chaussées*; *Documentation Routière pour Pays en Développement - Sélection de la DIRR - Informations générales*.

Société des Ingénieurs Civils de 9.20
France

[French Civil Engineering Society]
Address: 19 rue Blanche, 75009 Paris, France
Enquiries to: Director, Claude Herselin
Library holdings: 100 000 volumes.

Syndicat National des Industries 9.21
du Plâtre

– SNIP
[National Association of Gypsum Plaster Industries]
Address: 3 rue Alfred Roll, 75849 Paris Cedex 17, France
Telephone: (1) 47 66 03 64
Enquiries to: Secretary general, D. Daligand.
Centre d'Information du Plâtre
Founded: 1840
Subject coverage: Gypsum mining; gypsum plaster production; plaster board production; on site utilization.
Library facilities: Open to all users; no loan services; reprographic services (c).
Visits on request.
Information services: Bibliographic services (nc); no translation services; no literature searches; no access to on-line information retrieval systems.
Consultancy: Yes (nc).

Union Technique Interprofessionnelle du Bâtiment et des Travaux Publics, Centre de Documentation *

9.22

– UTIBTP

[Interprofessional Technical Union of Construction and Public Works, Documentation Centre]

Address: BP 1, Domaine de St-Paul, 78470 Saint Rémy les Chevreuse, France

Telephone: (1) 30 52 92 00

Telex: 695-527

Affiliation: Fédération National du Bâtiment

Founded: 1933

Subject coverage: Civil engineering; energy; soil mechanics.

Library facilities: Open to all users; loan services (nc); reprographic services (c).

Library holdings: 120 000 books; 400 periodicals.

Information services: Bibliographic services (c); literature searches (c); access to on-line information retrieval systems (c).

Pascal (CNRS); EDF (Doc.); CIS; NTIS; Compendex; Nasa.

Consultancy: Yes (nc).

GERMAN DEMOCRATIC REPUBLIC

Bauakademie der Deutschen Demokratischen Republik, Bauinformation

9.23

[Building Academy of the German Democratic Republic, Building Information Centre]

Address: Plauener Strasse 16, 1092 Berlin, German Democratic Republic

Telephone: 2007 240

Telex: 0112141 REGDD

Affiliation: Ministry of Construction

Enquiries to: Director, Professor M. Schimpfermann.

Central Technical Library for Building and Construction

Founded: 1953

Subject coverage: Fundamentals of civil engineering; research on structural engineering; architectural and structural design; architecture; building materials; technology of construction and assembling work, including reconstruction and upkeep; regional and town planning; technology of building material production; construction machinery and equipment; buildings and structures; building constructions and foundations; building equipment and technical services; economy, organization, management and planning in the construction industry.

Library facilities: Loan services; reprographic services (c).

The library is open to all citizens of the GDR from the age of sixteen (c).

Library holdings: 108 000 bound volumes; 900 current periodicals.

Information services: Bibliographic services (c); translation services (c); literature searches (c); access to on-line information retrieval systems (for planning and design instructions and documentation only - c).

Consultancy: No consultancy.

Courses: Advanced training courses for members of the staff of information facilities and building and construction libraries in the GDR, irregular.

Publications: Catalogue available on request.

GERMAN FEDERAL REPUBLIC

Deutscher Beton-Verein eV

9.24

– DBV

[German Concrete Association]

Address: PO Box 2126, Bahnhofstrasse 61, D-6200 Wiesbaden, German Federal Republic

Telephone: (06121) 1403-31

Enquiries to: Dr Helmut Fritz

Founded: 1898

Library facilities: Open to all users; no loan services; no reprographic services.

Information services: No information services.

Publications: List on request.

Forschungsinstitut der Zementindustrie

9.25

– FIZ

[Cement Industry Research Institute]

Address: Postfach 30 10 63, Tannenstrasse 2, D-4000 Düsseldorf 30, German Federal Republic

Telephone: (0211) 45781

Telex: 0858 4867

Affiliation: Verein Deutscher Zementwerke eV

Founded: 1948

Subject coverage: Cement and concrete research and technology; cement chemistry and mineralogy; cement manufacturing processes; environmental protection and works safety; safeguarding of the quality of cement.

Library facilities: Open to all users; no loan services; reprographic services (c - limited).

Services are primarily for the staff of the institute, but are also available to a limited extent to outside users.

Library holdings: 5500 monographs; 6300 volumes of journals; 150 German and foreign technical journals; patents; standards; research reports.
Information services: Bibliographic services; no translation services; literature searches; no access to on-line information retrieval systems.
Services are free to members and associate members of the association and, to a limited extent, to associated industries, universities, and students.
Consultancy: Yes.

Informationszentrum RAUM und BAU der Fraunhofer-Gesellschaft 9.26
– IRB
[Information Centre for Regional Planning and Building Construction]
Address: Nobelstrasse 12, D-7000 Stuttgart 80, German Federal Republic
Telephone: (49711) 6868 500
Telex: 7 255 167
Facsimile: (49711) 6868 399
Affiliation: Fraunhofer-Gesellschaft
Enquiries to: H. Degenkolbe
Founded: 1941
Subject coverage: Building construction; regional planning; town planning; housing; civil engineering.
Library facilities: Open to all users (nc); no loan services; reprographic services (c).
Library holdings: 73 170 bound volumes; 1629 current periodicals; 3367 reports.
Information services: Bibliographic services (c); no translation services; literature searches (c); access to on-line information retrieval systems (c).
The most important data bases of the IRB are: RSWB (regional planning, urban planning, housing, civil engineering); BAUFO (research projects on building construction and housing); FORS (research projects on regional planning, urban planning, housing). The IRB is a member of the International Council for Building Research Studies and Documentation (CIB) and has implemented, as the CIBDOC agency, several international data bases of the CIBDOC-System.
Consultancy: No consultancy.
Courses: On-line seminars.
Publications: *Schrifttum Bauwesen,* monthly; *Schrifttum Raumordnung, Städtebau, Wohnungswesen,* monthly; *Schrifttum Wohnungswesen/Wohnungswirtschaft,* quarterly; *Fachbuch INFO RSWB* (new books), monthly; *Kurzberichte aus der Bauforschung* (civil engineering research results), monthly; *Forschungsdokumentation Raumordnung, Städtebau, Wohnungswesen* (research projects on regional planning, town planning, housing), annually; *Thermendokumentationen Literaturauslesen, Katalog der Bauforschungsberichte* (research reports on civil engineering); *Katalog Literaturhinweise*

(bibliographies); *Bulldoks* (information services concerning new publications on building damage and the use of wood and concrete in construction), quarterly.

Stiftung Institut für Härterei-Technik * 9.27
– IHT
[Heat Technology Institute]
Address: POB 77 02 07, D-2820 Bremen 77, German Federal Republic
Telephone: (0421) 630007/8
Enquiries to: Metallographer, Ina Lehnert
Founded: 1950
Subject coverage: Heat technology; thermochemical treatment; transformation and transformation structures; mechanical and physical properties and test methods; corrosion; furnaces and materials.
Library facilities: Not open to all users; no loan services; reprographic services (c).
Information services: Bibliographic services (c); no translation services; literature searches (c); access to on-line information retrieval systems (c).
University of Bremen data bank.
Consultancy: Yes (c).

GREECE

Technical Chamber of Greece Library 9.28
Address: Odos Lekka 23-25, 105 62 Athinai, Greece
Telephone: (01) 3254 590
Subject coverage: Engineering.
Library holdings: 40 000 volumes.

HUNGARY

Építésügyi Tájékoztatási Központ 9.29
– ETK
[Building Information Centre]
Address: POB 83, Hársfaútca 21, H-1400 Budapest VII, Hungary
Telephone: 117-317
Telex: 226564
Affiliation: Építésügyiés Városfejlesztési Miniszterium (Ministry of Building and Urban Development)
Enquiries to: Chief librarian, Árpád Böjtös
Founded: 1950
Subject coverage: Technical and economic building information.

Library facilities: Open to all users (nc); loan services (nc); reprographic services (c).
Library holdings: 51 000 bound volumes; 1032 current periodicals; 7100 reports.
The library stocks traditional documents (books, periodicals, standards, patents) and non-traditional documents (research, study tour reports, product information, photos).
Information services: Bibliographic services (c); translation services (c); literature searches (c); access to on-line information retrieval systems (c).
Consultancy: Telephone answers to technical queries (c); writing of technical reports (c); compilation of statistical trade data (c); market surveys in technical areas (c).
Publications: *Technical and Economic Information on Building*, monthly; information publications on hungarian building, monthly; *World News on Building and Urban Development*, monthly; summary catalogue of building research activities; complete list on request.

ICELAND

Vegagerð Ríkisins*   9.30
[Public Roads Administration]
Address: Borgartún 5-7, 105 Reykjavík, Iceland
Telephone: (91) 21000
Affiliation: Ministry of Communication
Enquiries to: Librarian, Gunnar Gunnarsson
Founded: 1917
Subject coverage: Planning, construction, and maintenance of all state and country roads and bridges.
Library facilities: Open to all users; loan services (nc); reprographic services (nc).
Information services: No information services.

IRELAND

An Foras Forbartha   9.31
[National Institute for Physical Planning and Construction Research]
Address: St Martin's House, Waterloo Road, Dublin 4, Ireland
Telephone: (01) 602511
Telex: 30846 El
Affiliation: Department of the Environment
Founded: 1965
Subject coverage: Environmental research; roads; planning; conservation; water quality; construction;

housing; education and information.
Library facilities: Open to all users; loan services (interlibrary only); reprographic services.
Library holdings: 1000 bibliographies.
Information services: Bibliographic services; literature searches (retrospective searches, manual and computerized); access to on-line information retrieval systems.
IRS; Euronet; current awareness services; SDI computerised services; international exchanges.
Consultancy: Yes.
Publications: *Environmental Information Bulletin; Irish Journal of Environmental Science*, annual; microfiche reports; list on request.

ITALY

Associazione Italiana Tecnico   9.32
Economica del Cemento *
– AITEC
[Italian Assocation of Technical Economics]
Address: Via di Santa Teresa 23, 00198 Roma, Italy
Telephone: (06) 858505; (06) 8441809
Telex: 611321 AITEC I
Enquiries to: Editor, Dr Gaetano Bologna
Founded: 1929
Subject coverage: Civil engineering; cement chemistry.
Library facilities: Open to all users; no loan services; reprographic services (c).
Information services: Bibliographic services (c); no translation services; no literature searches.
Consultancy: No consultancy.
Publications: *L'Industria Italiana del Cemento; Il Cemento.*

Istituto di Ricerca sulle Acque *   9.33
– IRSA
[Water Research Institute]
Address: Via Reno 1, 00198 Roma, Italy
Telephone: (06) 841451
Telex: 614588
Affiliation: Consiglio Nazionale delle Ricerche
Enquiries to: Director, Professor Roberto Passino
Founded: 1968
Subject coverage: Water supply and water resources management; water pollution and water quality; wastewater treatment; sludge disposal.
Library facilities: No library facilities.
Information services: No information services.
Consultancy: Yes (nc).
Publications: Bibliographic list on request. The institute's publications can be acquired at CNR-Ufficio Vendita Pubblicazioni, Piazzale Aldo Moro 7, 00185

Roma, Italy.

LUXEMBOURG

Association Luxembourgeoise des Ingénieurs et Industriels *   9.34
[Luxembourg Industrial Association of Engineers]
Address: 4 Boulevard Grande-Duchesse-Charlotte, Luxembourg
Telephone: 2 33 39
Subject coverage: Architecture; engineering.
Library facilities: Library.

NETHERLANDS

Keuringsinstituut voor Waterleidingartikelen *   9.35
– KIWA nv
[Netherlands Waterworks Testing and Research Institute]
Address: Postbus 70, 2280 AB Rijswijk, Netherlands
Telephone: (070) 902720
Telex: 32480
Enquiries to: Librarian, A.Th.M. Wijers-de Bruijn
Founded: 1948
Subject coverage: Drinking water; waterworks testing and research.
Library facilities: Open to all users; loan services (except periodicals); reprographic services.
Information services: Literature searches (only for the institute's own use and water companies in the Netherlands); access to on-line information retrieval systems (only for the institute's own use and water companies in the Netherlands).

Koninklijk Instituut van Ingenieurs *   9.36
[Royal Institute of Engineers]
Address: Postbus 30424, Prinsessegracht 23, 2500 GK s'Gravenhage, Netherlands
Telephone: (070) 64 68 00
Telex: 33641
Subject coverage: All branches of engineering.
Library facilities: Library.

Stichting Bouwcentrum *   9.37
[Building Centre Foundation]
Address: Weena 700, 3014 AG Rotterdam, Netherlands
Telephone: (010) 116181
Telex: 22530 bouwc nl
Enquiries to: Librarian, Edo Dekker
Founded: 1947
Subject coverage: Building, housing, and housing environment.
Library facilities: Open to all users; loan services; reprographic services (for members only).
Consultancy: Yes.

Technische Hogeschool Delft   9.38
[Delft University of Technology, Department of Building Materials Science]
Address: Mekelweg 2, 2628 CD Delft, Netherlands
Library facilities: Library.

Technische Hogeschool Delft   9.39
[Delft University of Technology, Department of Building Science]
Address: Berlageweg 1, 2628 CR Delft, Netherlands
Library facilities: Library.

Technische Hogeschool Delft   9.40
[Delft University of Technology, Department of Civil Engineering]
Address: Stevinweg 1, 2628 CN Delft, Netherlands
Library facilities: Library.

Waterloopkundig Laboratorium   9.41
– WL
[Delft Hydraulics Laboratory]
Address: Postbus 177, Rotterdamseweg 185, 2600 MH Delft, Netherlands
Telephone: (015) 569353
Telex: 38176 HYDEL NL
Enquiries to: Document information coordinator, J.D. van der Tuin.
Postbus 152, 8300 AD Emmeloord, Netherlands
Founded: 1927
Subject coverage: Hydraulic, marine, fluvial, coastal, and offshore engineering; water resources; pumps; pipelines; hydrology; environmental hydrodynamics; hydrography, dredging, navigation.
Library facilities: Open to all users; loan services (nc); reprographic services (c - limited).
Information services: Bibliographic services (c); no translation services; literature searches (c); access to on-line information retrieval systems.
IRS/ESA; Lockheed; Dialog; also own data base - Delft Hydro Database from which on-line literature searches may be made.
Consultancy: Yes (c).
Publications: *Delft Hydroscience Abstracts*, monthly abstracts journal; *Delft Hydraulics Communications*, exchange medium.

NORWAY

Norges Byggforskningsinstitutt * 9.42
– NBI
[Norwegian Building Research Institute]
Address: Forskningsveien 3 B, N-0371 Oslo 3, Norway
Telephone: 469880
Affiliation: Norges Teknisk-Naturvitenskapelige Forskningsråd
Enquiries to: Director, Øystein Bergersen; Librarian, Margareth Grini
Founded: 1953
Subject coverage: Building materials and construction; building climatology; production engineering; building services; information and management in the design and construction process; functional principles of planning and equipping buildings.
Library facilities: Open to all users; loan services (nc); reprographic services (c).
Information services: Bibliographic services (c); literature searches (c); access to on-line information retrieval systems (c).
Lockheed; SDC; IRS.
Consultancy: Consultancy (c).
Publications: Publications list.

Norges Geotekniske Institutt 9.43
– NGI
[Norwegian Geotechnical Institute]
Address: PO Box 40, Taasen, N-0801 Oslo 8, Norway
Telephone: (02) 230388
Telex: 19787 ngi n
Affiliation: Royal Norwegian Council for Scientific and Industrial Research
Enquiries to: Librarian
Founded: 1953
Subject coverage: Soil mechanics; foundation engineering; rock mechanics; engineering geology; snow mechanics; avalanche investigations; site investigations; earth- and rockfill dams; foundation of offshore structure.
Library facilities: Open to all users; loan services (nc); reprographic services (c).
Information services: Bibliographic services; literature searches; access to on-line information retrieval systems.
Access to Polydoc, Dialog, and InfoLine.
Consultancy: Yes.

Norsk Institutt for Vannforskning * 9.44
– NIVA
[Norwegian Institute for Water Research]
Address: Postboks 333, N-0314 Blindern, Oslo 3, Norway
Telephone: (02) 235280
Telex: 74190 NIVA n
Affiliation: Norges Teknisk-Naturvitenskapelige Forskningsråd
Enquiries to: Information officer, Knut Pedersen
Founded: 1958
Subject coverage: Water quality management; pollution, utilization, and water exchange processes in lakes, river systems, estuaries, fjords, and coastal waters; purification, transport, and management of drinking water and waste water; prevention and treatment of acidification and eutrophication.
Library facilities: Open to all users; loan services (nc); no reprographic services.
Consultancy: Yes (c).
Publications: Publications List and Supplement available on request.

SINTEF Avdeling FCB 9.45
[SINTEF, FCB Division/Cement and Concrete Research Institute]
Address: N-7034 Trondheim-NTH, Norway
Telephone: (07) 594530
Telex: 55620 sintf n
Affiliation: Norwegian Institute of Technology
Enquiries to: Division manager, Ivar Holand; Secretary, Marit Tamvakis.
SINTEF Information Office
Founded: 1965
Subject coverage: Cement and concrete technology; concrete structures (in particular marine); condition monitoring; corrosion; testing of materials.
Library facilities: Open to all users (nc); loan services (nc); reprographic services (c).
The library mainly consists of reports on projects carried out by the division. Technical books, etc, are obtained from the main library at the Norwegian Institute of Technology.
Information services: No information services.
Consultancy: Telephone answers to technical queries (nc); writing of technical reports (c); no compilation of statistical trade data; no market surveys in technical areas.

Veglaboratoriet 9.46
[Norwegian Road Research Laboratory]
Address: PO Box 6390 Etterstad, N-0604 Oslo 6, Norway
Telephone: (02) 639900
Affiliation: Vegdirektoratet (Public Roads Administration)
Enquiries to: Librarian, Grethe A. Winge

Subject coverage: Road research and construction; soil and rock mechanics; winter maintenance; permafrost; tunnelling.
Library facilities: Open to all users; loan services (nc); reprographic services (nc).
Information services: Bibliographic services (nc); no translation services; literature searches (nc); access to on-line information retrieval systems (nc).
ESA-IRS system, especially the IRRD data base (International Road Research Documentation).
Consultancy: Yes (nc).

POLAND

Centralny Ośrodek Informacji Budownictwa   9.47
– COIB
[Building Information Main Centre]
Address: 27 Senatorska Street, 00-950 Warszawa, Poland
Telephone: 272449
Telex: WA 813736PL
Affiliation: Ministry of Building, Physical Planning and Local Economy
Enquiries to: Manager, Henryk Walcerz
Founded: 1960
Subject coverage: Scientific, economic, and technical information for the building sector and the building materials industry; automation; permanent building exhibition; professional training; international cooperation.
Library facilities: Open to all users (nc); loan services (nc); reprographic services (c).
Library holdings: 11 000 bound volumes; 477 current periodicals; 3900 reports.
Information services: Bibliographic services (c); no translation services; literature searches (nc); no access to on-line information retrieval systems.
Consultancy: Telephone answers to technical queries (nc); writing of technical reports (c); compilation of statistical trade data (c); market surveys in technical areas (nc).
Courses: Course on 'do it yourself' building work and house repairs, five times a month.
Publications: *Biuletyn Informacyjny o Budownictwie* (Building Information Bulletin), monthly; *Informacja Firmowa 'Busola'* - (Firms Literature 'Busola'), monthly; *Bibliografia Budownictwa* (Bibliography of Polish building literature), bimonthly; *Eksport Budownictwa; Problematyka Budownictwa* (Pamphlets on building problems), 16 issues per annum; catalogue and further information on request.

Instytut Meteoroliogii i Gospodarki Wodnej   9.48
– IMGW
[Meteorology and Water Management Institute]
Address: 61 Podlésna Street, 01-673 Warszawa, Poland
Telephone: 35 28 13
Telex: 81 43 31
Affiliation: Ministry for Environmental Protection and Nature Resources
Enquiries to: Head, Dr Hanna Mycielska. Information Centre
Founded: 1919
Subject coverage: Meteorology; hydrology; oceanology; water management; water engineering.
Library facilities: Open to all users (nc); loan services (nc); reprographic services (c).
Information services: Bibliographic services (c); no translation services; literature searches (c).
Consultancy: No consultancy.

Instytut Organizacji, Zarzadzania i Ekonomiki Przemystu Budowlanego *   9.49
– ORGBUD
[Institute for Organization, Management and Economics of the Building Industry]
Address: Ulica Filtrowa 1, 00-611 Warszawa, Poland
Telephone: 25 52 81
Telex: 813906
Enquiries to: Director, Professor H. Hajduk
Library holdings: 6477 volumes.

Instytut Techniki Budowlanej   9.50
– ITB
[Building Research Institute]
Address: Skryt poczt 998, Ulica Filtrowa 1, 00-950 Warszawa, Poland
Telephone: 25 04 71
Telex: 813022
Affiliation: Ministry of Building
Enquiries to: Director, Dr Marian Weglarz
Founded: 1945
Subject coverage: Building structures; building materials and elements; finishing techniques; acoustics; thermal physics; fire tests; building protection; soil and foundations; building on mining areas.
Library facilities: Open to all users; loan services (nc); reprographic services (c).
The information centre possesses: author, title, and systematic catalogues; microfilm reader; microfiche reader; xerographic printer.
Information services: Bibliographic services (c); translation services (c); literature searches (c); no access to on-line information retrieval systems.
The information centre is included in the automated information system on building (System Informacji

Naukowo-Technicznej i Organizacyjnej Budownicta), which is organized by the Building Information Main Centre.
Consultancy: Yes (c).
Publications: Publications list.

PORTUGAL

Centro de Informação Técnica para a Indústria 9.51
– CITI
[Technical Information Centre for Industry]
Address: Azinhaga dos Lameirosà Estrada do Paço do Lumiar, 1699 Lisboa Codex, Portugal
Telephone: 758 6141
Telex: 42 486
Affiliation: Laboratório Nacional de Engenharia e Tecnologia Industrial (LNETI)
Enquiries to: Director Dr Ana Maria Ramalho Correia
Founded: 1979
Subject coverage: Industrial technology (chemical, biotechnology, food, materials technology, electrical and electronics); energy (conventional, renewable, nuclear, and nuclear safety).
Library facilities: Open to all users; loan services; reprographic services (c).
Library holdings: 30 000 volumes; 800 current periodicals.
Information services: Bibliographic services (c); no translation services; literature searches (c); access to on-line information retrieval systems (c).
Access to Echo, ESA-IRS, Dialog, InfoLine, Questel, STN.
Consultancy: Yes (nc).

Laboratório Nacional de Engenharia Civil * 9.52
– LNEC
[National Laboratory of Civil Engineering]
Address: Avenida do Brasil 101, 1799 Lisboa Codex, Portugal
Telephone: 882131
Telex: 16 760 LNEC P
Affiliation: Ministério do Equipamento Social
Enquiries to: Director, A. Ravara
Founded: 1947
Subject coverage: Civil engineering and related subjects.
Library facilities: Open to all users; loan services (nc); reprographic services.
Library holdings: 100 000 volumes.

Information services: Bibliographic services (nc); literature searches (c); access to on-line information retrieval systems (c).
Questel; Lockheed.
Consultancy: Yes (nc).

ROMANIA

Consiliul Naţional al Inginerilor şi Tehnicienilor din RSR * 9.53
[National Council of Engineers and Technicians of the SRR]
Address: Calea Victoriei 118, 70179 Bucureşti, Romania
Telephone: 59 41 60
Enquiries to: General secretary, I.C. Ursu
Library holdings: 27 000 volumes.

Institutul de Cercetări şi Proiectări pentru Gospodărirea Apelor * 9.54
– ICPGA
[Water Resources and Engineering Research and Design Institute]
Address: 78-95 Splaiul Independentei 294, 77703 Bucureşti Sector 6, Romania
Telephone: (90) 492037
Affiliation: Consiliul National al Apelor
Enquiries to: Director, Călin Popescu
Founded: 1957
Subject coverage: Water resources systems operation; water allocation and use; water development schemes; water resources quality management; automatic water quality monitoring; aquatic toxicology and hydrology; water supply and treatment; wastewater and sludge treatment; flood control works; storage dams.
Library facilities: Open to all users; loan services (nc); reprographic services (c).
Exchange of publications.
Information services: Bibliographic services (c); no translation services; literature searches (c); no access to on-line information retrieval systems.
Consultancy: No consultancy.
Publications: Bibliographies in the following areas: water resources engineering; water quality protection; water supply; wastewater treatment; hydrology; hydraulics; hydrotechnical construction; river improvement works.

Oficiul de Informare 9.55
Documentară pentru Construcţii, Arhitectură şi Sistematizare *
– ODCAS
[Documentary Information Office for Building, Architecture and Town Planning]
Address: Căs. poştală 1-139, Bulevardul 1848 nr 10, 70058, Bucureşti, Romania
Telephone: 15 22 21
Affiliation: Institutul Central de Cercetare, Proiectare si Directivare în Construcţii
Enquiries to: Head, Aurelia Dobrescu
Founded: 1957
Library facilities: Not open to all users; loan services; reprographic services.
Library holdings: 65 680 volumes; 160 current periodicals.
Information services: Bibliographic services; literature searches.
Consultancy: Yes.

SPAIN

Centro de Estudios, Investigación 9.56 y Aplicaciones del Agua *
[Water Research and Investigation Centre]
Address: Paseo San Juan 39, Barcelona 9, Spain
Telephone: (93) 231 80 11
Telex: 54180 Aqua E
Affiliation: Sociedad General de Aguas de Barcelona SA
Enquiries to: General secretary, I. Aparicio
Founded: 1960
Subject coverage: Water supply; water treatment; sanitation; environmental protection.
Library facilities: Not open to all users; no loan services; no reprographic services.
Access to the library's facilities is provided at the secretary's discretion.
Information services: Bibliographic services; no translation services; no literature searches; no access to on-line information retrieval systems.
Consultancy: Yes (nc).

Centro de Estudios y 9.57 Experimentación de Obras Públicas, Centro de Información y Documentación
[Public Works Study and Experiment Centre, Information and Documentation Centre]
Address: Alfonso XII 3, 28014 Madrid 7, Spain
Telephone: (91) 467 37 08
Telex: 45022 CDX E

Enquiries to: Juan I. Cuesta
Founded: 1957
Subject coverage: Civil engineering; chemistry; physics.
Library facilities: Open to all users; loan services (internal only); reprographic services.
Information services: Bibliographic services; translation services; literature searches.
The centre is in the process of automating its five libraries and combining their publications.
Consultancy: Yes.
Publications: BoletínBibliográfico de IngenieríaCivil; Novedades Bibliográficas; documents; monographs and investigation notes.

Instituto de la Ingeñiería de 9.58 España *
[Engineering Institute of Spain]
Address: General Arrando 38, Madrid 4, Spain
Telephone: (91) 419 74 17
Library facilities: Library.

Instituto Eduardo Torroja de la 9.59 Construcción y del Cemento *
– IETCC
[Eduardo Torroja Building and Cement Institute]
Address: PO Box 19.002, Serrano Galvache s/n, Madrid 33, Spain
Telephone: (01) 202 04 40
Affiliation: Consejo Superior de Investigaciones Científicas
Enquiries to: Head, Antonio Comyn.
Documentation and Publications Group
Founded: 1934
Subject coverage: Building and its materials, particularly cement.
Library facilities: Open to all users; no loan services; reprographic services.
Information services: Bibliographic services; literature searches.
Consultancy: Yes. .

Laboratorio Central de 9.60 Estructuras y Materiales
[Central Laboratory for Structures and Materials]
Address: Alfonso XII 3, Madrid, Spain
Affiliation: Centro de Estudios y Experimentación
Enquiries to: Librarian, Margarita Pérez Tribaldos
Founded: 1898
Subject coverage: Civil engineering; building materials (including concrete, steel, and paint); structures (bridges, buildings).
Library facilities: Open to all users (c); no loan services; reprographic services (c).
Library holdings: 6000 bound volumes; 110 current periodicals.

Information services: Bibliographic services (c); literature searches (c); access to on-line information retrieval systems (c).
Publications: Catalogue of publications.

SWEDEN

Byggdok - Institutet för Byggdokumentation   9.61
[Swedish Institute of Building Documentation]
Address: Hälsingegatan 49, S-113 31 Stockholm, Sweden
Telephone: 08-34 01 70
Telex: 125 63
Enquiries to: Director, Bengt Eresund
Founded: 1966
Subject coverage: Housing; building; civil engineering; energy saving; installations; town planning; landscaping.
Library facilities: Open to all users; loan services (c); reprographic services (c).
Information services: Bibliographic services (c); translation services (c); literature searches (c); access to on-line information retrieval systems (c).
Information brokerage.
Quest, Dialog, Orbit, INKA, IRB, Pergamon InfoLine. Byggdok is the host of the data bases BODIL and BYGGFO which are used all over Scandinavia.
Consultancy: Telephone answers to technical queries (c); writing of technical reports (c); compilation of statistical trade data (c); market surveys in technical areas (c).
Overseas consultancy - building up documentation centres in, for example, developing countries (c).
Courses: Courses four times per annum on the use of the BODIL and BYGGFO data bases; biannual courses on information retrieval for the construction industry.
Publications: *Byggreferat*, 12 issues per year; *Bygginstitutioner 1986; Nya Byggregler*, four issues per year; *Exportservice; Reducerad UDK; Sökordlista till databasen BODIL; Bruksanvisning till databasen BODIL; Sökordlista till databasen BYGGFO; Bruksanvisning till databasen BYGGFO.*

Cement- och Betonginstitutet *   9.62
– CBI
[Cement and Concrete Research]
Address: S-100 44 Stockholm, Sweden
Telephone: 08-14 42 20
Telex: 12442 FOTEX S CONCRETE
Enquiries to: Librarian
Founded: 1942

Subject coverage: All aspects of concrete technology from basic research in materials science to production and control. (The structural behaviour of concrete buildings and components is not within the scope of the institute.).
Library facilities: Open to all users; loan services (c); reprographic services (c).
Information services: No information services.

Ingenjörsvetenskapsakademien *   9.63
[Royal Swedish Academy of Engineering Sciences]
Address: Box 5073, S-102 42 Stockholm, Sweden
Enquiries to: Director, Professor H.G. Forsberg
Subject coverage: Clearing house for scientific information.

Statens Institut för Byggnadsforskning *   9.64
– SIB
[National Swedish Institute for Building Research]
Address: PO Box 785, S-801 29 Gävle, Sweden
Telephone: 026-10 02 20
Telex: 47396 Byggfo
Enquiries to: Head of library, Lena Berntler
Founded: 1960
Library facilities: Not open to all users; no loan services.
Services are only provided for the staff. Other enquiries should be directed to Byggdok.

Svensk Byggtjänst *   9.65
[Swedish Building Centre]
Address: Box 7853, S-103 99 Stockholm, Sweden
Telephone: 08-730 51 00
Telex: 12442 fotexs
Enquiries to: Information officer, Elisabeth Sedig
Founded: 1934
Library facilities: No library facilities.
The literature service of the centre is not primarily a library. The centre furnishes information on and sells literature of significance to the work of those in the building trade. The building literature is recorded in a computer file together with a brief synopsis. This enables questions on literature to be answered properly.
Information services: Bibliographic services (nc); no translation services; literature searches (nc); access to on-line information retrieval systems (c).
Building Commodity File (Byggvaruregistret), is a computer register covering all building materials and components available on the Swedish market. This information is available in the form of computer printouts and on video display terminals. BESSY is a computer register for building specification texts linked to the Swedish AMA, General Specifications of Material and Workmanship. The AMA is used in Sweden as a reference document in compiling building

descriptions.
Consultancy: Yes (nc).
Publications: Journal of information on recently published books, six times a year; *Catalogue of Building Literature*, biennially.

Svenska Värmeverksföreningen * 9.66
[Swedish District Heating Association]
Address: Kammakargatan 62, S-111 24 Stockholm, Sweden
Telephone: 08-14 24 75
Enquiries to: Information secretary
Founded: 1949
Subject coverage: The association is an umbrella organization for district heating utilities - a collaboration body for Swedish municipal district heating utilities and other companies with interests in heat distribution, particularly in combination with power generation. The association aims to promote developments in the field of district heating, endeavours to bring about standardization, follow and support research, and look after the interests of members in relation to the authorities, etc.
Library facilities: No library facilities.
Publications: Technical reports (mostly in Swedish).

SWITZERLAND

École Polytechnique Fédérale de 9.67
Lausanne, Laboratoire
d'Hydraulique *
[Swiss Federal Institute of Technology of Lausanne, Hydraulics Laboratory]
Address: CH-1015 Lausanne, Switzerland
Telephone: (021) 47 23 75; 47 11 11
Telex: 24 478
Enquiries to: Professor Walter H. Graf
Founded: 1928
Library facilities: Open to all users; no reprographic services.
Information services: No information services.

Eidgenössische Anstalt für 9.68
Wasserversorgung,
Abwasserreinigung und
Gewässerschutz
– EAWAG
[Swiss Federal Institute for Water Resources and Water Pollution Control]
Address: Ueberlandstrasse 133, CH-8600 Dübendorf, Switzerland
Telephone: (01) 823 55 11

Telex: 56287 EAWA CH
Affiliation: Annexanstalt der Eidgenössische Technischen Hochschulen
Enquiries to: Library
Founded: 1936
Library facilities: Open to all users; loan services; reprographic services.

Schweizer Baudokumentation 9.69
– DOCU
[Swiss Building Documentation Service]
Address: CH-4249 Blauen, Switzerland
Telephone: (61) 89 41 41
Telex: 62681 docu ch
Enquiries to: Martin Bornand
Founded: 1967
Subject coverage: Information to architects and industry, especially on building products.
Library facilities: Open to all users; no loan services; no reprographic services.
The library contains documentation on building products.
Information services: Translation services; literature searches (connected to IRB data base).
Enquiry service based on DOCU's own data base.
Consultancy: Consultancy (on introducing building products into the Swiss market).
Publications: *DOCU Bulletin*, monthly.

TURKEY

Devlet Su Isleri Genel 9.70
Müdürlüğü *
[General Directorate of State Hydraulic Works]
Address: Yücetepe, Ankara, Turkey
Telephone: 90-41-33 92 30
Telex: 42305 DSIM TR
Affiliation: Ministry of Energy and Natural Resources
Enquiries to: DSI OET Dairesi Başkanliği
Founded: 1953
Subject coverage: All aspects of waterworks.
Library facilities: Open to all users; no loan services; no reprographic services.
Information services: No information services.

Yapi Arastirma Enstitüsü * 9.71
– YAE
[Building Research Institute]
Address: Bilir Sokak 17, Kavaklidere, Ankara, Turkey
Telephone: (41) 27 81 50
Telex: 43186 btak tr
Affiliation: Türkiye Bilimsel vel Teknik Arastirma Kurumu

Enquiries to: Director, Dr Mustafa Pultar
Founded: 1970
Subject coverage: Building research in general, specifically: environmental research; building structures; planning process in buildings; building construction activities.
Library facilities: Open to all users; no loan services; reprographic services (c).
Library holdings: The library holds about 8000 volumes plus about 100 titles in mainly primary periodicals. However, the parent organization keeps a larger library and a developed documentation centre (TÜRDOK) open for extra-organization use and information activities.
Information services: No information services.
Consultancy: Yes (c).

UNITED KINGDOM

Building Centre * 9.72
Address: 26 Store Street, London WC1E 7BT, UK
Telephone: (01) 637 1022
Telex: 261507 Ref. No. 3324
Enquiries to: Information services manager
Founded: 1931
Subject coverage: Building products.
Library facilities: Open to all users; no loan services; reprographic services.
All trade and non-trade information filed under CI/SFB.
Information services: Translation services; literature searches (only for members of the Centre's 'Find' Research Service).
Comprehensive range of manufacturers' literature available to visitors without charge.

Building Research Establishment 9.73
– BRE
Address: Garston, Watford WD2 7JR, UK
Telephone: (0923) 674040
Telex: 923220
Affiliation: Department of the Environment
Enquiries to: Building Research Advisory Service
Founded: 1921
Library facilities: Open to all users (by arrangement); no loan services; reprographic services (c).
Information services: No bibliographic services; no translation services; literature searches (c); no access to on-line information retrieval systems.
Data base of references to literature of building science, not yet available on-line. Searches made on request and a charge made for references found.
Consultancy: Consultancy (c).

Publications: BRE publications are available from the Publications Sales Office, lists available on request.

Building Research Establishment, 9.74
Princes Risborough Laboratory *
Address: Princes Risborough, Aylesbury, Buckinghamshire, UK
Telephone: (084 44) 3101
Telex: 83559 PRLRIS
Enquiries to: Librarian
Founded: 1925
Subject coverage: Building component performance; structural use of timber and boards; preservation of timber and stone, including biodeterioration and environmental aspects.
Library facilities: Open to all users (by appointment); loan services (to libraries, not to individuals); reprographic services (c).
Information services: Literature searches; access to on-line information retrieval systems (library staff will use own in-house system on behalf of enquirers - c).
Consultancy: Consultancy services provided by Advisory Service.

Building Services Research and 9.75
Information Association
– BSRIA
Address: Old Bracknell Lane West, Bracknell, Berkshire RG12 4AH, UK
Telephone: (0344) 426511
Telex: 848288 BSRIAC G
Enquiries to: Information officer
Founded: 1958
Subject coverage: Mechanical and electrical services of the built environment, including heating, ventilating, air conditioning, plumbing and sanitation, light and power, fire protection.
Library facilities: Not open to all users (BSRIA members only); loan services (BSRIA members only); reprographic services (no charge to BSRIA members).
Library holdings: 6000 bound volumes; 250 current periodicals; 15 000 reports.
Information services: Bibliographic services (c); no translation services; literature searches (c); access to on-line information retrieval systems (c).
No charge is made to BSRIA members for bibliographic services or literature searches.
Consultancy: Telephone answers to technical queries (charge by arrangement); writing of technical reports (c); compilation of statistical trade data (c); market surveys in technical areas (c).

Cement and Concrete Association * 9.76

Address: Wexham Springs, Slough SL3 6PL, UK
Telephone: (02816) 2727
Telex: 848352 CCA G
Affiliation: Cement manufacturing industry
Enquiries to: Head of library information group, H.A. Stoddart
Founded: 1935
Subject coverage: Portland cement application (ie concrete, etc); architecture; engineering; planning; building practice; material technology.
Library facilities: Open to all users (visit by appointment, otherwise by letter, phone or telex - service available to potential UK Portland cement users); loan services (mainly interlibrary); reprographic services (c).
Information services: No bibliographic services (special bibliographies made but regular services not currently provided); no translation services (but translations are available of many papers of direct interest to the association); literature searches.
Consultancy: Yes. A charge is sometimes made, but a quotation would be provided if there were to be a charge - the criterion is the service contribution made to the industry.
Publications: *Concrete Quarterly; Magazine of Concrete Research*, quarterly; complete list on request.

Chartered Institute of Building 9.77

Address: Englemere, King's Ride, Ascot, Berkshire SL5 8BJ, UK
Telephone: (0990) 23355
Enquiries to: Head of information, P.A. Harlow
Founded: 1834
Subject coverage: Building management.
Library facilities: Open to all users; loan services (nc); reprographic services (c).
Information services: Bibliographic services (nc); no translation services; literature searches (nc); no access to on-line information retrieval systems.
Consultancy: No consultancy.
Publications: Catalogue of publications.

Construction Industry Research and Information Association 9.78

– CIRIA
Address: 6 Storey's Gate, London SW1P 3AU, UK
Telephone: (01) 222 8891
Telex: 24224 (ref 2063)
Enquiries to: Membership and conference manager, F.G. Murray
Founded: 1960 (as CERA; name changed to CIRIA in 1967)
Subject coverage: Civil engineering; building; underwater and offshore engineering.
Library facilities: Not open to all users (members only); loan services (members only); no reprographic

services.
Library holdings: 170 reports.
Information services: Access to on-line information retrieval systems (underwater and offshore information - c).
Consultancy: Telephone answers to technical queries (members only).
Courses: About ten technical seminars each year on the results of recent CIRIA research projects.

Constructional Steel Research and Development Organisation * 9.79

– CONSTRADO
Address: 12 Addiscombe Road, Croydon CR9 3JH, UK
Telephone: (01) 688 2688
Telex: 946372
Affiliation: British Steel Corporation
Enquiries to: Information officer or Librarian
Founded: 1971
Subject coverage: CONSTRADO exists to ensure optimum efficiency from the use of steel in construction.
Library facilities: Open to all users (preferably by appointment); loan services (nc); reprographic services (in certain circumstances - nominal charge).
The library can provide bibliographies from an existing subject list.
Information services: No information services.
Consultancy: Yes (c).

Hydraulics Research Limited 9.80

– HR
Address: Wallingford, Oxfordshire OX10 8BA, UK
Telephone: (0491) 35381
Telex: 848552 HRS WAL
Enquiries to: Manager director, Dr T.J. Weare
Founded: 1952
Subject coverage: Civil engineering; open channel hydraulics.
Library facilities: Not open to all users (approved visitors may use library by prior permission - nc); loan services (usually available at discretion of library - nc); no reprographic services.
Library holdings: 3500 bound volumes; 200 current periodicals; 3000 reports.
Information services: No information services apart from literature searches, which could be considered in particular cases - nc.
Consultancy: Telephone answers to technical queries; writing of technical reports (c - frequently incorporated in package charge for work done at HR); no compilation of statistical trade data; market surveys in technical areas (c).
No charge is made for on-the-phone answers to technical queries for the first thirty minutes. After this, a consultancy fee is charged.

The prime service of HR is the investigation and research of civil engineering hydraulic problems.

Courses: Wallingford Storm Sewer Package (WASSP), about eleven per annum; seminars and courses on hydraulic subjects, including tideway, road drainage, wave prediction, and rock for shore protection, about seven per annum.

Institute of Hydrology 9.81

Address: Maclean Building, Crowmarsh Gifford, Wallingford, Oxfordshire OX10 8BB, UK
Telephone: (0491) 38800
Telex: 849365
Affiliation: Natural Environment Research Council
Enquiries to: Information officer, C. Kirby
Founded: 1965
Subject coverage: Mathematical, physical and applied hydrology, meteorology, climatology, geomorphology, soil science, agricultural hydrology, water resources development, water quality.
Library facilities: Loan services (BLDSC forms to cover costs); reprographic services (c).
The library is open to bona fide researchers at no charge; users are requested to telephone before arrival.
Library holdings: 7000 bound volumes; 230 current periodicals; 1500 reports.
Information services: Bibliographic services (when possible - nc); no translation services; no literature searches; no access to on-line information retrieval systems.
Consultancy: Telephone answers to technical queries (occasional charge); no compilation of statistical trade data; no market surveys in technical areas.
Courses: Occasional courses.
Publications: Technical reports - list available on request.

Institute of Offshore Engineering 9.82
– IOE
Address: Heriot-Watt University, Riccarton, Edinburgh EH14 4AS, UK
Telephone: (031) 449 3393; 449 3794
Telex: 727918 IOE HWU G
Affiliation: Heriot-Watt University
Enquiries to: Information officer, Arnold Myers
Founded: 1972
Subject coverage: Marine technology, including oceanography, petroleum technology, naval architecture, materials and corosion science, pollution control, underwater construction, diving, safety engineering, marine biology, instrumentation, navigation, geotechnics, law, and economics.
Library facilities: Open to all users (by appointment - c); no loan services; reprographic services (c).
Library holdings: Comprehensive reference library of offshore technology.
Information services: Bibliographic services (c); no translation services; literature searches (c); access to

on-line information retrieval systems (c).
Searches using in-house and commercial on-line retrieval systems as appropriate.
Consultancy: No telephone answers to technical queries; writing of technical reports (c); no compilation of statistical trade data; market surveys in technical areas (c).
Environmental impact assessments; contingency plans.
Courses: Courses provided by arrangement to meet requirements of clients.
Publications: *IOE Library Bulletin*, monthly; *Guide to Information Services in Marine Technology; Offshore Information Conference Papers*, reports; list on request.

Waste Management Information Bureau * 9.83
– WMIB
Address: Building 7.12, Harwell Laboratory, Didcot, Oxfordshire OX11 0RA, UK
Telephone: (0235) 24141
Telex: 83135 Atomha G
Affiliation: United Kingdom Atomic Energy Authority
Enquiries to: Manager, M.A. Lund
Founded: 1973
Subject coverage: Production, treatment, disposal and recycling of wastes, and related environmental problems.
Library facilities: Open to all users; reprographic services (c).
Visitors to the library must give one full working day's notice.
Information services: Bibliographic services (c); literature searches (c); access to on-line information retrieval systems (c).
The library has access to the following on-line systems: Dialtech; Dialog; Inis; Euronet.
The bureau is the UK national referral centre on waste management; it operates a computer-based bibliographic data base, and holds printed copies of all documents entered in the data base, at present numbering 27 000.
Consultancy: Yes (c).

Water Research Centre 9.84
Address: PO Box 16, Henley Road, Medmenham, Marlow, Buckinghamshire SL7 2HD, UK
Telephone: (0491) 571531
Telex: 848632
Enquiries to: Information officer.
Technical Information Section
Founded: 1974
Subject coverage: Water and wastewater treatment.
Library facilities: Open to all users; loan services (members only); reprographic services (c).
Information services: Bibliographic services; translation services; literature searches; access to on-line information retrieval systems.

WRC Aqualine data base.
Consultancy: Yes.
Publications: Reports series; publications list.

YUGOSLAVIA

Institut za Ispitivanje Materijala sr Srbije * 9.85
– IMS
[Materials Testing Institute Library]
Address: Bulevar Vojvode Mišića 43, 11000 Beograd, Yugoslavia
Telephone: 011-650 322
Telex: 12403
Enquiries to: Chief librarian, E. Janković
Subject coverage: Design; foundation; soil mechanics; building research (housing, environment, acoustics, insulation, concrete pre-stressed elements, prestressing, prefabrication); bridge design; test-loading; roads; architecture; metallurgy.
Library facilities: Not open to all users; loan services (nc); no reprographic services.
The library is open to similar institutions, universities, and other technical or special libraries.
Information services: No bibliographic services; translation services; literature searches.
Consultancy: No consultancy.

Jugoslovenski Gradjevinski Centar * 9.86
– JGC
[Yugoslav Building Centre]
Address: Bulevar revolucije 84, 11000 Beograd, Yugoslavia
Telephone: 011-436 122
Telex: 12007 YU JGC
Affiliation: Federal Economic Council
Enquiries to: Head of documentation and publications section, Mirjana Vučković
Founded: 1963
Subject coverage: Improvement of the building industry.
Library facilities: Open to all users; no loan services; reprographic services (c).
Information services: Bibliographic services; translation services; literature searches; access to on-line information retrieval systems.
Scientific technical documentation from the field of architecture and civil engineering in Yugoslavia and abroad.
Consultancy: Yes (occasional charge).

10 ENGINEERING, ELECTRICAL AND ELECTRONIC

AUSTRIA

Bundesversuchs- und Forschungsanstalt Arsenal * 10.1

[Federal Testing and Research Establishment Arsenal]
Address: POB 8, Franz Grill-Strasse 3, A-1031 Wien, Austria
Telephone: (0222) 782531-0
Telex: 1/36677
Affiliation: Ministry for Construction and Technology
Enquiries to: E. Böhm
Founded: 1950
Subject coverage: Electrotechnology; electrical machinery and materials; electronics; systems research.
Library facilities: Not open to all users (generally staff users only).
Information services: Literature searches (c).
Consultancy: Yes (c).

Staatlich Autorisierte Prüf- und Versuchsanstalt der Elektrizitätswerke Österreichs 10.2

[Government-Authorized Test Centre of the Austrian Electricity Supply Works]
Address: Obere Augartenstrasse 14 A, A-1020 Wien, Austria
Telephone: (0222) 336538-0
Telex: 61-3222536
Affiliation: Vienna Municipal Electricity Supply Works
Enquiries to: Head: F. Zankel
Founded: 1950
Subject coverage: Safety of electrical household appliances and installation equipment; testing; international cooperation; national and international certification procedures.
Library facilities: Not open to all users.
Information services: Information on existing systems in Austrian national, regional, and international certification procedures is given free of charge.
Consultancy: Yes (mainly for applicants for the Austrian Safety Mark for electrical equipment or for regional or international certification procedures).

BELGIUM

Comité Électrotechnique Belge 10.3
– CEB

[Belgian Electrotechnical Committee]
Address: Galérie Ravenstein 3, Boîte 11, B-1000 Bruxelles, Belgium
Telephone: (02) 51 20 028
Affiliation: Commission Électrotechnique Internationale; Cerelec
Founded: 1909
Library facilities: Open to all users; no loan services; no reprographic services.
Information services: Sale of IEC publications, European standards and Belgian standards.
Consultancy: Yes.

BULGARIA

Central Institute for Computing Technique * 10.4

Address: boulevard Lenin 7 km, Sofia 1113, Bulgaria
Telephone: (02) 71 251
Telex: 22645
Affiliation: State Economic Board
Enquiries to: Senior researcher, Zhivko Paskalev
Founded: 1966
Subject coverage: Computers; microcomputers; minicomputers; telecommunications; disk drives;

magnetic tape units; software.

Library facilities: Open to all users; loan services (nc) reprographic services (c).

Information services: Bibliographic services (nc); translation services (c); literature searches (nc); access to on-line information retrieval systems (c).

The institute is developing a document and term data base.

Consultancy: Yes (nc).

Publications: *Proceedings of CIIT; Express Information; Signal Information.*

DENMARK

EDB-Rådet 10.5

[Danish Data Processing Council]

Address: Bredgade 58, DK-1260 København K, Denmark

Telephone: (01) 11 35 00

Founded: 1966

Subject coverage: Information on data processing questions to associated bodies; promotion of the appropriate use of data processing via the educational system.

Library facilities: Open to all users; loan services; no reprographic services.

Information services: Bibliographic services; literature searches.

Consultancy: Yes.

Lysteknisk Laboratorium 10.6
– LTL

[Danish Illuminating Engineering Laboratory]

Address: Lundtoftevej 100, Bygning 325, DK-2800 Lyngby, Denmark

Telephone: (02) 87 39 11

Affiliation: Danish Academy of Technical Sciences

Enquiries to: Director, Bjarne Nielsen

Founded: 1964

Subject coverage: Illuminating engineering: measurements, calculations, development of lighting fittings, quality criteria.

Library facilities: Not open to all users.

The library is mainly for the use of the laboratory staff, but if material is available only at LTL's library it may be borrowed or copied. A charge will normally be made for this.

Library holdings: 1000 bound volumes; 8 current periodicals; 100 reports.

Information services: No bibliographic services; translation services (c); literature searches (mainly by

means of external data bases - c); no access to on-line information retrieval systems.

Consultancy: Telephone answers to technical queries (no charge for brief answers); writing of technical reports (c); no compilation of statistical trade data; no market surveys in technical areas.

Courses: Ten courses about once or twice per annum, on topics that include the following: road lighting, interior lighting, lighting fittings.

Publications: Technical reports, two to five per year.

FINLAND

Teknillinen Korkeakoulu 10.7

[Helsinki University of Technology, Department of Electrical Engineering]

Address: Otakaari 5, SF-02150 Espoo, Finland

Telephone: 90-4512340

Telex: 125161

Enquiries to: Librarian, Jarmo Rinne

Subject coverage: Acoustics; digital electronics; electron physics and semiconductors; telecommunication and information technology; radio engineering; applied electronics; systems theory; power systems; electrical machinery; theoretical electrical engineering; measuring techniques; power electronics and illuminating engineering; control engineering; communication engineering.

Library facilities: Open to all users.

loan and reprographic services are available through the main library of the university.

Publications: List on request.

FRANCE

Centre National d'Études des 10.8
Télécommunications
– CNET

[National Telecommunications Research Centre]

Address: 38-40 rue du Général Leclerc, 92131 Issy les Moulineaux, France

Telephone: (1) 45 29 44 44

Telex: 250317F

Affiliation: Direction Générale des Télécommunications

Enquiries to: Directeur de l'information, de la coopération et des échanges techniques, J.P. Bloch

Founded: 1945

Subject coverage: Telecommunications and related science and technology.

Library facilities: Open to all users (nc); loan services (nc); reprographic services (c).
The library is open all the year round, from Monday to Friday.
Library holdings: 30 000 bound volumes; 900 current periodicals.
Information services: No bibliographic services; no translation services; no literature searches; access to on-line information retrieval systems (c).
The data base Teledoc can be reached by Transpac, Euronet, and Tymnet. The information vendor is Télésystèmes-Questel. Teledoc is a bibliographical data base devoted to telecommunications and to related science and technology.
Consultancy: No consultancy.
Publications: *Annales des Télécommunications*, 6 issues per year; *Bulletin Signalétique des Télécommunications*, 12 issues per year; *L'Écho des Recherches*, 4 issues per year; *Collection Technique et Scientifique des Télécommunications*.

Électricité de France, Direction des Études et Recherches 10.9
– EdF
[French Electricity, Study and Research Section]
Address: 1 avenue du Général de Gaulle, 92141 Clamart, France
Telephone: (1) 47 65 41 37
Telex: 20434 FEDFNORM
Facsimile: (1) 47 65 31 24
Enquiries to: Head, J.P. Allard
Founded: 1946
Subject coverage: Generation transmission, and distribution of electric power.
Library facilities: Open to all users (by special arrangement); no loan services; reprographic services (only original documents referenced in EDF-DOC data base - c).
Library holdings: About 68 000 bound volumes; 1500 current periodicals; 100 000 reports.
Information services: No bibliographic services; no translation services; no literature searches; access to on-line information retrieval systems.
The EDF data base is available through two hosts: ESA/IRS; Questel.
Consultancy: Yes (occasionally).
Publications: *Documentation Technique*, monthly; *Thesaurus EDF*.

Société des Electriciens, des Electroniciens et des Radioélectriciens 10.10
[Electrical, Electronic, and Radiocommunications Technologists' Society]
Address: 48 rue de la Procession, 75724 Paris Cedex 15, France

Telephone: (1) 45 67 07 70
Telex: 200 565
Library facilities: Library.

GERMAN FEDERAL REPUBLIC

Verband Deutscher Elektrotechniker eV 10.11
– VDE
[German Association of Electrical Engineers]
Address: Stresemann Allee 21, D-6000 Frankfurt am Main 70, German Federal Republic
Enquiries to: Secretary, Professor P. Dietrich

HUNGARY

Automatislási Kutató Intézet MTA 10.12
[Automation Research Institute]
Address: Egri J. útca 18, Budapest XI, Hungary
Affiliation: Magyar Tudományos Akadémia (Hungarian Academy of Sciences)

Magyar Elektrotechnikai Egyesület 10.13
[Hungarian Electrotechnical Association]
Address: Kossuth Lajos tér 6/8, 1055 Budapest, Hungary
Library facilities: Library.

PRODINFORM Müszaki Tanácsadó Vállalat 10.14
[PRODINFORM Technical Consulting Company]
Address: PO Box 453, Munkácsy M u 16, Budapest 1372, Hungary
Telephone: 323 770
Telex: prod-4 227750
Enquiries to: Library manager, József Bauer
Founded: 1951
Subject coverage: Metallurgy; machine industry; electrotechnology; electronics; mining.
Library facilities: Not open to all users (nc); loan services (nc); reprographic services (c).
Library holdings: 37 355 bound volumes; 503 current periodicals; 20 000 reports.
Information services: Bibliographic services (c); translation services (c); literature searches (c); no access to on-line information retrieval systems.

Consultancy: Telephone answers to technical queries (nc); no writing of technical reports; no compilation of statistical trade data; market surveys in technical areas (c).

Publications: *Iparjogvédelmi tájékoztató* (Information on Industrial Rights); *Ipari Szabványosítás* (Industrial Standardization); *Korszerü technológiák* (Modern Technologies); *Minőség és megbizhatóság* (Quality and Reliability); *Kohászati uj külföldi könyvbeszerzések*, metallurgy bibliography; *Hiradástechnikai és müszeripari uj külföldi könyvbeszerzések*, communications, electronics and measurements bibliography; *Szerszámgépipari uj külföldi könyvbeszerzések*, machine tools industry bibliography.

IRELAND

Electricity Supply Board 10.15
– ESB

Address: Lower Fitzwilliam Street, Dublin 2, Ireland
Telephone: (01) 765831
Telex: 25313
Enquiries to: Head of Library and Information Services, James C. O'Reilly
Founded: 1926
Subject coverage: Electricity generation and supply.
Library facilities: Open to all users (by appointment only); loan services (to libraries only - nc); reprographic services (nc).
Information services: No bibliographic services; translation services (c); no literature searches; no access to on-line information retrieval systems.
Consultancy: Telephone answers to technical queries (nc); no writing of technical reports; no compilation of statistical trade data; no market surveys in technical areas.

ITALY

Associazione Elettrotècnica ed 10.16
Elettrònica Italiana
[Italian Association for Electrical and Electronic Engineering]

Address: Viale Monza 259, 20126 Milano, Italy
Telephone: (02) 25 50 641
Library facilities: Library.

Fiat TTG SpA, Documentazione 10.17
e Biblioteca *
[Fiat TTG SpA, Documentation Office and Library]

Address: PO Box 599, Via Cuneo 20, 10152 Torino, Italy
Telephone: (011) 2600290
Telex: 221050
Enquiries to: Librarian and Information Officer, Dr F. Ferrero
Founded: 1906
Subject coverage: Manufacture of electric energy and cogeneration plants; energy in general.
Library facilities: Not open to all users; loan services; reprographic services (c).
The library is not available to non-staff users (some particular cases excepted). Cooperation with technical and scientific organizations; liaison with national professional associations.
Information services: No bibliographic services; no translation services; literature searches; access to on-line information retrieval systems.
Bibliographic data base organization in progress; bibliographic and subject-based computerized searches.
Consultancy: Yes (nc).

NETHERLANDS

NV Kema Library * 10.18

Address: Postbus 9035, 6800 ET Arnhem, Netherlands
Telephone: (085) 45 70 57
Telex: 450 16
Enquiries to: A.J. te Grotenhuis
Founded: 1920
Subject coverage: Electrical engineering; nuclear science; environmental pollution; materials testing.
Library facilities: Not open to all users; loan services; reprographic services.
Information services: Literature searches; access to on-line information retrieval systems.
The library has on-line access to the following: SDC; ESA; Inis.

Technische Hogeschool Delft 10.19
[Delft University of Technology, Department of Electrotechnology]

Address: Mekelweg 4, 2628 CD Delft, Netherlands
Library facilities: Library.

NORWAY

Norges Vassdrags- og Energiverk   10.20
– NVE
[Norwegian Water Resources and Energy Office]
Address: Postboks 5091, Majorstua 0301, Oslo 3, Norway
Telephone: (02) 46 98 00
Telex: 71912 nveso n
Enquiries to: Head, Public Relations Office, Øystein Skarheim
Founded: 1920
Subject coverage: Water resources; electricity.
Library facilities: Open to all users; loan services (inter-library); reprographic services.
Information services: Literature searches.
Consultancy: Yes.

Norsk Regnesentral *   10.21
[Norwegian Computing Centre]
Address: PO Box 335, Forskningsvn 1 B, Blindern, 0314 Oslo 3, Norway
Telephone: (02) 466930
Telex: 16518 ncc n
Affiliation: Royal Norwegian Council for Scientific and Industrial Research
Enquiries to: Librarian, Tove Loe
Founded: 1958
Library facilities: Not open to all users; loan services (nc); no reprographic services.
The library primarily provides services to the staff, but also participates in the interlibrary loan cooperation.
Information services: No information services.
Consultancy: No consultancy.

POLAND

Instytut Elektotechniki *   10.22
[Eletrical Engineering Institute]
Address: Ulica Pozaryskiego 28, 04-703 Warszawa, Poland
Telephone: 12 20 68
Telex: 813279
Enquiries to: Director, Zdzisław Czajczyński
Founded: 1951
Subject coverage: Electrical engineering.
Library facilities: Open to all users; loan services; reprographic services.
International interlibrary loan services.
Information services: Bibliographic services; literature searches.

The centre is engaged in international cooperation on the Comecon information retrieval system.
Publications: *Informacja Ekspresowa*, bimonthly; *Przeglad Dokumentacyjny Elektotechniki*, monthly.

Instytut Maszyn Matematycznych *   10.23
– IMM
[Mathematical Machines Institute]
Address: Ulica Krzywickiego 34, 02-078 Warszawa, Poland
Telephone: 21 50 96; 21 84 41; 29 92 71
Telex: 813517
Enquiries to: Director, Dr Bronisław Piwowar
Founded: 1957
Subject coverage: Research and development work in the field of computer hardware and software: development bases of systems and subsystems design methods, chosen applications in form of problem-oriented control computers; standardization works; patent protection.
Library facilities: Open to all users; loan services; reprographic services.
National and international interlibrary loan services.
Publications: *Biuletyn Informacyjny - Nauki i Techniki Komputerowe* (Information Bulletin of Science and Technology), bimonthly; *Przeglad Dokumentacyjny* (Documentation Review), bimonthly; *Ekspress Informacja* (Express Information), monthly.

Przemysłowy Instytut Elektroniki *   10.24
– PIE
[Industrial Electronics Institute]
Address: Ulica Długa 44/50, 00-241 Warszawa, Poland
Telephone: 31 38 39; 31 52 21
Telex: 813260 pie pl
Enquiries to: Head, Zbigniew Stolarski
Scientific-Technical and Economics Information Centre
Founded: 1956
Subject coverage: Microelectronics; integrated circuits; optoelectronic; products; technology; applications, automatic testing; measuring and control equipment; technological equipment for semiconductor manufacturing.
Library facilities: Open to all users; loan services; reprographic services.
Library holdings: The institute has three technical libraries, which hold 40 000 volumes and 300 current technical periodicals.
Information services: Bibliographic services; literature searches.
The institute's information services are controlled by its computer-based APIT system; print-out available includes SDI monthly service on 250 topics; subject,

country and patent bibliographies.

Publications: *Przeglad Dokumentacyjny* (Documentation Review), monthly series includes: journals, firms and commercial literature, new books, patents for inventions, and scarce materials; *Bibliografia wydawnictw PIE* (Index of PIE Publications), annual; *Katalogi i Listy Preferencyjne wyrobów Naukowo-Produkcyjnego Centrum Pólprzewodników CEMI* (Catalogue and Preference Lists of Semiconductor Research and Production Centre CEMI).

PORTUGAL

Electricidade de Portugal *   10.25
– EDP
[Electricity of Portugal]
Address: Avenida José Malhoa, A13, 1000 Lisboa, Portugal
Telephone: (01) 723013
Telex: 15563 EDP EC P
Enquiries to: Head of Information and Documentation Department, Olímpio Neves Gonçalves
Founded: 1976
Subject coverage: Electricity.
Library facilities: Not open to all users; loan services; reprographic services.
Exhibition and reproduction of microfilms: extracts of articles; compilation of bibliographies; systematic search; retrospective search; search through terminal.
Information services: Bibliographic services; translation services; literature searches; access to on-line information retrieval systems.
Access through terminal to the following data banks: NSBS - Economy of Energy; EDP data bank, Operational Body of Generation and Transmission Equipment.
Consultancy: Yes.

SPAIN

Asociación Electrotécnica   10.26
Española
– IEI
[Electrical Engineering Society of Spain]
Address: Avenida del Brasil 7-9, Madrid 20, Spain
Telephone: 456 76 64
Library facilities: Library.

SWEDEN

Svenska Elverksföreningen,   10.27
Föreningen för Elektricitetens
Rationella Användning
– SEF/FERA
[Swedish Association of Electricity Supply, Society for Rational Use of Electricity]
Address: Box 6405, S-113 82 Stockholm, Sweden
Telephone: 08-225890
Telex: 13593 gaselle s
Enquiries to: Editor, Bo Gustrin
Founded: 1903
Subject coverage: Production, distribution, and rational and safe use of electrical energy.
Library facilities: Open to all users; no loan services; reprographic services (c).
Information services: No information services.
Consultancy: Yes (c).

SWITZERLAND

International Telecommunication   10.28
Union
– ITU
Address: Place des Nations, CH-1211 Genève 20, Switzerland
Telephone: (022) 99 51 11
Telex: 421 000 uit ch
Affiliation: United Nations Organization
Enquiries to: Head of Central Library and Archives, A.G. El-Zanati
Founded: 1865
Library facilities: Open to all users; loan services; reprographic services.
The union has a film library (359 films) and a photo library (15 000 photographs) on telecommunications.
Information services: Bibliographic services; literature searches.
Consultancy: Yes.
Publications: List of Periodicals; List of Annuals; List of Recent Acquisitions; List of ITU Publications; Catalogue of Telecommunication and Electronics Films; Photo Library Catalogue.

Schweizerische PTT-Betriebe/ Entreprise des PTT Suisses/ Azienda Svizzera delle PTT *   10.29
[Swiss Post, Telephone, and Telegraphy, Library and Documentation]
Address: Viktoriastrasse 21, CH-3030 Bern, Switzerland
Telephone: (031) 622749
Telex: 32011ptt ch
Enquiries to: Library Chief, Walter Bruderer
Founded: 1893
Subject coverage: Post, telephone, and telegraphic services.
Library facilities: Open to all users; loan services (nc); no reprographic services.
Information services: Bibliographic services (nc); translation services; literature searches; access to on-line information retrieval systems (nc).
RADOS (information about home documentation): bibliographic references about books, titles of periodicals, articles, and files of cuttings.
Consultancy: Yes (nc).

Schweizerischer Elektrotechnischer Verein *   10.30
– SEV
[Swiss Electrotechnical Institution]
Address: Seefeldstrasse 301, CH-8034 Zürich, Switzerland
Telephone: (01) 384 92 95
Telex: 56047 SEV ch
Enquiries to: Librarian, Trudi Jaeger
Subject coverage: Electrotechnics; electrotechnology; electronics; standardization.
Library facilities: Open to all users; loan services (for members only - nc); reprographic services (c).
Information services: Bibliographic services (for members only); literature searches (for members only); no access to on-line information retrieval systems.
Consultancy: Yes (for members only).
Publications: Bibliographic list.

UNITED KINGDOM

British Telecom Research Laboratories   10.31
Address: Martlesham Heath, Ipswich, Suffolk IP5 7RE, UK
Telephone: (0473) 642839
Enquiries to: Manager, David Alsmeyer Information Services
Subject coverage: Telecommunications.

Library facilities: Not open to all users; loan services (interlibrary).
Library holdings: 15 000 bound volumes; 600 current periodicals.
Consultancy: No consultancy.

Central Electricity Generating Board   10.32
– CEGB
Address: 15 Newgate Street, London EC1A 7AU, UK
Telephone: (01) 248 1202
Telex: 883141
Affiliation: Electricity Council
Publications: *CEGB Research*, annual report.

Institution of Electrical Engineers   10.33
– IEE
Address: Savoy Place, London WC2R 0BL, UK
Telephone: (01) 240 1871
Telex: 261176
Facsimile: (01) 240 7735
Enquiries to: Information manager, J.P. Tomlinson
Founded: 1871
Subject coverage: Electrical engineering; electronics; computing; control engineering; physics; energy; safety regulations including those for domestic electrical wiring, industry, shipping, and offshore structures.
Library facilities: Open to all users; loan services (to members of the IEE and the British Computer Society and to other libraries only - nc); reprographic services (c).
The library also houses the British Computer Society library and the Advanced Manufacturing Information Service.
Library holdings: 200 000 bound volumes; 960 current periodicals; 15 000 reports.
Information services: Bibliographic services (nc); no translation services; literature searches (c); access to on-line information retrieval systems (c).
Access to external data base hosts includes Dialog, IRS, Euronet, Datasolve, Polis, SIA, INKA, BRS, EKOL, Echo, and Télésystèmes-Questel. Internal data base - Advanced Manufacturing Information.
Consultancy: Telephone answers to technical queries (brief queries free of charge); no writing of technical reports; compilation of statistical trade data (c); market surveys in technical areas (c).
Publications: *IEE Proceedings; Computer-Aided Engineering Journal*, bi-monthly; *IEE News*, monthly; *Electronics Letters;* fortnightly; *Electronics and Power*, monthly; *Software Engineering Journal*, bimonthly; publications catalogue; Inspec information services catalogue.

National Computing Centre * 10.34

Address: Oxford Road, Manchester M1 7ED, UK
Telephone: (061) 228 6333
Telex: 668962
Enquiries to: Information Services
Founded: 1966
Subject coverage: Computing; information technology; telecommunications; computer-aided engineering; computer hardware and software.
Library facilities: Not open to all users; no loan services; reprographic services (c).
Information services: No translation services; literature searches (c); no access to on-line information retrieval systems.
Data bases are maintained on nationally available computing hardware, software, service companies, engineering software, UK data sources, UK computing salaries and major UK computer users.
Consultancy: Yes (c).
Publications: Directories of computing suppliers, equipment, software, and computer-aided engineering; abstracts; regular and ad hoc surveys.

YUGOSLAVIA

Institut Mihailo Pupin 10.35
[Mihailo Pupin Institute]
Address: Volgina 15, 11060 Beograd, Yugoslavia
Telephone: 011-776 222
Telex: 11584 YU IMP BG
Enquiries to: Head of Library, Zorka Djordjević
Founded: 1948
Subject coverage: Automatic control; computer science; telecommunications; piezoelectric components; hydraulics and pneumatics.
Library facilities: Open to all users; loan services (nc); reprographic services (c).
Library holdings: Over 16 000 journals (408 Yugoslav and 15 892 foreign); 7500 books (500 Yugoslav and 7024 foreign).

11 ENGINEERING, MECHANICAL

AUSTRIA

Bundesversuchs- und Forschungsanstalt Arsenal * 11.1
[Federal Testing and Research Establishment Arsenal Engineering]
Address: Franz Grill-Strasse 3, Arsenal Obj 210, A-1030 Wien, Austria
Telephone: (0222) 78 25 31
Telex: 07/6677
Enquiries to: Head, A. Diemling
Founded: 1950
Subject coverage: Vehicle testing; acoustic and vibration measurements; fluid engineering; refrigeration and air-conditioning; thermal engineering.
Library facilities: Not open to all users.
Information services: Literature searches (c).
Consultancy: Yes (c).

Schiffbautechnische Versuchsanstalt in Wien * 11.2
[Vienna Model Basin]
Address: Brigittenauerlände 256, A-1200 Wien, Austria
Telephone: (0222) 33 43 18
Telex: 132619 svaiw a
Enquiries to: Superintendent, Dr Gerhard Strasser
Founded: 1912
Library facilities: Open to all users; no loan services; reprographic services.
Consultancy: Yes.

BELGIUM

Société Belge des Mécaniciens 11.3
[Belgian Society of Mechanical Engineers]
Address: Rue des Drapiers 21, B-1050 Bruxelles, Belgium
Telephone: (02) 511 82 86

Library facilities: Library.

BULGARIA

Mechanical Engineering Information Centre 11.4
Address: 12 Ho-Chi-Min boulevard, Sofia 157A, Bulgaria
Telephone: 72 38 21
Telex: 22744
Affiliation: Central Institute of Mechanical Engineering
Enquiries to: Director
Founded: 1978
Subject coverage: Mechanical engineering.
Library facilities: Not open to all users; loan services; reprographic services.
Information services: Bibliographic services (c); translation services (c); literature searches (c).
Documentary information is provided in the form of retrospective literature searches, selective dissemination of information, and through question-answer services.
Consultancy: Yes (c).
Publications: *Tehnologichna Informazija,* monthly; *Tehnologia na Mashinostroeneto,* 5 issues a year on engineering technology; *Informatzionen List,* bimonthly current awareness bulletin; monthly abstracts summaries.

CZECHOSLOVAKIA

Československá Společnost pro Mechaniku při ČSAV 11.5
[Czechoslovak Society of Mechanics]
Address: Vyšehradská 49, 128 00 Praha, Czechoslovakia
Telephone: 29 64 51

Výzkumný Ústav Zemědělské Techniky   11.6
– VÚZT
[Agricultural Engineering Research Institute]
Address: Kšancím 50, 163 07 Praha 6 - Řepy, Czechoslovakia
Telephone: 359 531
Enquiries to: Director.
Oddělení Vědeckotechnických Informací (Department of Scientific and Technical Information)
Founded: 1951
Subject coverage: Principles and technology of agricultural engineering.
Library facilities: Open to all users; loan services (through State Library of ČSR, Praha 1); no reprographic services.
Information services: Bibliographic services (c); literature searches (c).
Consultancy: Yes.
Publications: Annual report (summaries in Russian, English, French and German are available free or for exchange); scientific reprints.

DENMARK

Danmarks Tekniske Højskole, Afdeling for Mekanisk Teknologi   11.7
[Technical University of Denmark, Mechanical Technology Department]
Address: Byning 423, DK-2800 Lyngby, Denmark
Telephone: (02) 88 25 22
Enquiries to: Librarian, National Technical Library
Subject coverage: Mechanical technology.
Library facilities: Not open to all users; loan services (limited); no reprographic services.
Consultancy: Yes.

Danmarks Tekniske Højskole, Laboratoriet for Køleteknik   11.8
[Technical University of Denmark, Refrigeration Laboratory]
Address: Building 402, DK-2800 Lyngby, Denmark
Telephone: (02) 88 46 22
Telex: 37529 DTHDIA DK
Enquiries to: Librarian, Birger Rosendahl
Founded: 1939
Subject coverage: Refrigeration techniques; heat pump systems.
Library facilities: Open to all users; loan services (nc); reprographic services (c).
Loan services are available only through the National Technological Library of Denmark.
Information services: Not available.

Jordbrugsteknisk Institut *   11.9
[Agricultural Engineering Institute]
Address: Rolighedsvej 23, DK-1958 København V, Denmark
Telephone: (01) 351788
Parent: Royal Veterinary and Agricultural University
Enquiries to: Head, Professor T. Tougaard Pedersen
Subject coverage: Agricultural machinery; tractors; processing on the farm; alternative energy (wind and biomass).
Library facilities: Not open to all users; no loan services; reprographic services (c).
Information services: Bibliographic services (c); translation services (c); no literature searches; no access to on-line information retrieval systems.
Consultancy: Yes (nc).

Skibsteknisk Laboratorium *   11.10
– SL
[Danish Maritime Institute]
Address: Hjortekaersvej 99, DK-2800 Lyngby, Denmark
Telephone: (02) 87 99 33
Telex: 37223 shilab
Enquiries to: Librarian, Lisbert Wedendahl
Founded: 1959
Subject coverage: Marine engineering; hydrodynamics; aerodynamics.
Library facilities: Open to all users; loan services; reprographic services (c).
Information services: Bibliographic services; literature searches.
Consultancy: Yes.
Publications: List of published reports on request.

Statens jordbrugstekniske Forsøg   11.11
[Danish Agricultural Engineering Institute]
Address: Bygholm, DK-8700 Horsens, Denmark
Telephone: (05) 623199
Enquiries to: Head, F. Guul-Simonsen
Founded: 1978
Subject coverage: Machinery, buildings, and work studies for farms.
Library facilities: Open to all users (nc); no loan services; no reprographic services.
Library holdings: 550 bound volumes; 90 current periodicals; 75 reports.
Consultancy: Telephone answers to technical queries (nc); writing of technical reports (c); no compilation of statistical trade data; market surveys in technical areas (c).

FINLAND

Teknillinen Korkeakoulu   11.12
[Helsinki University of Technology, Department of Mechanical Engineering]
Address: Otakaari 4, SF-02150 Espoo, Finland
Telephone: (90) 451 2658
Telex: 125161
Enquiries to: Librarian, Kaisu Tapiola
Subject coverage: Hydraulic machinery, combustion engines, automotive and tool engineering, machine design, production engineering, materials technology, foundry technology, energy economy and power stations, industrial energy technology, steam engineering, applied thermodynamics and machine design, sanitary engineering, industrial economy, work psychology, information processing, ship-building, aerospace engineering, fluid flow, aerodynamics.
Library facilities: Open to all users.
Loan and reprographic services are available through the main library of the university.
Publications: List on request.

Valtion Maatalousteknologian   11.13
Tutkimuslaitos
– VAKOLA
[State Research Institute of Engineering in Agriculture and Forestry]
Address: PPA 1, SF-03400 Vihti, Finland
Telephone: (913) 46 211
Parent: Ministry of Agriculture and Forestry
Enquiries to: Inspector, P. Olkinuora
Founded: 1949
Subject coverage: Machinery and equipment used in agriculture, forestry, horticulture, dairies, home economics and industry; standardization and safety; promotion of research cooperation with regard to agricultural buildings.
Library facilities: Open to all users; no loan services; no reprographic services.
Information services: No information services.
Consultancy: Telephone answers to technical queries (nc); writing of technical reports; compilation of statistical trade data (nc); no market surveys in technical areas.
Publications: Test reports; study reports; test bulletins.

FRANCE

Centre National d'Études et   11.14
d'Expérimentation de
Machinisme Agricole *
– CNEEMA
[National Institute of Agricultural Engineering]
Address: Parc de Tourvoie, 92160 Antony, Hauts de Seine, France
Telephone: (1) 46 66 21 09
Telex: 204565 CNEEMA ANTY
Affiliation: Ministry of Agriculture
Enquiries to: Head of documentation, M Ganneau; M Dao
Founded: 1957
Subject coverage: Farm machinery; agricultural mechanization.
Library facilities: Open to all users; no loan services; reprographic services (c).
Information services: Bibliographic services; translation services; literature searches.
Consultancy: Yes.

Centre Technique de l'Industrie   11.15
Horlogère *
– CETEHOR
[French Technical Centre for Watch- and Clock-making]
Address: BP 1145, 39 avenue de l'Observatoire, 25003 Besançon Cedex, France
Telephone: 81 50 38 88
Telex: 360293 FO88
Founded: 1949
Subject coverage: Horology; microtechnology; biomedicine; establishing standards; technical assistance and inspection tests; horological and mechanical/electronic precision products. The centre acts as a regional standards office.
Library facilities: Open to all users; no loan services; reprographic services (c).
Information services: Bibliographic services (c); no translation services; literature searches (c); access to on-line information retrieval systems (c).
Access is available to ESA, Télésystème, Sligos, and INKA.
Consultancy: Yes (c).

Centre Technique des Industries   11.16
Mécaniques *
– CETIM
[French Mechanical Engineering Technical Centre]
Address: BP 67, 60304 Senlis, France
Telephone: (1) 44 53 32 66
Telex: 140006
Enquiries to: Jean-Noël Ostermann.
Centre de Documentation de la Mécanique
Founded: 1964

Subject coverage: Mechanical engineering; technical research and aid for the French industries.
Library facilities: Open to all users; no loan services; reprographic services (c).
Information services: Bibliographic services (c); translation services (c); no literature searches; access to on-line information retrieval systems.
The centre maintains the CETIM bibliographic data base on host ESA-IRS.
Consultancy: Yes (c).

Institut de Recherches de la Construction Navale   11.17

– IRCN

[Shipbuilding Research Institute]

Address: 47 rue de Monceau, 75008 Paris, France
Telephone: (1) 45 61 99 11
Telex: 280756 F Navir
Affiliation: Chambre Syndicale des Constructeurs de Navires
Enquiries to: Adjoint an director général, B. Diot
Founded: 1948
Subject coverage: Merchant ships and their construction.
Library facilities: Not open to all users.

International Institute of Refrigeration *   11.18

– IIR

Address: 177 boulevard Malesherbes, 75017 Paris, France
Telephone: (1) 42 27 32 35
Telex: 643259 F
Enquiries to: Director
Founded: 1908
Subject coverage: Cryology; thermodynamics and transport processes; refrigerating machinery; biology and food science; freeze drying; storage and transport; air conditioning and energy recovery.
Library facilities: Open to all users; no loan services.
Information services: Bibliographic services; translation services; literature searches.
The library carries out bibliographic searches in fields covered by the institute, and provides copies of abstracts, reports and regional articles. Translations are made for personal use only into French and English; in other languages, quotations are supplied on request.
The institute does not deal with any trade, commercial or professional matters.
Consultancy: Yes.
Publications: *Bulletin of the International Institute of Refrigeration*, bimonthly; *International Journal of Refrigeration*, bimonthly; proceedings of the 11 scientific commissions, codes of practice, recommendations and guides. Complete publications list on request.

Union Technique de l'Automobile, du Motocycle et du Cycle   11.19

– UTAC

[Technical Union of the Automotive, Motorcycle and Cycle Industries]

Address: 157 rue Lecourbe, 75015 Paris, France
Telephone: (1) 88 42 53 90
Telex: UTAC 692775 F
Affiliation: Chambre Syndicale des Constructeurs d'Automobiles Pascale Savory.
Service de Documentation de l'UTAC
Enquiries to: Head, documentation
Subject coverage: Automotive engineering.
Library facilities: Open to all users (nc); no loan services; reprographic services (c).
Library holdings: 3797 bound volumes; 226 current periodicals.
Information services: Bibliographic services (nc); translation services (c); literature searches (nc); no access to on-line information retrieval systems.
Consultancy: No consultancy.
Publications: *Bulletin de Documentation*, monthly.

GERMAN DEMOCRATIC REPUBLIC

VEB Kombinat Schiffbau   11.20

Address: Doberaner Strasse 110/111, Postfach 79, DDR-2500 Rostock 1, German Democratic Republic
Telephone: Rostock 3670
Telex: 031279
Enquiries to: Director, research and product development, Dr Rolf Michael.
Leitstelle für Information und Dokumentation (Information and Documentation Centre)
Founded: 1954
Subject coverage: Products of shipbuilding; shipbuilding techniques.
Library facilities: Not open to all users; loan services; reprographic services (c).
Library holdings: The centre holds approximately 32 000 monographs, 350 periodicals, and various publications of commercial organizations.
Information services: Bibliographic services (c); no translation services; literature searches (c).
Access to off-line information retrieval systems (c).
The centre provides literature searches on the basis of the computer-aided storage and retrieval system Idis for periodicals since 1969 (about 80 000 abstracts); conducts monthly SDI searches; collects about 6000 abstracts a year on microfiche on specific subjects.
Consultancy: Yes (nc).

GERMAN FEDERAL REPUBLIC

Dokumentation Kraftfahrwesen eV *   11.21

– DKF

[Automotive Engineering Information Service]

Address: Postfach 1508, Etzelstrasse 1, D-7120 Bietigheim-Bissingen, German Federal Republic

Telephone: (07142) 52081

Enquiries to: Marketing manager, Wolfram F. Schürmann.

Grönerstrasse 5, D-7140 Ludwigsburg

Founded: 1974

Subject coverage: Automotive engineering: literature information service; performance and design data; international rules and regulations.

Library facilities: Not open to all users; no loan services; reprographic services (c).

Information services: No bibliographic services; translation services (c); literature searches (c); access to on-line information retrieval systems (c).

The basis of the DKF's services is the *DKF Literature Information Service*, published every 3 weeks and containing 800 brief reports covering the broad spectrum of the automotive industry. It is published either as printed matter (paper or index card), or on magnetic tape, or on microfilm; on-line access is also available using customer's own terminal. The *Standard Information Service* appears regularly, giving specialist information in common problem areas. Individual user requests are also accepted.

The DKF also provides customized literature searches, press reviews, guides to rules and regulations, review of military technology, and design data.

Consultancy: No consultancy.

Dokumentationsstelle für Schiffstechnik *   11.22

– STG

[Ship Technology Documentation and Information Centre]

Address: Institut für Schiffbau, Lämmersieth 90, D-2000 Hamburg 60, German Federal Republic

Telephone: (040) 690 14 13

Parent: Schiffbautechnische Gesellschaft eV

Enquiries to: Manager, Professor Dr Siegfried Weiss

Founded: 1963

Subject coverage: Ship design, construction, operation and navigation; marine and ocean engineering; equipment; ports and shipping; management; economics; cargo handling; research; model testing; offshore technology; propulsion technology; hydrodynamics; vibration; propellers; cavitation; fuel.

Library facilities: Not open to all users; no loan services; reprographic services (c).

Information services: Bibliographic services (c); no translation services; literature searches (c); access to on-line information retrieval systems.

Consultancy: Yes (c).

Publications: 250 monthly abstracts from current technical literature, printed on cards; monthly list of new abstracts.

HUNGARY

Gépipari Tudományos Egyesület   11.23

[Mechanical Engineers' Society]

Address: PO Box 451, H-1372 Budapest V, Hungary

Library facilities: Library.

ITALY

Atti 'Fondazione Giorgio Ronchi'   11.24

Address: Largo E. Fermi 1, 50125 Firenze, Italy ·

Telephone: (055) 22 11 63

Enquiries to: President, Professor Vasco Ronchi

Founded: 1944

Subject coverage: Mechanical engineering.

Istituto Nazionale per Studi ed Esperienze di Architettura Navale   11.25

[National Institute of Naval Architecture Studies and Experiments]

Address: Via Corrado Segre 60, 00146 Roma, Italy

Telephone: 5563634

Telex: 612379

Enquiries to: President, A. Ferrauto

Library holdings: 1200 volumes.

NETHERLANDS

Instituut voor Mechanisatie, Arbeid en Gebouwen *   11.26

– IMAG

[Agricultural Engineering Institute]

Address: Postbus 43, 6700 AA Wageningen, Netherlands

Telephone: (08370) 19119

Telex: 45330 CTWAG

Enquiries to: Assistant librarian, A.F.C. Bonenberg

Founded: 1974

Subject coverage: Agricultural mechanization; horticultural mechanization; agricultural and horticultural buildings; labour.

Library facilities: Open to all users; loan services; reprographic services.
Information services: No information services.
Consultancy: Yes (nc).
Publications: *Research Reports*; publications list on request.

Instituut voor Wegtransportmiddelen TNO   11.27

– IW-TNO
[Road Vehicles Research Institute TNO]
Address: Postbus 237, 2600 AE Delft, Netherlands
Telephone: (015) 569330
Telex: 38071 zptno nl
Affiliation: Organisatie voor Toegepast Natuurwetenschappelijk Onderzoek
Enquiries to: Librarian, E.G. van Koperen
Founded: 1970
Subject coverage: Research, development and advice in the field of design and application of road vehicles and their parts. Special fields of interest include application of alternative fuels (LPG, LNG and alcohol fuels); safety, crash phenomena and biomechanics; optimization and application of restraint-systems for adults and children; general bicycle technology; approval testing according to national and international regulations.
Library facilities: Open to all users (nc); loan services (nc); reprographic services (nc).
Information services: No bibliographic services; no translation services; literature searches (via CID - TNO - c); access to on-line information retrieval systems (via CID - TNO - c).
Consultancy: No consultancy.
Publications: *By the Way*, newsletter.

Technische Hogeschool Delft   11.28
[Delft University of Technology, Department of Marine Engineering]
Address: Mekelweg 2, 2628 CD Delft, Netherlands
Library facilities: Library.

NORWAY

Landbruksteknisk Institutt   11.29
– LTI
[Norwegian Institute of Agricultural Engineering]
Address: Postboks 65, N-1432 Ås-NLH, Norway
Telephone: (02) 949370
Affiliation: Royal Ministry of Agriculture
Enquiries to: Professor Kristian Aas
Founded: 1947
Subject coverage: Research and testing activities in agricultural and horticultural engineering, extension work.

Library facilities: Not open to all users; no loan services; no reprographic services.
The library is available only to the staff at the institute and to students at the Agricultural University of Norway. The library is a special library consisting mainly of text books, periodicals, and magazines, concerned with agricultural and horticultural engineering. The library cooperates with the main library of the Agricultural University of Norway.
Information services: Bibliographic services (c); no translation services; no literature searches; no access to on-line information retrieval systems.
The International Association on Mechanization of Field Experiments (IAMFE) has its temporary secretariat and Information Centre at the Institute.
Consultancy: Yes (nc).
Publications: *Test and Research Reports; Orientation*; these publications are distributed abroad on an exchange basis, as is the annual report.

Norges Skipsforskningsinstitutt *   11.30
– NSFI
[Ship Research Institute of Norway]
Address: Postboks 4125, Valentinlyst, N-7001 Trondheim, Norway
Telephone: (02) 595500
Telex: 55146 NSFITN
Parent: Royal Norwegian Council for Industrial and Tehnical Research
Enquiries to: Librarian Vera Romberg.
MTS Library
Founded: 1952
Subject coverage: Design, testing and operation of ships, offshore structures and systems.
Library facilities: Open to all users; loan services (nc); reprographic services (c).
Library holdings: The library is operated as a joint venture with the NTHB (the library at the Norwegian Institute of Technology), specializing in marine technology, and has 13 000 titles and 360 subscriptions to technical journals and report series.
Information services: Bibliographic services (c); translation services (c); literature searches (c); access to on-line information retrieval systems (c).
The library maintains a shipbuilding abstracts data base; other data bases are available on request.
Consultancy: Yes (c).

Norsk Undervannsteknologisk Senter   11.31
– NUTEC
[Norwegian Underwater Technology Centre]
Address: Postboks 6, Gravdalsveien 255, N-5034 Ytre Laksevaag, Bergen, Norway
Telephone: 05-34 16 00
Telex: 42 892 nutec n
Facsimile: 05-34 16 00

Affiliation: Statoil, Norsk Hydro Saga
Enquiries to: Head, Irene Hunskaar
Document Administration
Founded: 1985 (as a limited company)
Subject coverage: Testing, research and development in underwater technology, hyperbaric medicine and physiology, underwater communication, underwater vehicles.
Library facilities: Open to all users (on request); loan services; reprographic services.
Library holdings: 2500 bound volumes; 200 current periodicals.
Information services: Bibliographic services; no translation services; literature searches; access to on-line information retrieval systems.
Consultancy: Telephone answers to technical queries; no writing of technical reports; no compilation of statistical trade data; no market surveys in technical areas.

POLAND

Centrum Techniki Okrętowej   11.32
– CTO
[Ship Design and Research Centre]
Address: Skryt poczt 270, Wica Waly Piastowskie 1, 80-958 Gdańsk, Poland
Telephone: 374-201
Telex: 0152474 CTO PL
Enquiries to: Director, Willi Fandrey
CTO Biblioteka Naukowo-Techniczna
Subject coverage: Advanced design of cargo ships and other specialized craft; developing new types of ship equipment; ship structure mechanics, vibration and noise; ship hydromechanics; materials technology, corrosion and ship environment protection; development of ship's propulsion machinery plant; development of deck machinery and fish processing plants; electrical and automation equipment; equipment testing standardization; developing the materials coding systems.
Library facilities: Open to all users (nc); loan services (nc); reprographic services (c).
Library holdings: 35 800 books; 8671 volumes of bound journals (332 titles); 285 current periodicals (including 104 from the West and 51 from the East); 28 850 reports and other special publications.
Information services: Bibliographic services (c); translation services (c); literature searches (c - computer-aided system APIS-4 for data searches).
Consultancy: Telephone answers to technical queries (nc); writing of technical reports (c).
Publications: *Informacja Ekspresowa Przemyslu Okrętowego* (Information Review of Shipbuilding Publications), monthly; *Przegląd Dokumentacyjny* (Journal of Abstracts on Shipbuilding), monthly;

Zeszyty Problemowe (Papers on Shipbuilding Technology Problems), quaterly; *Serwis Informacyjny Przemyslu Okrętowego - monotematyczny* (Information Review of Chosen Shipbuilding Problems), quarterly; *Przegląd Dokumentacyjny Patentów* (Documentary Review of Patents), quarterly; *Komunikat Normalizacyjny* (Standarization Bulletin), monthly.

Instytut Podstawow   11.33
Podstawowrych Problemów
Techniki PAN
[Fundamental Technological Research Institute]
Address: Ulica Świętokrzyska 21, 00 049 Warszawa, Poland
Telephone: 26 12 81
Telex: 815638
Enquiries to: Director, Professor H. Frackiewicz
Subject coverage: Topics covered include mechanical systems and acoustics.
Library holdings: 80 000 volumes.

SPAIN

Instituto de la Ingeñiería   11.34
de España
[Spanish Institute of Engineering]
Address: General Arrando 38, 28010 Madrid, Spain
Enquiries to: General secretary, J. Tornos

SWITZERLAND

Centre Suisse de Documentation   11.35
dans le Domaine de la
Microtechnique
– CENTREDOC
Address: Case postale 27, rue Breguet 2, CH-2000 Neuchâtel 7, Switzerland
Telephone: (038) 25 41 81
Telex: 952 655 CSEM
Enquiries to: Manager, Bernard Chapuis; on-line services, J.P. Häring, R. Chopard
Founded: 1964
Subject coverage: Science and technology, particularly microtechnology and watch-making; economic forecasts and market analysis; technology monitoring; patents.
Library facilities: Not open to all users (c); loan services; reprographic services.
Library holdings: 3400 bound volumes; 160 current periodicals.
Information services: Bibliographic services (c); no translation services; literature searches (c); access to

on-line information retrieval systems (c).

CENTREDOC is connected to the following hosts: Datastar; INKA; ESA-IRS; Lis-Dialog; SDC Orbit; Télésystèmes-Questel; G-Cam; INPADOC; Pergamon.

Consultancy: Telephone answers to technical queries (c); compilation of statistical trade data (c); market surveys in technical areas (c).

Publications: *La Revue des Inventions Horlogères - RIH* a summary of patent applications in precision engineering in the principal industrial nations; *Bulletin Centredoc,* a weekly summary of scientific periodicals covering precision engineering, electronics, metallurgical chemistry, physics and mechanical engineering; *Centredoc Online.*

Verein Schweizerischer Maschinen-Industrieller 11.36

– VSM
[Swiss Association of Machinery Manufacturers]

Address: Kirchenweg 4, CH-8032 Zürich, Switzerland
Telephone: (01) 47 84 00
Telex: 816519 vsm ch
Enquiries to: Dr K. Meier
Founded: 1883
Library facilities: Not available.
Consultancy: Yes (usually nc).
Publications: Annual report; lists of companies and products.

UNITED KINGDOM

BHRA Fluid Engineering Centre * 11.37

– BHRA
Address: Cranfield, Bedford MK43 0AJ, UK
Telephone: (0234) 750422
Telex: 82505AG
Enquiries to: Head of information services, G.A. Watts
Founded: 1947
Subject coverage: All aspects of fluid engineering, specializing in: physical and mathematical modelling of flow problems; high-pressure technology; pumps and transporting solids by pipeline; mixing and sealing technology.
Library facilities: Not open to all users; loan services (c); reprographic services (c).
The library is not open to all users, but enquiries are accepted at the discretion of the Information Group. Loan services are free for members of BHRA.
Information services: Bibliographic services (c); translation services (c); literature searches (c); access to on-line information retrieval systems (c).
Specialist literature searches are free to BHRA members. The centre offers on-line access to the

Fluidex data base.
Consultancy: Yes (free to BHRA members only).
Publications: Quarterly newsletter; nine specialist abstracts journals, free to BHRA members; publications list on request.

British Internal Combustion Engine Research Institute Limited 11.38

– BICERI Ltd
Address: 111-112 Buckingham Avenue, Trading Estate, Slough, Berkshire SL1 4PH, UK
Telephone: (0753) 27371
Telex: CHAMCOM SLOUGH 848314 (BICERI)
Enquiries to: Librarian
Founded: 1943
Subject coverage: Internal combustion engines: lubrication; fuel injection; noise; torsional vibration and stress; alternative fuels; thermodynamics.
Library facilities: Not open to all users; loan services (c); reprographic services (c).
The library contains a classified card index on references to internal combustion engines.
Information services: Literature searches (c).
Consultancy: Yes (c).
Publications: BICERI Abstracts, weekly, from technical and patent publications; list of reports on request.

British Maritime Technology Limited 11.39

– BMT
Address: Wallsend Research Station, Wallsend, Tyne and Wear NE28 6UY, UK
Telephone: (091) 262 5242
Telex: 53476
Enquiries to: Manager, J.G. Kerr.
Technical Information
Founded: 1985
Subject coverage: Design, construction and operation of ships and marine vehicles; offshore engineering.
Library facilities: Not open to all users; no loan services; reprographic services (c).
Library holdings: 15 000 bound volumes; 250 current periodicals; 50 000 reports.
Information services: Bibliographic services (c); translation services (c); literature searches (c); access to on-line information retrieval systems (c).
The association maintains the BSRA Abstracts on-line data base.
Consultancy: No consultancy.

Industrial Unit of Tribology, Leeds University 11.40

Address: Woodhouse Lane, Leeds LS2 9JT, UK
Telephone: (0532) 431751
Telex: 51311 Relays G
Enquiries to: Manager, Dr C.N. March
Founded: 1969
Subject coverage: Tribology; lubrication; bearings; wear; friction.
Library facilities: No library facilities.
Information services: No bibliographic services; no translation services; literature searches (c); access to on-line information retrieval systems (c).
Consultancy: Telephone answers to technical queries (free to membership scheme subscribers only); writing of technical reports (c); no compilation of statistical trade data; market surveys in technical areas (c). Design appraisal - c.

Institution of Mechanical Engineers 11.41

– IMechE
Address: 1 Birdcage Walk, Westminster, London SW1H 9JJ, UK
Telephone: (01) 222 7899
Telex: 917944
Enquiries to: Information services manager, J. Ollerton.
Information and Library Services
Founded: 1847
Subject coverage: Mechanical engineering: pressure vessels; materials strength; mechanisms; heat exchangers.
Library facilities: Open to all users (by appointment - priority given to members and members of other engineering institutions); loan services (charge for inter-library loans); reprographic services (c).
Library holdings: 150 000 bound volumes, including reports; 500 current periodicals.
Information services: Bibliographic services; no translation services; literature searches (c); access to on-line information retrieval systems (c - ESA/IRS; InfoLine; Dialog; Datasolve).
Manual searches up to half an hour are free to members.
Consultancy: Telephone answers to technical queries (charge after 30 minutes); no writing of technical reports; no compilation of statistical trade data; no market surveys in technical areas.
Publications: *Mechanical Engineering Publications* annual; subject bibliographies; *Current Periodicals List*; information packs; awareness bulletins; additions list.

Machine Tool Industry Research Association 11.42

– MTIRA
Address: Hulley Road, Macclesfield, Cheshire, UK
Telephone: (0625) 25421
Enquiries to: Chief information officer, J.O. Cookson
Founded: 1961
Subject coverage: Machine tool technology including computer-aided design and manufacture and production engineering systems.
Library facilities: Not open to all users (non-members by special arrangement); loan services (members only); reprographic services (c).
Library holdings: 4000 bound volumes; 150 current periodicals; 1000 reports.
Information services: Bibliographic services; no translation services; literature searches; access to on-line information retrieval systems (c).
Bibliographic services and literature searches are free to members.
Consultancy: Telephone answers to technical queries (nc); writing of technical reports (c); compilation of statistical trade data (c); market surveys in technical areas (c).
Courses: Courses held irregularly.

Motor Industry Research Association * 11.43

– MIRA
Address: Watling Street, Nuneaton, Warwickshire CV11 6AJ, UK
Telephone: (0682) 348541
Telex: 311277 MIRA G
Enquiries to: Chief information officer, M.J. Shields
Founded: 1946
Subject coverage: Motor vehicle engineering.
Library facilities: Open to all users (c); no loan services; reprographic services (c).
Information services: Bibliographic services (c); translation services (c); literature searches (c).
Advisory service on international legal regulations for motor vehicle construction.
Consultancy: Yes (c).
Publications: The association is the European agent for publications of the Society of Automotive Engineers (SAE), USA.

National Engineering Laboratory 11.44

– NEL
Address: East Kilbride, Glasgow G75 0QU, UK
Telephone: (03552) 20222
Telex: 777888
Affiliation: Department of Trade and Industry
Enquiries to: Head, T. Archbold.
Library and Information Services
Founded: 1947

Subject coverage: Mechanical engineering. Topics covered include flow measurement and distribution, turbomachinery, offshore engineering, alternative energy (wave and wind), materials technology, and power systems engineering. NEL is the custodian of the national standards of flow.

Library facilities: Open to all users (at librarian's discretion); loan services (through British Library Document supply Centre).

Information services: Bibliographic services; no translation services; literature searches.

Discretionary charges are made for bibliographic services and literature searches.

The library has on-line access to the following data bases: Orbit (SDC); Dialog (Lockheed); IRS (ESA); INKA GRIPS (Fachinformationszentrum Technik eV); Pergamon InfoLine; Télésystèmes-Questel.

Consultancy: Telephone answers to technical queries; no writing of technical reports; no compilation of statistical trade data; no market surveys in technical areas.

NEL undertakes consultancy for a wide range of United Kingdom manufacturing industry.

Courses: A wide range of conferences, seminars and courses.

Publications: *Index of NEL Publications; A Resource for Industry; Introducing NEL.*

National Institute of Agricultural Engineering   11.45

– NIAE

Address: Wrest Park, Silsoe, Bedford MK45 4HS, UK

Telephone: (0525) 60000

Telex: 825808 NIAE-WP-G

Affiliation: British Society for Research in Agricultural Engineering

Enquiries to: Head, scientific information department

Founded: 1924

Subject coverage: Agricultural and horticultural engineering research and development.

Library facilities: Open to all users (by appointment); loan services; reprographic services (c - plus postage).

Information services: Limited literature searches.

Production Engineering Research Association *   11.46

– PERA

Address: Melton Mowbray, Leicestershire LE13 0PB, UK

Telephone: (0664) 4133

Telex: 34684 PERAMM

Enquiries to: Head, V.C. Watts. Library and Information Services

Founded: 1946

Subject coverage: Manufacturing engineering.

Library facilities: Loan services; reprographic services (c).

The library is open only to PERA staff, member firms, and members of the Institute of Production Engineers.

Information services: Bibliographic services (c); translation services (c); literature searches (c); access to on-line information retrieval systems (c).

The library has access to the following on-line information retrieval systems: ESA; Dialog; SRC; Blaise; and Euronet.

Consultancy: Yes (c).

Publications: *PERA Bulletin*, current awareness service for members only.

Spring Research and Manufacturers' Association   11.47

– SRAMA

Address: Henry Street, Sheffield S3 7EQ, UK

Telephone: (0742) 760771

Telex: 547676

Enquiries to: Director, J.A. Bennett

Founded: 1946

Subject coverage: All aspects of springs, spring materials, and spring manufacturing technology.

Library facilities: Open to all users; no loan services (but volumes may be loaned informally to members); reprographic services (c - but no charge to members). Library charges vary according to the extent of use, but most services are free to SRAMA members.

Library holdings: About 550 bound volumes; 40 current periodicals; 5000 reports.

Information services: Bibliographic services (c); no translation services; literature searches (c); access to on-line information retrieval systems (c).

Consultancy: Telephone answers to technical queries (c); writing of technical reports (as cooperative research or contract - c); compilation of statistical trade data (limited by availability of non-confidential statistics - c); market surveys in technical areas (as part of research programme - may be undertaken as contract work).

Other technical and professional consultancies: quality assurance, materials selection etc.

A charge may be made for complex technical queries answered by phone; non-members may be required to write in and be billed.

Courses: Courses on technical subjects - spring design, material selection, quality assurance, SPC, heat treatment and inspection, etc - about three times per annum.

Publications: Annual report; list on request.

YUGOSLAVIA

Tovarna Avtomobilov in Motorjev
11.48

– TAM

[Automobile and Engine Factory]

Address: Ptujska cesta 184, 62000 Maribor, Slovenija, Yugoslavia

Telephone: (062) 32 321

Telex: 33 111 YU AUTOMA

Enquiries to: Head, Lidija Podbevŝek. Information and Documentation Department and Library

Founded: 1946

Subject coverage: Manufacturing of buses, lorries, special-purpose vehicles and engines.

Library facilities: Open to all users; loan services; reprographic services.

Library holdings: 33 672 bound volumes.

Information services: Bibliographic services (nc); translation services (nc); literature searches (nc) no access to on-line information retrieval systems.

Consultancy: No consultancy.

12 ENVIRONMENTAL STUDIES

AUSTRIA

Bundesanstalt für Wassergüte 12.1
[Federal Institute for Water Quality]
Address: Schiffmühlenstrasse 120, POB 52, Kaisermühlen, A-1223 Wien, Austria
Telephone: (0222) 234591
Telex: 136 946 bawas a
Affiliation: Ministry of Agriculture and Forestry
Enquiries to: Librarian, Dr Ilse Ottendorfer
Founded: 1947
Subject coverage: Water management, monitoring, and quality; limnology; wastewater treatment; hydrobiology.
Library facilities: Open to all users; loan services; no reprographic services.
Information services: Bibliographic services (occasional).

Institut für Umweltforschung 12.2
– ifu
[Environmental Research Institute]
Address: Elisabethstrasse 11, Graz, Styria, Austria
Telephone: (316) 36030
Affiliation: Forschungsgesellschaft Joanneum - Graz
Enquiries to: Director, Dr Mert König
Founded: 1970
Subject coverage: Environmental research; regional planning; energy systems; energy technologies; biotechnology; housing research.
Consultancy: Telephone answers to technical queries (nc); writing of technical reports (c).

Österreichischer Naturschutzbund 12.3
[Austrian Nature Conservation Association]
Address: Arenbergstrasse 10, A-5020 Salzburg, Austria
Telephone: (0662) 74 3 71
Library facilities: Library.

BELGIUM

Centre Belge d'Étude et de Documentation de l'Eau 12.4
– CEBEDEAU
[Belgian Centre for Water Research and Documentation]

Address: 2 Rue Armand Stévart, B-4000 Liège, Belgium
Telephone: (041) 52 12 33
Enquiries to: D. van den Ackerveken
Founded: 1947
Subject coverage: Water and corrosion technology: analyses; wastewater treatment; industrial and domestic uses of water; corrosion measurement; self purification of rivers; waste treatment; valorization.
Library facilities: Open to all users; no loan services; reprographic services.
Information services: Bibliographic services; literature searches.
Consultancy: Yes.

CYPRUS

Nature Conservation Service 12.5
Address: Nicosia, Cyprus
Affiliation: Ministry of Agriculture and Natural Resources

CZECHOSLOVAKIA

Státní Ústav Památkové Péče a Ochrany Přírody, Praha

12.6

[State Institute for the Care of Historical Monuments and Nature Conservation, Prague]
Address: Valdštejnské nám 1, 118 01 Praha 1, Czechoslovakia
Telephone: 513, line 335
Affiliation: Ministry of Culture
Enquiries to: Chief librarian, Jaroslav Dušek
OBIS VTEI (Scientific, Technological and Economic Information Centre)
Founded: 1958
Subject coverage: Documentation of historical monuments in the Czech Socialist Republic; architectonic and historical research and urban evaluation of historical centres of towns, optimization of their preservation methods and of methods of their functional regeneration; biological and geological science; register of protected plant and animal species, register of protected areas of nature.
Library facilities: Open to all users (nc); loan services (nc); no reprographic services.
Priority in the use of the library is given to specialists in nature conservation and care of monuments.
Library holdings: 28 000 bound volumes; 204 current periodicals.
Information services: Bibliographic services (nc); no translation services; literature searches (nc); no access to on-line information retrieval systems.
Consultancy: No consultancy.
Publications: Publications list on request.

Výskumný Ústav Vodného Hospodárstva

12.7

– VÚVH
[Water Research Institute]
Address: nabr gen L. Svobodu 5, 881 01 Bratislava, Czechoslovakia
Telephone: 3351-32
Telex: 093213
Enquiries to: Director
Scientific and Technical Information Centre
Founded: 1951
Subject coverage: All aspects of water management and water resources development, and water pollution control.
Library facilities: Not open to all users; loan services; reprographic services.
Information services: Bibliographic services; translation services; literature searches.
The institute serves as the Water Management Investment Construction Evaluating Centre for Czechoslovakia.

Consultancy: Yes.
Publications: *Prace a Studie* (Works and Studies); *Veda a Výskum Praxi* (Science and Research for Practice); *Informácie* (Information).

DENMARK

Danmarks Naturfredningsforening

12.8

[Danish Society for Nature Conservation]
Address: Frederiksberg Runddel 1, DK-2000 København F, Denmark
Telephone: (01) 15 41 55
Enquiries to: Director, D. Rehling
Founded: 1911
Subject coverage: Nature conservation and environment planning in Denmark.
Library holdings: 5000 volumes.
Consultancy: Yes.

Danmarks Tekniske Højskole, Laboratoriet for Økologi og Miljølaere *

12.9

[Technical University of Denmark, Ecology and Environmental Science Laboratory]
Address: Building 224, DK-2800 Lyngby, Denmark
Telephone: (02) 88 40 66
Telex: 37529 dthdia dk
Enquiries to: Director, Finn Bro-Rasmussen
Founded: 1977
Subject coverage: Environmental chemistry and ecological impact.
Library facilities: Not open to all users (only to students attached to the laboratory).
The library is linked to the DTB Danish Technical Library of the DTH, situated nearby.
Information services: No information services.
Consultancy: Yes (nc).

Miljøstyrelsen *

12.10

– NAEP
[National Agency of Environmental Protection]
Address: Strandgade 29, DK-1401 København K, Denmark
Telephone: (01) 57 83 10
Telex: 31209
Affiliation: Miljøministeriet (Ministry for the Environment)
Enquiries to: Documentation officer, Mike Robson
Founded: 1972
Subject coverage: Environmental protection; pollution control.

Library facilities: Open to all users; loan services (nc); no reprographic services.
Facilities are available to personal callers only.
Information services: Bibliographic services (c); no translation services; literature searches (c); access to on-line information retrieval systems (c).
The library has access to the following on-line information retrieval systems: Blaise, CIS, DIMDI, ESA-IRS, Datacentralen, Dialog.
The agency is the Danish national enquiry point for UNEP/Infoterra.
Consultancy: No consultancy.

Vandkvalitetsinstitutett   12.11
[Water Quality Institute]
Address: 11 Agern Allé, DK-2970 Hørsholm, Denmark
Telephone: (02) 86 52 11
Telex: 37874 vkicph
Affiliation: Akademiet for de Tekniske Videnskaber (ATV)
Subject coverage: Water quality: emissions; surveys; planning; waste water treatment technology; restoration and aquaculture; data processing; ecology; hygiene.
Information services: The institute provides various information-related services, including publications, preparation of summaries, collecting and disseminating rare information, and organizing conferences and lectures.
Publications: *Vand,* quarterly.

FINLAND

Suomen Luonnonsuojeluliitto ry   12.12
[Finnish Nature Protection Society]
Address: Perämiehenkatu 11A, 00150 Helsinki 15, Finland
Telephone: (90) 642 881

FRANCE

Association Française pour   12.13
l'Étude des Eaux
– AFEE
[French Water Study Association]
Address: 21 rue de Madrid, 75008 Paris, France
Telephone: (1) 45 22 14 67 (Paris); 93 74 22 23 (Valbonne)
Enquiries to: Secretary

Alternative address: Sophia-Antipolis, 06560 Valbonne, France
Founded: 1949
Subject coverage: Documentation on water resources; water treatment; water supply; sewage treatment; water analysis; hydrobiology; toxicology.
Library facilities: Open to all users; no loan services; reprographic services (c).
Information services: Bibliographic services (bulletin and paperfiles); translation services (c); literature searches (c); access to on-line information retrieval systems.
Spidel, Paris.
Consultancy: Yes.
Publications: *Information Eaux,* bulletin, eleven issues per year.

Centre d'Information et de   12.14
Recherche sur les Nuisances
– CIRN
[Environmental Information and Research Centre]
Address: Tour Elf, Cedex 45, 92078 Paris La Défense, France
Telephone: (1) 47 44 45 46
Telex: ELFA 615400 F
Affiliation: Société Nationale Elf Aquitaine
Enquiries to: Environment manager, Bernard Tramier
Founded: 1971
Subject coverage: Air water and waste pollution in relation to the oil and petrochemical industries.
Library facilities: No library facilities.
Information services: The centre has a legal and technical information service.

Conseil de l'Europe   12.15
[Council of Europe]
Address: BP 431, 67007 Strasbourg Cedex, France
Enquiries to: European Information Centre for Nature Conservation
Subject coverage: Information and advice to Europeans through documentation, publications and campaigns about the natural environment.

Fédération Française des Sociétés   12.16
de Protection de la Nature
[French Federation of Nature Protection Societies]
Address: 57 rue Cuvier, 75231 Paris Cedex 05, France
Telephone: (1) 43 36 79 95

Institut National de Recherche   12.17
Chimique Appliquée
– IRCHA
[National Institute of Applied Chemical Research]
Address: BP 1, 91710 Vert le Petit, France
Telephone: (1) 64 93 24 75
Telex: 600820F

Affiliation: Ministry of Industry
Enquiries to: External relations officer, Alain Prats
Founded: 1959
Subject coverage: Fine chemistry, biotechnology, new materials and environment. The institute undertakes research and development, and provides advice and technical assistance.
Library facilities: Open to all users (normally - nc); loan services (normally - nc); reprographic services (nc).
Library holdings: About 10 000 bound volumes; about 500 current periodicals.
Information services: Bibliographic services (c); no translation services; literature searches (c); access to on-line information retrieval systems.
Consultancy: Telephone answers to technical queries; no writing of technical reports; no compilation of statistical trade data; no market surveys in technical areas.
Courses: Training courses, held two or three times per annum, on the following topics: exotoxicology; good laboratory practice; French and foreign chemical regulations.

Intergovernmental Oceanographic Commission * 12.18

– IOC
Address: Unesco, 7 Place de Fontenoy, 75007 Paris, France
Parent: United Nations
Enquiries to: Secretary
Subject coverage: Marine environment; marine biology; marine geology; marine geophysics; fishery statistics; marine pollution (organic/inorganic/radioactive substances); marine meteorology; bathymetry; oceanography (physical, chemical, biological, etc); hydrography; remotely sensed observations of the sea.
Information services: On-line machine searching; on-line machine retrospective searching; manual searching; manual retrospective searching; data referral; specialized searches on request.
Data base - *Marine Environmental Data Information Referral System* (MEDI). MEDI is searched and maintained via on-line access to the Unesco computer facility. However, as the facility is not registered on any of the international telecommunications networks, on-line access to MEDI is not available through public on-line systems.
Publications: *Marine Environmental Data Information Referral Catalogue* (periodic publication in the IOC manuals and guides series); brochure (in English, French, Spanish and Russian).

Secrétariat Faune-Flore * 12.19
[Fauna and Flora Secretariat]
Address: Muséum National d'Histoire Naturelle, 57 rue Cuvier, 75231 Paris Cedex 05, France
Telephone: (1) 43 36 54 32
Enquiries to: Director, F. de Beaufort
Founded: 1979
Subject coverage: Ecology, fauna, flora, natural habitats: inventories, mapping scientific bibliography.
Library facilities: Not open to all users; loan services; reprographic services.
Information services: Bibliographic services (c); No translation services; no literature searches; no access to on-line information retrieval systems.
The secretariat has a data base, FAUNA-FLORA.
Consultancy: Yes (c).
Publications: List of publications available on request.

Service de la Documentation et des Publications 12.20
– SDP
[Documentation and Publications Service]
Address: Boite Postal 337, 29273 Brest Cedex, France
Telephone: 98 22 40 13
Telex: OCEANEX 940627 F
Affiliation: Institut de Recherche pour l'Exploitation de la Mer
Enquiries to: Head, Raoul Piboubes
Founded: 1984
Subject coverage: Oceanology: marine geology, biology, ecology, geochemistry, geophysics, physics and chemistry; marine technology; marine pollution; fisheries; aquaculture; law of the sea; economics.
Library facilities: Open to all users (nc); loan services (serials only); reprographic services (c).
Information services: Bibliographic services (c); no translation services; literature searches (c); access to on-line information retrieval systems (c).

Société Nationale de Protection de la Nature et d'Acclimatation de France 12.21
[National Society for Nature Protection and Zoology in France]
Address: 57 rue Cuvier, 75231 Paris Cedex 05, France
Telephone: (1) 47 07 31 95
Library facilities: Library.

GERMAN FEDERAL REPUBLIC

Bundesanstalt für Gewässerkunde   12.22
[Federal Institute of Hydrology]
Address: Kaiserin-Augusta-Anlagen 15-17, D-5400 Koblenz, German Federal Republic
Telephone: (0261) 12431
Telex: 8-62 499
Affiliation: Bundesminister für Verkehr
Founded: 1948
Subject coverage: Hydrology, water resources, water pollution control and related fields.
Library facilities: Not open to all users; reprographic services (c - if more than 20 pages).
Library holdings: 40 000 bound volumes.
Information services: No information services.

Bundesforschungsanstalt für   12.23
Naturschutz und
Landschaftsökologie
[Federal Research Centre for Nature Conservation and Landscape Ecology]
Address: Konstantinstrasse 110, D-5300 Bonn, German Federal Republic
Telephone: (0228) 84910
Affiliation: Bundesministerium für Ernährung, Landwirtschaft und Forsten
Enquiries to: Centre for Agricultural Documentation and Information
Subject coverage: Ecology of plants, nature conservation, animal ecology, landscape management, landscape and recreational planning, and related fields.
Library facilities: The centre holds records of national nature reserves and wetlands, endangered species, nature parks and national parks.
Publications: List on request.

Deutscher Naturschutzring eV,   12.24
Bundesverband für Umweltschutz
[German Federation for Environmental Protection]
Address: Kalkuhlstrasse 24, 5300 Bonn-Oberkassel, German Federal Republic
Telephone: (0228) 44 15 05

Deutsches Hydrographisches   12.25
Institut
[German Hydrographic Institute]
Address: Postfach 220, Bernhard-Nocht-Strasse 78, D-2000 Hamburg 4, German Federal Republic
Telephone: (040) 3190-1
Telex: 211 138 bmvhh d
Facsimile: (040) 3190 5150
Affiliation: Federal Ministry for Transport
Enquiries to: Information officer, G. Heise

Bibliothek im Deutschen Hydrographischen Institut
Founded: 1945
Subject coverage: Sea surveying; oceanography; pollution monitoring.
Library facilities: Open to all users (nc); loan services (nc); reprographic services (nc).
Library holdings: 113 000 bound volumes, including reports; 1500 current periodicals.
Information services: Bibliographic services (c - Hydrographische Dokumentation); no translation services; literature searches (nc); access to on-line information retrieval systems (c).
Consultancy: Telephone answers to technical queries (c); writing of technical reports (nc); compilation of statistical trade data (nc); no market surveys in technical areas.
Publications: Navigational charts; *Seehandbücher* (Pilots); notices to mariners; (list of lights); *Deutsche Hydrographische Zeitschrift*, a scientific journal; *Meereskundliche Beobachtungen* (Oceanographic Observations); annual report; *Ozeanographie* (collected reprints on oceanography); *Überwachung des Meeres* (Monitoring the Seas).

GKSS-Forschungszentrum   12.26
Geesthacht GmbH
[GKSS Research Centre Geesthacht GmbH]
Address: Postfach 11 60, Max-Planck-Strasse, D-2054 Geesthacht, German Federal Republic
Telephone: (04152) 12-1
Telex: 0218712 gkssg
Enquiries to: Siegfried Otto
Founded: 1956
Subject coverage: Materials technology; underwater technology; environmental research, meteorological research, environmental techniques.
Library facilities: Open to all users; no loan services; no reprographic services.
Information services: Bibliographic services (nc); no translation services; no literature searches.
On-line information retrieval systems for staff users only.
Consultancy: Yes (nc).
Publications: *Wissenschaftlich-technische Berichte der GKSS*, annually.

Institut für Gewerbliche   12.27
Wasserwirtschaft und
Luftreinhaltung eV
[Industrial Water Management and Air Pollution Control Institute]
Address: Gustav-Heinemann-Ufer 84-86, D-5000 Köln 51, German Federal Republic
Telephone: (0221) 3708 397
Library facilities: Library.

Institut für Wasser-, Boden- und Lufthygiene 12.28
[Institute for Water, Soil and Air Hygiene]
Address: Corrensplatz 1, D-1000 Berlin 33, German Federal Republic
Telephone: (030) 8308-0
Telex: 0184016 bgesa d
Affiliation: Bundesgesundheitsamt
Enquiries to: Librarian
Founded: 1901
Subject coverage: Environmental hygiene, including human ecology, sanitary engineering, drinking-water, waste water and water pollution control, air and soil hygiene, water catchment.
Library facilities: Open to all users; loan services; reprographic services.
Information services: Bibliographic services; literature searches; access to on-line information retrieval systems (via Euronet).
Consultancy: Yes.
Publications: *Literaturberichte über Wasser, Abwasser, Luft und feste Abfallstoffe*, reference journal; annual publications list; current contents list, weekly.

Verein für Wasser, Boden - und Lufthygiene eV 12.29
[Society for Water, Soil and Air Hygiene]
Address: Postfach, Corrensplatz 1, D-1000 Berlin 33, German Federal Republic
Subject coverage: The society supports the Institut für Wasser-, Boden- und Lufthygiene.
Library facilities: Library.

GREECE

Hellenic Environmental Pollution Association 12.30

Address: Xenofontos 14, Athinai 118, Greece
Telephone: 32 43 534

HUNGARY

Országos Környezet- és Természetvédelmi Hiyatal * 12.31
– OKTH
[National Authority for Environment and Nature Protection]
Address: PO Box 33, H-1531 Budapest, Hungary
Telephone: (01) 166 600

Telex: 226115 OKTH H
Enquiries to: Manager of environmental information system, I. Juhász; Librarian, M. Kuczka
Founded: 1979
Subject coverage: Environmental and nature protection; land protection; water protection; air quality protection; flora and fauna protection; landscape protection; human settlement protection.
Library facilities: Not open to all users; loan services (nc); reprographic services (c).
Information services: No bibliographic services; no translation services; no literature searches; access to on-line information retrieval systems (nc).
An environmental information system contains integrated data from 9 subsystems and user programs.
Consultancy: Yes (c).

ICELAND

Náttúruverndarrád 12.32
[Nature Conservation Council]
Address: Hverfigötu 26, Reykjavík, Iceland
Telephone: (91) 2 25 20

IRELAND

An Foras Forbartha 12.33
[National Institute for Physical Planning and Construction Research]
Address: St Martin's House, Waterloo Road, Dublin 4, Ireland
Telephone: (01) 602511
Telex: 30846 El
Affiliation: Department of the Environment
Founded: 1965
Subject coverage: Environmental research; roads; planning; conservation; water quality; construction; housing; education and information.
Library facilities: Open to all users; loan services (interlibrary only); reprographic services.
Library holdings: 1000 bibliographies.
Information services: Bibliographic services; literature searches (retrospective searches, manual and computerized); access to on-line information retrieval systems.
IRS; Euronet; current awareness services; SDI computerised services; international exchanges.
Consultancy: Yes.
Publications: *Environmental Information Bulletin; Irish Journal of Environmental Science*, annual;

microfiche reports; list on request.

ITALY

Istituto Inquinamento   12.34
Atmosferico
– IIA
[Air Pollution Institute]
Address: Casella Postale 10, Via Salaria Km 29 300, 00016 Monterotondo Scalo, Italy
Telephone: (06) 9005349
Telex: 610076 CNR RMI
Parent: Consiglio Nazionale delle Ricerche
Enquiries to: Director, Professor A. Liberti
Founded: 1968
Subject coverage: Air pollution; industrial hygiene; analytical chemistry.
Library facilities: Open to all users (nc); loan services (nc); reprographic services (nc).
Library holdings: 400 bound volumes; 10 current periodicals.
Information services: Bibliographic services (nc); no translation services; literature searches (nc); access to on-line information retrieval systems (c).
Consultancy: Telephone answers to technical queries (c); writing of technical reports (c); no compilation of statistical trade data; market surveys in technical areas (c).

LUXEMBOURG

Ligue Luxembourgeoise pour la   12.35
Protection de la Nature et des
Oiseaux
[Luxembourg League for the Protection of Nature and Birds]
Address: BP 709, 6 boulevard F.D. Roosevelt, 2017 Luxembourg
Library facilities: Library.

MALTA

Regional Oil Combating Centre   12.36
for the Mediterranean Sea *
– ROCC
Address: Manoel Island, Malta
Telephone: 37296/7/8
Telex: 1464 UNROCC; 1396 UNROCC
Affiliation: IMO, United Nations Environmental Programme
Enquiries to: Director, P. Le Lourd
Founded: 1976
Subject coverage: Prevention and control of oil pollution in the Mediterranean. Exchange of information, technical cooperation, training, contingency planning assistance in cases of emergency.
Library facilities: Open to all users; no loan services; no reprographic services.
Information services: Bibliographic services (nc); no translation services; literature searches (nc); no access to on-line information retrieval systems.
The centre's information services are free of charge for contracting parties to the Barcelona Convention.
Consultancy: Yes (nc).
Publications: A quarterly publication *ROCC News* (English and French versions) is available free of charge.

NETHERLANDS

Hoofdgroep Maatschappelijke   12.37
Technologie TNO
– HMT-TNO
[TNO Technology for Society Division]
Address: Postbus 342, 7300 AH Apeldoorn, Netherlands
Telephone: (055) 773344
Telex: 36395 tnoap
Facsimile: (055) 419837
Parent: Netherlands Organization for Applied Scientific Research (TNO)
Enquiries to: Managing director, C.J. Duyverman
Founded: 1976
Subject coverage: Chemical analysis; biological environmental research; air pollution; environmental technology; wind nuisance; living and working atmosphere; biotechnology; organic chemistry and synthesis; chemical engineering; energy saving and combustion; safety in industry and coal technology.
Library facilities: Open to all users; loan services; reprographic services.

The library has microfiche and microfilm reader/printer facilities.

Information services: Bibliographic services; no translation services; literature searches; access to on-line information retrieval systems.

The library has access to the following information retrieval systems: Euronet, SDC, Dialog, Datastairs, Inis.

Internationaal Bodemreferentie en Informatie Centrum 12.38
– ISRIC

[International Soil Reference and Information Centre]

Address: Postbus 353, 6700 AJ Wageningen, Netherlands

Telephone: (08370) 19063

Enquiries to: Director, Dr W.G. Sombroek

Subject coverage: Soil science.

Library facilities: Open to all users (nc); no loan services; reprographic services (c).

Library holdings: 10 500 bound volumes; 50 current periodicals.

Information services: Bibliographic services (usually nc); no translation services; no literature searches; no access to on-line information retrieval systems.

Consultancy: Telephone answers to technical queries (nc); no writing of technical reports; no compilation of statistical trade data; no market surveys in technical areas.

Rijksinstituut voor Natuurbeheer * 12.39
– RIN

[Nature Management Research Institute]

Address: Postbus 9201, Kemperbergerweg 67, 6800 HB Arnhem, Netherlands

Telephone: (085) 452991

Affiliation: Ministry of Agriculture and Fisheries

Enquiries to: Librarian, L.P. van der Veen-Heins

Founded: 1970

Subject coverage: Nature management; environmental pollution; estuarine ecology; ornithology, zoology and botany; chemical analysis.

Library facilities: Not open to all users; loan services (nc); reprographic services (c).

Information services: No bibliographic services; no translation services; no literature searches; access to on-line information retrieval systems.

The library has on-line access to the Central Agriculture Catalogue (CLC) of the Agricultural University of Wageningen, Netherlands. It also maintains a list of professional ecologists available for overseas assignments. .

Consultancy: Yes.

Studie- en Informatiecentrum TNO voor Milieu-Onderzoek * 12.40
– SCMO-TNO

[Environmental Research and Information Centre TNO]

Address: Postbus 186, 2600 AD Delft, Netherlands

Telephone: (015) 569330

Telex: 38071 zptno

Affiliation: Organisatie voor Toegepast Natuurwetenschappelijk Onderzoek

Enquiries to: Head, information section, L. de Lavieter

Founded: 1970

Subject coverage: Environmental research: water, air and soil pollution; noise; wastes; nature and landscape protection.

Library facilities: Open to all users; loan services (nc); reprographic services (c - in some cases).

Information services: Bibliographic services (c); no translation services; literature searches (c); access to on-line information retrieval systems (c).

All charges are discretionary.

The library maintains its own data base of 4500 research projects and 10 000 documents resulting from these projects.

Consultancy: Yes (c).

NORWAY

Norsk Institutt for Luftforskning * 12.41
– NILU

[Norwegian Institute for Air Research]

Address: Postboks 130, N-2001 Lillestrøm, Norway

Telephone: (02) 714170

Affiliation: Royal Norwegian Council for Scientific and Industrial Research

Enquiries to: Assistant director, Odd F. Skogvold; Librarian, Kari M. Kvamsdal

Founded: 1969

Subject coverage: Air pollution; atmospheric corrosion; instrumentation and chemical analysis; meteorological measurements and analysis.

Library facilities: Open to all users; loan services (nc); reprographic services (nc).

Information services: Bibliographic services (nc); no translation services; literature searches (c); access to on-line information retrieval systems (c).

The library has on-line access to the Dialog information service.

Consultancy: Yes (nc).

Norsk Institutt for Vannforskning *   12.42
– NIVA
[Norwegian Institute for Water Research]
Address: Postboks 333, N-0314 Blindern, Oslo 3, Norway
Telephone: (02) 235280
Telex: 74190 NIVA n
Affiliation: Norges Teknisk-Naturvitenskapelige Forskningsråd
Enquiries to: Information officer, Knut Pedersen
Founded: 1958
Subject coverage: Water quality management; pollution, utilization, and water exchange processes in lakes, river systems, estuaries, fjords, and coastal waters; purification, transport, and management of drinking water and waste water; prevention and treatment of acidification and eutrophication.
Library facilities: Open to all users; loan services (nc); no reprographic services.
Consultancy: Yes (c).
Publications: *Publications List* and *Supplement* available on request.

POLAND

Biuro Projektów Ochrony Atmosfery   12.43
– PROAT
[Air Pollution Control Design Office]
Address: Plac Orła Białego 1, 70-562 Szczecin, Poland
Telephone: 443 66; 371 25 (information department)
Telex: 0422491
Enquiries to: Head, information department, Jacek Scheibe
Founded: 1948
Library facilities: Open to all users; loan services; reprographic services.
Information services: Bibliographic services; translation services; literature searches.
Consultancy: Yes.

Centrum Ochrony Srodowiska *   12.44
– COS
[Environmental Pollution Abatement Centre]
Address: Kossutha 6, 40-832 Katowice, Poland
Telephone: (032) 540164
Telex: 0312532
Affiliation: Instytut Kształtowania Srodowiska (Institute for Environmental Development)
Enquiries to: Scientific director, Dr M.J. Łaczny
Founded: 1973

Subject coverage: Protection of air, water and soil against pollution; technology of waste water treatment and solid waste disposal; ecological effects of air pollution; physical-chemical methods of analysis connected with the environment; radioecology; toxicology; environmental chemistry; computer science.
Library facilities: Open to all users; loan services (nc); no reprographic services.
Information services: Bibliographic services (c); translation services (c); literature searches (c); no access to on-line information retrieval systems.
The centre's own bibliographic bases are small; use is made of other Polish information centres.
Consultancy: Yes (c).

Instytut Ekologii PAN   12.45
[Ecology Institute, Polish Academy of Sciences]
Address: Dziekanów Leśny, 05-092 Łomianki, Poland
Telex: 81 73 78
Affiliation: Polska Akademia Nauk (Polish Academy of Sciences)
Enquiries to: Librarian
Founded: 1963
Subject coverage: Ecology; environmental sciences; limnology; bioenergetics.
Library facilities: Open to all users; loan services; reprographic services.
Library holdings: 60 000 volumes; 1616 periodicals.
Information services: Bibliographic services; literature searches.
Publications: The library conducts an exchange of four ecological journals published by the institute.

Zakład Ochrony Przyrody i Zasobów Naturalnych *   12.46
[Nature and Natural Resources Protection Research Centre]
Address: Ariańska 1, 31-505 Kraków, Poland
Telephone: (094) 21 51 44
Affiliation: Polska Akademia Nauk (Polish Academy of Sciences)
Enquiries to: Chief librarian, Maria Słupik
Founded: 1920
Subject coverage: Conservation of nature and natural resources; environmental and landscape conservation; botany; zoology; forestry; geography; geology; hydrology; agriculture; spatial planning.
Library facilities: Open to all users; loan services (nc); no reprographic services.
Library holdings: Besides books and periodicals, the library collects maps, photographs and slides in the field of nature conservation, and makes them available to users. The library also prepares and makes available to users card indices concerning national parks, nature reserves, areas of protected landscape, and natural

monuments in Poland.

Information services: Bibliographic services (nc); translation services (nc); literature searches (nc).

Consultancy: Yes (nc).

Publications: Bibliography on the Protection of Nature in Poland, available in index card form.

SPAIN

Centro de Estudios, Investigación y Aplicaciones del Agua * 12.47
[Water Research and Investigation Centre]

Address: Paseo San Juan 39, Barcelona 9, Spain

Telephone: (93) 231 80 11

Telex: 54180 Aqua E

Affiliation: Sociedad General de Aguas de Barcelona SA

Enquiries to: General secretary, I. Aparicio

Founded: 1960

Subject coverage: Water supply; water treatment; sanitation; environmental protection.

Library facilities: Not open to all users; no loan services; no reprographic services.

Access to the library's facilities is provided at the secretary's discretion.

Information services: Bibliographic services; no translation services; no literature searches; no access to on-line information retrieval systems.

Consultancy: Yes (nc).

Instituto Nacional de Investigaciones Agrarias, Departamento Nacional de Protección Vegetal' 12.48
[National Institute of Agronomical Research, Plant Protection Department]

Address: Puerta de Hierro 1, Carretera de la Coruña Km 7, Apdo 8.111, Madrid, Spain

Enquiries to: Director, Manuel Arroyo Varela

Library holdings: The institute has a library of 40 000 volumes.

The address of the institute is: Ministerio de Agricultura, José Abascal 56, 28003 Madrid.

Instituto Nacional para la Conservación de la Naturaleza 12.49
– ICONA

[National Institute for the Conservation of Nature]

Address: Gran Via de San Francisco 35, Madrid 5, Spain

Telephone: (91) 266 82 00

Telex: 27422

Affiliation: Ministry of Agriculture

Founded: 1972

Library facilities: Open to all users; loan services; reprographic services (c).

Information services: Bibliographic services; translation services; literature searches.

Consultancy: Yes.

SWEDEN

Institutet för Vatten- och Luftvårdsforskning 12.50
– IVL

[Water and Air Pollution Research Institute]

Address: Hälsingegatan 43, Box 21060, S-100 31 Stockholm, Sweden

Telephone: (08) 24 96 80

Telex: 21400

Affiliation: Department of Agriculture; Swedish industry

Enquiries to: Librarian

Founded: 1966

Subject coverage: Industrial pollution of air and water; handling and conversion of solid waste; sampling and analysis; characterization of risk chemicals; forecasting of stress in the ecosystem; recycling liquid and solid wastes.

Library facilities: Not open to all users; no loan services; reprographic services (c).

Publications: List on request.

Miljövårdscentrum 12.51
[Environmental Sciences Centre]

Address: S-100 44 Stockholm, Sweden

Telephone: (08) 78 79 118

Affiliation: Royal Institute of Technology

Founded: 1970

Subject coverage: Environmental science, education and technology.

Library facilities: Open to all users; loan services no reprographic services.

Information services: Bibliographic services; literature searches.

Consultancy: Yes.

Riksförbundet för Hembygdsvård 12.52
[National Association for the Preservation of Swedish Culture and Nature]

Address: Box 20031, Rutger Fuchsgatan 4, S-104 60 Stockholm, Sweden

Telephone: (08) 23 31 50

Subject coverage: Preservation of the environment.

Library facilities: Library.

Svenska Naturskyddsföreningen   12.53
[Swedish Society for Nature Conservation]
Address: Box 6400, S-113 82 Stockholm, Sweden
Telephone: (08) 15 15 50
Library facilities: Library.

SWITZERLAND

Bundesamt für Umweltschutz   12.54
– BUS
[Federal Office of Environmental Protection]
Address: CH-3003 Bern, Switzerland
Telephone: (031) 61 93 11
Telex: 912 304
Facsimile: (031) 61 99 81
Affiliation: Eidgenössisches Departement des Innern
(Ministry for Home Affairs)
Enquiries to: Information officer, E. Gysin.
Dokumentationsdienst BUS
Founded: 1971
Subject coverage: Environmental protection: water,
air, soil, noise, chemical substances, wastes, fisheries.
Library facilities: Open to all users (nc); no loan
services; no reprographic services.
Library holdings: 10 000 bound volumes; 200 current
periodicals.
Information services: No information services.
Consultancy: Telephone answers to technical
queries; no writing of technical reports; no compilation
of statistical trade data; no market surveys in technical
areas.
Publications: *Schriftenreihe Umweltschutz*, reports
on environmental issues; *Umweltschutz in der Schweiz*,
periodical; recommendations and regulations; reports
on fisheries issues; publications list on request.

Eidgenössische Anstalt für   12.55
Wasserversorgung,
Abwasserreinigung und
Gewässerschutz
– EAWAG
[Swiss Federal Institute for Water Resources and Water
Pollution Control]
Address: Ueberlandstrasse 133, CH-8600 Dübendorf,
Switzerland
Telephone: (01) 823 55 11
Telex: 56287 EAWA CH
Affiliation: Annexanstalt der Eidgenössische
Technischen Hochschulen
Enquiries to: Library
Founded: 1936
Library facilities: Open to all users; loan services;
reprographic services.

Schweizerischer Bund für   12.56
Naturschutz/Ligue Suisse pour la
Protection de la Nature
– SBN/LSPN
[Swiss League for Nature Protection]
Address: Wartenbergstrasse 22, Postfach 73, CH-
4020 Basel, Switzerland
Telephone: (061) 42 74 42
Enquiries to: Secretariat
Founded: 1909
Subject coverage: Protection of nature in Switzerland.
Library facilities: Open to all users (nc); loan services
(nc); reprographic services (c).
Loan services are only provided within Switzerland,
and there are no periodicals that can be borrowed.
Information services: No information services.
Consultancy: Telephone answers to technical queries
(when staff are available); no writing of technical
reports; no compilation of statistical trade data; no
market surveys in technical areas.
Courses: The league has two centres for field studies.
Publications: *Schweizer Naturschutz*, 6 issues a year.

TURKEY

Türkiye Tabiatini Koruma   12.57
Cemiyeti
[Turkish Association for Nature and Natural Resources
Conservation]
Address: Menekşe Sok 29/4, Ankara, Turkey
Telephone: 17 18 70

UNITED KINGDOM

Commonwealth Human Ecology   12.58
Council *
– CHEC
Address: 63 Cromwell Road, London SW7, UK
Telephone: (01) 373 6761
Affiliation: Commonwealth Foundation
Enquiries to: Executive vice chairman, Zena Daysh
Founded: 1969
Subject coverage: CHEC's functions are to educate
and to coordinate and initiate action in the field of
human ecology. The council's services are available
to governments, government departments, national
and international development agencies, professional
bodies, universities, educational institutions and
industrial and business corporations and companies.

Information services: Access to on-line information retrieval systems.

Retrieval facilities for information related to human ecology.

Publications: *CHEC Journal; CHEC Points,* periodically; lists of books, conference reports, bibliographies and other CHEC publications available from the Membership Secretary.

Department of the Environment and Transport Library 12.59

Address: Marsham Street, London SW1P 3EB, UK
Telephone: (01) 212 4847
Telex: 22221
Affiliation: Department of the Environment; Department of Transport
Enquiries to: Librarian
Subject coverage: Housing; town and country planning; local government; pollution; ports; transport (including shipping and civil aviation); water supply; roads; bridges; mineral workings; regional planning.
Library facilities: Not open to all users; loan services; no reprographic services.

Loan services are available only to libraries with BL forms. Library facilities are restricted primarily to officers of both departments; researchers only with prior permission.

Information services: Bibliographic services (c). Published bibliographic items only.

Publications: *Library Bulletin,* fortnightly abstracts series; *Annual List of Publications; Monthly Supplements*; bibliographies in planning, housing, transport and local government.

Environmental Data Services 12.60
– ENDS

Address: Unit 24, Finsbury Business Centre, 40 Bowling Green Lane, London EC1R 0NE, UK
Telephone: (01) 278 4745
Subject coverage: Pollution control; conservation; waste management; environmental policy.
Library facilities: Not open to all users (informal access is sometimes available); no loan services; reprographic services.
Publications: Monthly journal.

Institute of Hydrology 12.61

Address: Maclean Building, Crowmarsh Gifford, Wallingford, Oxfordshire OX10 8BB, UK
Telephone: (0491) 38800
Telex: 849365
Affiliation: Natural Environment Research Council
Enquiries to: Information officer, C. Kirby
Founded: 1965
Subject coverage: Mathematical, physical and applied hydrology, meteorology, climatology, geomorphology, soil science, agricultural hydrology, water resources

development, water quality.
Library facilities: Loan services (BLDSC forms to cover costs); reprographic services (c).

The library is open to bona fide researchers at no charge; users are requested to telephone before arrival.

Library holdings: 7000 bound volumes; 230 current periodicals; 1500 reports.
Information services: Bibliographic services (when possible - nc); no translation services; no literature searches; no access to on-line information retrieval systems.
Consultancy: Telephone answers to technical queries (occasional charge); no compilation of statistical trade data; no market surveys in technical areas.
Courses: Occasional courses.
Publications: Technical reports - list available on request.

Institute of Offshore Engineering 12.62
– IOE

Address: Heriot-Watt University, Riccarton, Edinburgh EH14 4AS, UK
Telephone: (031) 449 3393; 449 3794
Telex: 727918 IOE HWU G
Affiliation: Heriot-Watt University
Enquiries to: Information officer, Arnold Myers
Founded: 1972
Subject coverage: Marine technology, including oceanography, petroleum technology, naval architecture, materials and corosion science, pollution control, underwater construction, diving, safety engineering, marine biology, instrumentation, navigation, geotechnics, law, and economics.
Library facilities: Open to all users (by appointment - c); no loan services; reprographic services (c).
Library holdings: Comprehensive reference library of offshore technology.
Information services: Bibliographic services (c); no translation services; literature searches (c); access to on-line information retrieval systems (c).

Searches using in-house and commercial on-line retrieval systems as appropriate.

Consultancy: No telephone answers to technical queries; writing of technical reports (c); no compilation of statistical trade data; market surveys in technical areas (c).

Environmental impact assessments; contingency plans.

Courses: Courses provided by arrangement to meet requirements of clients.
Publications: *IOE Library Bulletin,* monthly; *Guide to Information Services in Marine Technology; Offshore Information Conference Papers,* reports; list on request.

Institute of Terrestrial Ecology * 12.63
– ITE

Address: Merlewood Research Station, Grange-over-Sands, Cumbria LA11 6JU, UK
Telephone: (044 84) 2264

Telex: 65102 MERITE
Affiliation: Natural Environment Research Council
Enquiries to: Chief Librarian, J. Beckett
Founded: 1973
Subject coverage: Biology applied to the study of natural ecosystems and environmental management.
Library facilities: Open to all users; loan services; reprographic services.
Library facilities are open to genuine researchers only; loan and reprographic services are made through the interlibrary network.
There are libraries at all research stations within the ITE: Monks Wood Experimental Station (Huntingdon); Merlewood Research Station (Grange-over-Sands); Furzebrook Research Station (Wareham); Edinburgh; Banchory; Bangor Research Station; and Culture Centre of Algae and Protozoa (Cambridge).
Information services: No information services.
Consultancy: No consultancy.

Marine Biological Association of the United Kingdom   12.64

– MBA
Address: Citadel Hill, Plymouth PL1 2PB, UK
Telephone: (0752) 221761
Affiliation: Natural Environment Research Council
Enquiries to: Head, Library and Information Services
Founded: 1888
Subject coverage: Marine sciences: marine and estuarine biology, ecology, pollution; oceanography; fisheries and related subjects.
Library facilities: Not open to all users; loan services; reprographic services.
The library is open only to marine scientists and loans only material not available from the British Library; reprographic services are limited also to this material.
Library holdings: 60 000 bound volumes; 1400 current periodicals; 65 000 pamphlets and reprints.
There is an extensive collection of expedition reports, together with a rare books section, archives, colour slides, films, charts, microfilm and microfiche material.

The library includes the Marine Pollution Information Centre, with a collection of 32 000 documents on the scientific aspects of marine pollution.
Information services: Bibliographic services (c); no translation services; literature searches (c); access to on-line information retrieval systems (c).
The library is the UK focal point and main input centre for the UN/FAO Aquatic Sciences and Fisheries Information System (ASFIS), contributing abstracts to the computer-searchable data base and monthly publication *Aquatic Sciences and Fisheries Abstracts CASFAD*. It also provides data and tests formats for the Environmental Chemicals Data and

Information Network of the EEC (ECDIN), and collaborates along similar lines with the UNEP International Register of Potentially Toxic Chemicals (IRPTC).
Data analysis is undertaken of the effects of chemicals in marine and estuarine environments.
The library makes a charge for bibliographic services only if appreciable work is required.
Consultancy: Telephone answers to technical queries (no charge if queries are brief); no writing of technical reports; no compilation of statistical trade data; no market surveys in technical areas.
Courses: Work experience and practical training are provided for overseas marine librarians and information specialists.
Publications: *Marine Pollution Research Titles*, monthly reference bulletin; bibliographies of marine and estuarine pollution and of estuary and coastal waters biology; library catalogue; publications list on request; booklet, *A Guide to the Library*, available on request.

National Centre for Alternative Technology *   12.65

– NCAT
Address: Llwyngwern Quarry, Machynlleth, Powys, UK
Telephone: (0654) 2400
Affiliation: Society for Environmental Improvement
Enquiries to: Information officer, Felicity Shooter
Founded: 1974
Subject coverage: Alternative sources of energy; organic gardening; sewage disposal; building energy-conserving structures.
Library facilities: Not open to all users; no loan services; reprographic services (c).
The library is a fairly small part of the information service; much of the information is in files, accessible through enquiries only at present. There is a large alternative technology bookshop with international mail order service.
Information services: An enquiry answering service is available by telephone or post.
Consultancy: Yes (c).
Publications: Bibliographic list.

Nature Conservancy Council   12.66

– NCC
Address: Northminster House, Peterborough PE1 1UA, UK
Telephone: (0733) 40345
Affiliation: Department of the Environment
Enquiries to: Librarian
Founded: 1973
Subject coverage: Conservation of fauna, flora, and geological and physiographical features.

Library facilities: Open to all users (nc); loan services (nc); reprographic services (c).

EP The library is open to visitors by appointment only, and loan requests must be made on BLDSC forms.

Information services: Bibliographic services (nc); no translation services; literature searches (a charge may be levied for long searches); access to on-line information retrieval systems (c).

Access to NCC WILDSCAPE data base via Datasolve.

Consultancy: Telephone answers to technical queries (nc); no writing of technical reports; no compilation of statistical trade data; no market surveys in technical areas.

Publications: Reading lists; bibliographies; information sheets.

Waste Management Information Bureau *   12.67

– WMIB

Address: Building 7.12, Harwell Laboratory, Didcot, Oxfordshire OX11 0RA, UK

Telephone: (0235) 24141

Telex: 83135 Atomha G

Affiliation: United Kingdom Atomic Energy Authority

Enquiries to: Manager, M.A. Lund

Founded: 1973

Subject coverage: Production, treatment, disposal and recycling of wastes, and related environmental problems.

Library facilities: Open to all users; reprographic services (c).

Visitors to the library must give one full working day's notice.

Information services: Bibliographic services (c); literature searches (c); access to on-line information retrieval systems (c).

The library has access to the following on-line systems: Dialtech; Dialog; Inis; Euronet.

The bureau is the UK national referral centre on waste management; it operates a computer-based bibliographic data base, and holds printed copies of all documents entered in the data base, at present numbering 27 000.

Consultancy: Yes (c).

Water Research Centre   12.68

Address: PO Box 16, Henley Road, Medmenham, Marlow, Buckinghamshire SL7 2HD, UK

Telephone: (0491) 571531

Telex: 848632

Enquiries to: Information officer

Technical Information Section

Founded: 1974

Subject coverage: Water and wastewater treatment.

Library facilities: Open to all users; loan services (members only); reprographic services (c).

Information services: Bibliographic services; translation services; literature searches; access to on-line information retrieval systems.

WRC Aqualine data base.

Consultancy: Yes.

Publications: Reports series; publications list.

YUGOSLAVIA

Plant Protection Institute   12.69

Address: POB 936, 11000 Beograd, Yugoslavia

Telephone: 660 049

Enquiries to: Director, Dr T. Stamenković

Library holdings: 7000 books; 12 650 periodicals.

13 FOOD AND DRINK

AUSTRIA

Bundesanstalt für Lebensmitteluntersuchung und - Forschung 13.1

[Federal Institute of Food Inspection and Research]
Address: Kinderspitalgasse 15, A-1090 Wien, Austria
Telephone: (0222) 42 76 61
Telex: 7/6000
Affiliation: Federal Ministry of Health and Environmental Protection
Enquiries to: Director
Founded: 1897
Subject coverage: General food research.
Library facilities: Available to non-staff users only with prior permission of the director.
Consultancy: Yes (limited).

Lebensmittel-Versuchsanstalt Forschungsinstitut der Ernährungswirtschaft 13.2

[Food Research Institute]
Address: Blaasstrasse 29, A-1190 Wien, Austria
Telephone: (0222) 36 22 55
Enquiries to: Head, Professor Herbert Woidich.
Information Centre of Food and Nutrition Sciences
Founded: 1927
Subject coverage: Food research; food analysis; food technology; cosmetics; environmental analysis.
Library facilities: Open to all users (nc); no loan services; reprographic services (c).
Information services: Bibliographic services (nc); no translation services; literature searches (c); access to on-line information retrieval systems (c).
Access to the following on-line information retrieval systems: Dialog; SDC Orbit; ESA/IRS; Cornell University PBM/STIRS; STN.
Consultancy: Telephone answers to technical queries (occasional charge); writing of technical reports; no compilation of statistical trade data; no market surveys in technical areas.

BELGIUM

Institut Belge pour l'Amélioration de la Betterave 13.3

– IBAB
[Belgian Beet Research Institute]
Address: Molenstraat 45, B-3300 Tienen, Belgium
Telephone: (016) 815171
Affiliation: Sugar industry
Enquiries to: Librarian
Founded: 1932
Subject coverage: Sugar beet and by-products.
Library facilities: Open to all users; loan services; reprographic services.
Information services: Bibliographic services; translation services; literature searches.
Consultancy: Yes.

Institut des Industries de Fermentation - Institut Meurice Chimie 13.4

[Fermentation Institute - Meurice Chemistry Institute]
Address: CERIA, Avenue Émile Gryzon, B-1070 Bruxelles, Belgium
Telephone: (02) 523 20 80
Affiliation: Centre d'Enseignement et de Recherches des Industries Alimentaires et Chimiques
Enquiries to: Director
Subject coverage: Chemistry and biochemistry.
Library facilities: Open to all users (students only); loan services (students only); reprographic services.

BULGARIA

Canning Research Institute * 13.5

Address: Ul. Mizia 10, BG-4000 Plovdiv, Bulgaria
Telephone: (032) 5 21 09
Telex: 44463
Affiliation: National Agro-Industrial Union

Enquiries to: Director, Boris Michov
Founded: 1962
Subject coverage: Fruit and vegetable storage and processing.
Library facilities: Open to all users; no loan services; reprographic services (nc).
Information services: Bibliographic services (nc); translation services (c); literature searches (c); access to on-line information retrieval systems.
Access is available to the on-line information system AGRIS.
Consultancy: Yes (nc). .
Publications: *Naucni Trudove* (Scientific Works), in Bulgarian, with abstracts in English and Russian.

Food Industry Scientific and Technical Union 13.6

Address: Rakovski 108, 1000 Sofia, Bulgaria
Telex: 22 185 nts bg

CZECHOSLOVAKIA

Středisko Technických Informaci Potravinářského Průmyslu 13.7
– STI
[Technical Information Centre for the Food Industry]
Address: Londýnská 55, 120 21 Praha 2, Czechoslovakia
Telephone: 25 73 00
Telex: 122348 VÚPP C
Affiliation: VÚPP-Výzkumný ústav potravinářského průmyslu (Food Research Institute)
Enquiries to: Manager, F. Kastl
Founded: 1958
Subject coverage: Collection, dissemination and retrieval of scientific, technical, and economic information in food industries area, including detergents, cosmetics and tobacco industries.
Library facilities: Open to all users; loan services (nc); reprographic services (c).
The library also conducts international loan services and interchange of information materials (books, journals, and photocopies).
Information services: Bibliographic services (nc); translation services (c); literature searches (c); access to on-line information retrieval systems (c).
The centre maintains the data base FSTA - Food Service Technology Abstracts - on magnetic tapes, which is used for SDI and information search services.
Consultancy: Yes (nc).

DENMARK

Slagteriernes Forskningsinstitut * 13.8
[Danish Meat Research Institute]
Address: Maglegaardsvej 2, DK-4000 Roskilde, Denmark
Telephone: (02) 36 12 00
Telex: 43241 meatre dk
Enquiries to: Librarian, Anja Møller Rasmussen
Founded: 1954
Subject coverage: Meat research.
Library facilities: Not open to all users; loan services (nc); reprographic services (nc).
Information services: Bibliographic services (nc); no translation services; literature searches (c); access to on-line information retrieval systems (c).
The institute acts as a host for the following data bases: Dialog, DIMDI, InfoLine, ESA-IRS.
Consultancy: No consultancy.

Statens Levnedsmiddelinstitut * 13.9
– SL
[National Food Institute]
Address: Mørkhøj Bygade 19, DK-2860 Søborg, Denmark
Telephone: (01) 69 66 00
Telex: 16289
Affiliation: Ministry of Environmental Protection
Enquiries to: Head of division, Søren C. Hansen
Founded: 1968
Subject coverage: Food legislation; food control; toxicology.
Library facilities: Open to all users; no loan services; reprographic services (nc).
Information services: No bibliographic services; no translation services; no literature searches; access to on-line information retrieval systems (nc).
The institute has access to about 15 international data bases.
Consultancy: Yes (nc).
Publications: Publications list on request.

FINLAND

Teknillinen Korkeakoulu 13.10
[Helsinki University of Technology, Department of Chemistry]
Address: Kemistintie 1A, SF-02150 Espoo, Finland
Telephone: (90) 451 2743
Telex: 125161
Enquiries to: Librarian, Marjukka Patrakka
Subject coverage: Organic chemistry; biochemistry and food technology; physical chemistry; inorganic and analytical chemistry; chemical technology; chemical engineering.

Library facilities: Open to all users.
Loan and reprographic services are available through the main library of Helsinki University of Technology.
Publications: List on request.

FRANCE

Centre de Documentation Internationale des Industries Utilisatrices de Produits Agricoles
13.11

– CDIUPA

[International Documentation Centre for Industries using Agricultural Products]

Address: 1 avenue des Olympiades, 91300 Massy, France
Telephone: (1) 69 20 97 38
Affiliation: Association pour la Promotion Industrie - Agriculture
Enquiries to: Director, M. Carra
Founded: 1967
Subject coverage: Food science, technology and economics; food products.
Library facilities: Open to all users (c); loan services (c); reprographic services.
Information services: Bibliographic services; translation services; literature searches.
Access to on-line information retrieval systems: data base IALINE through Télésystèmes.
Charge made for all services.
Consultancy: No consultancy.

CIE (Centre, Institut, École) *
13.12

Address: 42-44 rue d'Alésia, 75014 Paris, France
Telephone: (1) 43 27 16 74
Telex: CIEALE 20 40 60 F
Affiliation: Confédération Française de la Conserve
Enquiries to: Information officers, Pierre Lott, Anne-Marie Mekkaoui.
Service de Documentation et d'Information du CIE
Subject coverage: CIE is a grouping of the following three organizations, active in the following areas: Centre Technique des Conserves de Produits Agricoles (CTCPA) standardization and control in the area of preservation of agricultural products; Institut Appert (IA), research, documentation and laboratory control of canned foods; École d'Application des Techniques de Conservation des Produits Alimentaires (ETCPA), professional education in canning. The information service is common to all three.
Library facilities: Open to all users; no loan services; reprographic services (c).
Information services: Bibliographic services (c); literature searches (c); access to on-line information

retrieval systems (c).
The service has access to the following on-line information retrieval systems: Dialog; ESA Quest; Questel. Information services are available in subjects other than canning; consultancy services are available on national food and labelling regulations.
Consultancy: Yes (c).
Publications: *Bulletin analytique du CIE*, monthly; *Nouvelles de la Rue d'Alésia*, monthly, limited to French professional canners affiliated to national organizations; *Barèmes de stérilisation pour aliments appertisés*.

Institut des Corps Gras
13.13

– ITERG

[Fats and Oils Research Institute]

Address: 10a rue de la Paix, 75002 Paris, France
Telephone: (1) 42 96 50 29
Telex: 230905 STABILI
Enquiries to: Director, Service de Documentation
Founded: 1943
Subject coverage: Fats and oils and their derivatives: production methods, properties, applications in foodstuffs (oils, proteins, margarine) and in chemicals industry (soaps, detergents, paints, cosmetics). Chemistry, biochemistry, catalysis and analysis of oil-based products.
Library facilities: Open to all users; no loan services; reprographic services (for exchange only).
Information services: For members only.
Publications: *Revue Française des Corps Gras*, monthly; list on request.

Institut Technique de la Vigne et du Vin
13.14

– ITV

[Technical Institute of Viticulture and Oenology]

Address: 21 rue François Ier, 75008 Paris, France
Telephone: (1) 45 22 31 68
Founded: 1949
Subject coverage: Viticulture and oenology.
Library facilities: Not open to all users; loan services (films and photographs only); no reprographic services.
Consultancy: Yes.
Publications: *Vignes et Vin*; annual report.

GERMAN DEMOCRATIC REPUBLIC

Institut für Getreideverarbeitung der DDR   13.15
[Processed Grain Institute of the GDR]
Address: Arthur-Scheunert-Allee 40/41, DDR-1505 Bergholz-Rehbrücke, German Democratic Republic
Telephone: (033) 252
Telex: 015 241
Library facilities: Library.

Zentralinstitut für Ernährung *   13.16
[Central Institute of Nutrition]
Address: Arthur-Scheunert-Allee 114-116, DDR-1505 Bergholz-Rehbrücke, German Democratic Republic
Telephone: (033) 321
Affiliation: Academy of Sciences of the German Democratic Republic
Enquiries to: Director, Dr Joachim Voight. Scientific Information Service
Founded: 1946
Subject coverage: Food and human nutrition.
Library facilities: Open to all users; loan services; reprographic services.
The institute provides international interlibrary loans.
Information services: Bibliographic services (c); no translation services; literature searches (c); no access to on-line information retrieval systems.
The institute provides literature search services as follows: DSI (monthly); information retrieval service (retrospective to 1970).
Publications: Bibliographies: *Nahrung und Ernährung des Menschen: Series FO-Lebensmittelwissenschaft; Series F9-Ernährung* (Human Nutrition: Series FO Food Science; series F9 Nutrition). Annual subject indexes and author indexes.

GERMAN FEDERAL REPUBLIC

Bundesanstalt für Fettforschung *   13.17
[Federal Centre for Lipid Research]
Address: Piusallee 76, D-4400 Münster, German Federal Republic
Telephone: (0251) 43510
Affiliation: Bundesministerium für Ernährung, Landwirtschaft und Forsten
Enquiries to: Director, Professor H.K. Mangold
Founded: 1943
Subject coverage: Lipid research.
Library facilities: Open to all users; no loan services; no reprographic services.
Information services: No information services.

Bundesanstalt für Milchforschung   13.18
[Federal Dairy Research Centre]
Address: Hermann Weigmann Strasse 1, D-2300 Kiel, German Federal Republic
Telephone: (0431) 609 323
Telex: 29 29 66
Enquiries to: Director, Professor H.W. Kay. Department of Data Processing and Information Services
Founded: 1877
Subject coverage: Dairy science, food science, dairy industry, nutrition.
Library facilities: Open to all users; no loan services; reprographic services (c).
Information services: Bibliographic services (c); no translation services; literature searches (c); access to on-line information retrieval systems (c).
The centre offers on-line information retrieval via DIMDI from the following data banks: AGRIS; ASFA; Biosis prev AB; CAB Animal; CAB Plant; Cancerlit-1; Embase; FSTA; ISI/Biomed; ISI/ISTPB; ISI/Multisci; Medlars-2; Toxline.
Consultancy: No consultancy.

Bundesforschungsanstalt für Ernährung   13.19
– BFE
[Federal Research Centre for Nutrition]
Address: Engesserstrasse 20, D-7500 Karlsruhe 1, German Federal Republic
Telephone: (0721) 60114 6
Enquiries to: Informationszentrum und Bibliotheken
Founded: 1974
Subject coverage: Food sciences (technology, chemistry, toxicology, microbiology); nutrition; home economics; catering.
Library facilities: Open to all users (nc); no loan services; reprographic services (nc - in-house publications only).
Library holdings: About 10 000 bound volumes; 450 current periodicals.
Information services: Bibliographic services (nc); no translation services; literature searches (c); access to on-line information retrieval systems.
DIMDI, INKA, STN, ESA, Datastar.
Consultancy: Telephone answers to technical queries (nc); no writing of technical reports; no compilation of statistical trade data; no market surveys in technical areas.
Publications: Annual report; publications list on request.

Bundesforschungsanstalt für Getreide- und Kartoffelverarbeitung 13.20

[Federal Research Centre for Grain and Potato Processing]

Address: Postfach 23, Schützenberg 12, D-4930 Detmold, German Federal Republic
Telephone: (05231) 28042
Telex: 09-35851
Affiliation: Senate of Berlin
Enquiries to: Head, Magda Klüver. Information/Documentation Department
Founded: 1907
Subject coverage: Research in grain and potato processing; baking technology; milling technology; starch and potato technology; biochemistry and analysis.
Library facilities: Open to all users; no loan services; reprographic services (c).
Library holdings: 45 000 bound volumes; 480 current periodicals.
Information services: Bibliographic services (c); no translation services; literature searches (c); access to on-line information retrieval systems (c).
The centre provides an information service on cereal science and processing (printed version issued fornightly).
Consultancy: Yes (nc)..
Publications: *Bibliography of Cereals Processing*, annual; *Bibliography of Publications of the Federal Research Centre for Grain and Potato Processing*, every five years.

Institut für Ernährungswissenschaft, Dokumentationsstelle 13.21

[Nutrition Institute, Documentation Centre]

Address: Goethestrasse 55, D-6300 Giessen, German Federal Republic
Telephone: (0641) 702 6022; 702 6021
Affiliation: Justus-Liebig-Unversität, Giessen
Enquiries to: Documentation staff, Frau Powilleit, Frau Scheer
Founded: 1963
Subject coverage: Human nutrition; food science and technology.
Library facilities: Not open to all users; no loan services; reprographic services (c).
Information services: No bibliographic services; no translation services; literature searches (c); access to on-line information retrieval systems (c).
Consultancy: No consultancy.

Universitätsbibliothek der TU Berlin, Abteilung Zuckertechnologie mit Fachdokumentation 13.22

[University Library of the TU Berlin, Department of Sugar Technology and Documentation Centre]

Address: Amrumerstrasse 32, D-1000 Berlin 65, German Federal Republic
Telephone: (030) 3147540
Telex: 01-83872
Affiliation: Technische Universität Berlin (TU)
Enquiries to: Information officer, Dr Elmar W. Krause.
Central Sugar Library
Founded: 1865
Subject coverage: All aspects of sugar technology including history, economics, technology, analysis, plants, agriculture (beet and cane), food engineering, and medicine. Formerly part of the Sugar Institute Berlin, the library was incorporated in the TU Berlin in 1978.
Library facilities: Open to all users; loan services; reprographic services.
Library holdings: The library holds about 30 000 volumes including 1000 periodicals (530 current). Books and photocopies may be ordered through the international interlibrary loans service.
Information services: Bibliographic services; no translation services; literature searches; access to on-line information retrieval systems.
The library has access to the following on-line data bases: FSTA; AGRIS; CAB.
Consultancy: Yes.
Publications: *New Sugar Titles*, published in cooperation with: BC Sugar, Vancouver, Canada; British Sugar Corporation, Norwich, UK; Sugar Industry Research Institute, Reduit, Mauritius; Tate and Lyle Limited, Reading, UK.

Zentralstelle für Agrardokumentation und - Information 13.23

– ZADI

[Centre for Agricultural Documentation and Information]

Address: Postfach 20 05 69, Villichgasse 17, D-5300 Bonn 2, German Federal Republic
Telephone: (0228) 357097
Affiliation: Bundesministerium für Ernährung, Landwirtschaft und Forsten
Enquiries to: Director, Dr Eugen Müller
Founded: 1969
Subject coverage: Food, agriculture, forestry, including viticulture, plant protection, animal nutrition and foodstuffs, animal production and husbandry, fisheries, food hygiene.

Library facilities: No library facilities. The centre uses the facilities of the Zentralbibliothek der Landbauwissenschaft und Abteilungsbibliothek für Naturwissenschaft und vorklinische Medizin, Postfach 2460, Nussallee 15a, D-5300 Bonn 1, German FR.

Information services: Bibliographic services (c); no translation services; literature searches; access to online information retrieval systems.

The centre has access to the following:.

CAB; AGRIS; AGRICOLA; ASFA; FSTA; PSTA; Biosis; Agrep; PHYTOMED; VITIS.

Consultancy: No consultancy.

Publications: Kongresse und Tagungen der Ernährungswissenschaften, Landbauwissenschaften, Forstwissenschaften, Holzwirtschaftswissenschaften, Veterinärmedizin (Conferences in Nutrition, Agricultural, Forestry and Veterinary Sciences), annual; *Forschungsvorhaben im Bereich der Landbau-, Ernährungs- Forst- und Holzwirtschaftswissenschaften sowie der Veterinärmedizin; Teil 1: Landbauwissenschaften; Teil 2: Tierische Produktion/Veterinärmedizin; Teil 3: Ernährungswissenschaften; Teil 4: Forst- und Holzwirtschaftswissenschaften*, research reports in agriculture, animal production and veterinary science, nutrition and forestry; *Nachweise von Literatur und Forschungsvorhaben; Informationsbereitstellung und Datenverarbeitung im Agrarbereich*, quarterly review of literature and research, information science in agriculture; *Nachweise von Literatur und Forschungsvorhaben: Alternativen im Landbau*, quarterly review of literature and research, alternatives in agriculture. Publications list available on request.

HUNGARY

Központi Élelmiszeripari Kutató Intézet 13.24
– KÉKI

[Central Food Research Institute]

Address: Postafiók 76, Herman Ottó utca 15, H-1022 Budapest II, Hungary

Telephone: 361 558 928

Telex: 224709 OSzBK

Enquiries to: Chief librarian, Erzsébet Pethő

Founded: 1959

Subject coverage: Food industry; biology; enzymology; microbiology; fermentation and protein technology; chemistry; measuring and control engineering in the processing and preservation of foods; biotechnology.

Library facilities: Open to all users (nc); loan services (nc); reprographic services (c).

Library holdings: 19 000 bound volumes; 200 current periodicals.

Information services: No information services.

Consultancy: No consultancy.

Publications: Publications list on request.

Magyar Élelmezésipari Tudományos Egyesület 13.25
– MÉTE

[Hungarian Scientific Society for the Food Industry]

Address: V Akadémia utca 1-3, H-1361 Budapest, Hungary

Telephone: (1) 122 859

Telex: 225792 mtesz h

Affiliation: Müszaki és Természettudományi Egyesületek Szövetsége

Enquiries to: Secretary-general, Professor István Tóth-Zsiga

Founded: 1949

Subject coverage: Food chemistry; chemical, physical and biological methods; food sciences.

Library facilities: Not open to all users; loan services; no reprographic services.

Consultancy: Yes.

Országos Húsipari Kutatointézet 13.26
– OHKI

[Hungarian Meat Research Institute]

Address: Gubacsi utca 6/b, H-1097 Budapest IX, Hungary

Telephone: (1) 337 350

Telex: 22-4980

Affiliation: Ministry of Agriculture and Food Affairs

Enquiries to: Head librarian, K. Molnár.

Information Centre of the Meat Industry

Founded: 1959

Subject coverage: Research and development in the meat industry.

Library facilities: Open to all users; loan services (nc); reprographic services (c).

The library maintains a register of translations and research reports, and a register of motion pictures available for loan.

Information services: Bibliographic services (c); translation services (c); literature searches (nc); access to on-line information retrieval systems (c).

The library has on-line connection, through the Information Centre of the Ministry for Agriculture and Food Affairs (MÉM Agroinform), to Lockheed Dialog.

Consultancy: Consultancy (nc).

Publications: Guide to the Technical Literature of the Meat Industry, abstracts journal.

ICELAND

Rannsóknastofnun Fiskiðnaðarins   13.27
[Icelandic Fisheries Laboratories]
Address: PO Box 1390, Skúlagata 4, 121 Reykjavík, Iceland
Telephone: (91) 20240
Telex: Simtex IS 3000 - Fishlab
Enquiries to: Librarian, Eiríkur T. Einarsson
Founded: 1965
Subject coverage: Sea food research; bacteriology; microbiology; biotechnology; mechanical engineering concerning the fish industry.
Library facilities: Open to all users (nc); loan services (nc); reprographic services.
Books only, not periodicals, may be borrowed.
Library holdings: About 5000 bound volumes; 145 current periodicals.
Information services: Bibliographic services (c); no translation services; literature searches (c); access to on-line information retrieval systems (c).
Data banks used are Dialog, Pergamon InfoLine, and the data bases of Food RA in Leatherhead, United Kingdom.
Consultancy: Telephone answers to technical queries (nc); no writing of technical reports; no compilation of statistical trade data; no market surveys in technical areas.
Publications: Annual report; technical reports.

ITALY

Istituto Sperimentale per la   13.28
Valorizzazione Tecnologica dei
Prodotti Agricoli *
– IVTPA
[Experimental Institute for the Technological Improvement of Agricultural Products]
Address: Via G. Venezian 26, 20133 Milano, Italy
Telephone: (02) 29 37 41
Enquiries to: Director, Professor Andrea Monzini
Founded: 1967
Subject coverage: Cool storage; freezing technology; quality control of vegetables, fish and meat.
Library facilities: Open to all users; loan services (limited); reprographic services.
Publications: Annual report; catalogue on request.

Stazione Sperimentale per   13.29
l'Industrie degli Oli e dei Grassi
– SSOG
[Oils and Fats Industry Research Station]
Address: Via Giuseppe Colombo 79, 20133 Milano, Italy

Telephone: (02) 236 10 51
Telex: 340129 SSOG I
Founded: 1923
Subject coverage: Olive and seed oils, edible and industrial fats, vegetable proteins, paints and varnishes, soaps, detergents and surfactants, mineral oils.
Library facilities: Open to all users; no loan services; reprographic services.
Information services: Bibliographic services (c); literature searches (c).
Consultancy: Yes (c).
Publications: *La Rivista Italiana delle Sostanze Grasse*, monthly; *Tribologie e Lubrificazione*, quarterly.

NETHERLANDS

Hoofdgroep Voeding en   13.30
Voedingsmiddelen TNO
– VV-TNO
[Food and Nutrition Research Division TNO]
Address: Postbus 360, 3700 AJ Zeist, Netherlands
Telephone: (03404) 52244
Telex: 40022 civo nl
Affiliation: Nederlandse Organisatie voor Toegepast-Natuurwetenschappelijk Onderzoek TNO
Enquiries to: Director
Founded: 1940
Library facilities: Open to all users; loan services; reprographic services (c).
Information services: Bibliographic services; translation services; literature searches.
Charge made for all services.
Consultancy: Yes.

Instituut voor Graan, Meel en   13.31
Brood TNO
– IGMB - TNO
[Cereals, Flour and Bread Institute TNO]
Address: Postbus 15, 6700 AA Wageningen, Netherlands
Telephone: (08370) 19051
Telex: 4022 civo nl
Affiliation: Nederlandse Organisatie voor Toegepast-Natuurwetenschappelijk Onderzoek TNO
Enquiries to: Director
Founded: 1941
Subject coverage: Food products and mixed feed from cereals, seeds and pulses.
Library facilities: Open to all users; loan services; reprographic services (c).
Library, including card catalogue on cereal chemistry and technology, in German.
Publications: List on request.

Instituut voor Visserijprodukten *   13.32
– IV-TNO
[Fishery Products Institute TNO]
Address: Postbus 183, 1970 AD IJmuiden, Netherlands
Telephone: (02550) 19023
Affiliation: CIVO Technologie TNO
Enquiries to: Director, G.M. Straatsburg-Braam
Subject coverage: Technology, chemistry and bacteriology of fishery products.
Library facilities: Open to all users; loan services (nc); reprographic services (c).
Information services: Bibliographic services (nc); no translation services; literature searches (c); no access to on-line information retrieval systems.
Consultancy: Consultancy (nc).

Nederlands Instituut voor   13.33
Zuivelonderzoek *
– NIZO
[Netherlands Institute for Dairy Research]
Address: Postbus 20, 6710 BA Ede, Netherlands
Telephone: (08380) 19013
Telex: 37205 NL
Enquiries to: Head, E. Otter
Founded: 1948
Subject coverage: Dairy technology; dairy microbiology; analytical chemistry; process technology; biochemistry; nutrition.
Library facilities: Not open to all users; loan services (within the Netherlands); reprographic services (c).
Information services: No information services.

NORWAY

Bryggeriindustriens   13.34
Forskningslaboratorium
[Brewing Industry Research Laboratory]
Address: Forskningsveien 1, Blindern, Oslo 3, Norway
Telephone: (02) 45 20 10
Affiliation: Bryggeriernes Servicekontor
Enquiries to: Information officer, Sturla Lie
Founded: 1946
Subject coverage: Brewing: chemistry, biochemistry, microbiology.
Library facilities: Available to Norwegian breweries staff only.
The laboratory is a part of the NAERINFO joint information project, organized through the Norwegian Centre for Informatics (NSI).

Norges Slakterilaboratorium *   13.35
– NSL
[Norwegian Meat Research Laboratory]
Address: Lörenveien 37, PO Box 96, Refstad, N-0513 Oslo 5, Norway
Telephone: (02) 15 05 10
Telex: 71302
Enquiries to: Director, Karl Martin Anthonsen
Founded: 1961
Library facilities: Open to all users; no loan services; reprographic services.
Information services: Bibliographic services; translation services; literature searches; no access to on-line information retrieval systems.
Consultancy: Yes.

Norsk Institutt for   13.36
Naeringsmiddelforskning
– NINF
[Norwegian Food Research Institute]
Address: Postboks 50, N-1432 Ås-NLH, Norway
Telephone: (02) 94 08 60
Telex: 72400 fotex n Attn. NINF, Ås
Affiliation: Selskapet for Landbrukets Naeringsmiddelforskning
Enquiries to: Librarian, Line Arneberg
Founded: 1971
Subject coverage: Food science and technology.
Library facilities: Open to all users (nc); loan services (nc); reprographic services (nc).
The library is primarily for internal use, but provides external services providing it does not inconvenience the institute's own research.
Library holdings: 6000 bound volumes; 213 current periodicals.
Information services: No translation services; literature searches; access to on-line information retrieval systems (c).
The library is connected to Dialog and Pergamon InfoLine.
Consultancy: Telephone answers to technical queries (nc); writing of technical reports (free or reduced price to meat, vegetables, fruit and berries industry - c) no compilation of statistical trade data; no market surveys in technical areas.
Courses: Sensory analysis (twice yearly), food hygiene in the food industry (yearly), detergents and disinfectants (yearly).

Sildolje- og Sildemelindustriens   13.37
Forskningsinstitutt
– SSF
[Norwegian Herring Oil and Meal Industry Research Institute]
Address: Bjørgeveien 220, N-5033 Fyllingsdalen, Bergen, Norway
Telephone: (05) 12 31 00
Telex: 40087 forskn

Affiliation: Norwegian fish meal industry
Enquiries to: Librarian, Berit B. Gurvin
Founded: 1948
Subject coverage: Chemistry; chemical engineering; biochemistry; microbiology; nutrition; animal husbandry; food science; fisheries technology (aquaculture).
Library facilities: Open to all users (nc); loan services (nc); reprographic services (nc).
Library holdings: About 100 current periodicals; 548 external reports; about 1000 books.
Information services: No information services.

POLAND

Centralny Ośrodek Informacji Naukowej przy 'Społem', Centralnyn Ośrodku Badawczo-Rozwojowym Przemysłu Gastronomicznego *   13.38

– COINTE COBRPGiAS
['Społem' Scientific Information Centre, Central Research and Development Institute for the Gastronomic and Foodstuffs Industry]
Address: Ulica Kopernika 15/17, 90-503 Łódź, Poland
Telephone: 689 20
Telex: 886765
Enquiries to: Director, Wojciech Nowak
Founded: 1970
Subject coverage: Scientific research and development studies, experiments and practical applications in gastronomy. Studies concerning the production of delicatessen, non-alcoholic beverages, and the foodstuff market.
Library facilities: Open to all users; loan services; reprographic services (from library holdings only). International and national interlibrary loan services.
Information services: Patent agency, problems of normalization, trade literature.
Publications: *Biuletyn Informacyjny* (Information Bulletin), quarterly.

Instytut Przemysłu Mleczarskiego *   13.39

[Dairy Industry Institute]
Address: Hoza 66/68, 00-682 Warszawa, Poland
Telephone: (022) 28 58 12
Affiliation: Central Union of Dairy Cooperatives
Enquiries to: Chief librarian
Subject coverage: Dairy processing.
Library facilities: Not open to all users; loan services (inter-library only); reprographic services (c).

Information services: Bibliographic services; translation services (institute staff and staff of the Central Union of Dairy Cooperatives only); literature searches.
Consultancy: Yes.

PORTUGAL

Empresa Pública de Abastecimento de Cereais *   13.40

– EPAC
[Cereals Supply Public Enterprise]
Address: Avenida Gago Coutinho 26, Apartado 5129, 1753 Lisboa Codex, Portugal
Telephone: 802322 PPCA
Telex: 13380 EPAC P
Enquiries to: Head, Documentation and Information Department
Founded: 1977
Subject coverage: Cereals: storage, technology, chemistry, trade, import.
Library facilities: Open to all users; loan services; reprographic services.
Information services: Bibliographic services; translation services; literature searches.
Consultancy: Yes.
Publications: List on request.

Instituto do Azeite e Produtos Oleaginosos *   13.41

– IAPO
[Olive Oil and Oil Seeds Board]
Address: Avenida António Augusto de Aguiar 23-2, 1098 Lisboa Codex, Portugal
Telephone: 572566
Affiliation: Secretaria de Estado do Comércio Externo.
Enquiries to: Librarian, Dr Octávio Henrique Pinto Faustino
Founded: 1973
Subject coverage: Coordination of production, import, export and trade in olive oil, oil seeds, oil meals, margarines, soaps, and some animal fats; technical and economic studies concerning these products, and checking of their origin and quality.
Library facilities: Open to all users; loan services (c); reprographic services (c).
Information services: Bibliographic services (c); translation services (c).
Consultancy: Yes (c).

Instituto do Vinho do Porto   13.42

– IVP
[Port Wine Institute]
Address: Rua de Ferreira Borges, 4000 Porto, Portugal
Telephone: (02) 26522

Telex: 25337
Affiliation: Secretaria de Estado da Alimentação, Ministério da Agricultura, Pescas e Alimentação
Enquiries to: Director, Dr Maria Lúcia M.C.C. Alves Moreira
Founded: 1933
Subject coverage: Protection and control of the genuineness and quality of port wine.
Library facilities: No loan services; no reprographic services.
The library is open to researchers only.
Library holdings: 14 098 bound volumes; 260 current periodicals.
Information services: Bibliographic services (nc); no translation services; literature searches (nc).
The institute maintains information on the following: port wine in general; demarcated area of the Douro port wine producing region; Entrepôt (V.N. de Gaia); Portuguese and foreign literature on port; legislation on port; oenology and viticulture.
Consultancy: No telephone answers to technical queries; writing of technical reports; compilation of statistical trade data; market surveys in technical areas.
Publications: *Anais do Instituto do Vinho do Porto*, three yearly; *Cadernos Mensais de Estatistica e Informação*, bimonthly statistical report; *O Vinho do Porto*, annual port business survey.

SPAIN

Instituto de Agroquímica y Tecnología de Alimentos   13.43
– IATA
[Food Technology and Agrochemistry Institute]
Address: Jaime Roig J1, 46010 Valencia, Spain
Telephone: (96) 369 0800
Telex: 64197 AYTVE
Affiliation: Consejo Superior de Investigaciones Científicas
Enquiries to: Director, Luis Duran Hidalgo
Founded: 1950
Subject coverage: Food science and technology; agricultural chemistry.
Library facilities: Open to all users (nc); no loan services; reprographic services.
Library holdings: 6000 books; 4000 bound journals; 250 current periodicals.
Information services: No information services.
Consultancy: Telephone answers to technical queries (nc); writing of technical reports; compilation of statistical trade data; market surveys in technical areas.
Courses: Curso de alta especialización en tecnología de alimentos, annual course on food technology.

Instituto de Fermentaciones Industriales   13.44
[Industrial Fermentation Institute]
Address: Juan de la Cierva 3, Madrid 6, Spain
Affiliation: Consejo Superior de Investigaciones Científicas
Enquiries to: Librarian
Founded: 1940
Subject coverage: Food, beverages and brewing.
Library facilities: Open to all users; loan services; reprographic services.
The library is coordinated with the other libraries of the National Research Council and therefore can provide material available in these libraries.
Information services: Bibliographic services; literature searches; access to on-line information retrieval systems (through the Information and Documentation Service - ICYT).
Consultancy: Yes.

Instituto de la Grasa y sus Derivados   13.45
[Fats and Derivatives Institute]
Address: Avenida P. García Tejero 4, Sevilla 12, Spain
Telephone: (954) 61 15 50
Affiliation: Consejo Superior de Investigaciones Científicas
Enquiries to: Librarian/information officer.
Servicio de Información y Documentación
Founded: 1947
Subject coverage: Vegetable oils, in particular olive oil; vegetable proteins; table olives; soaps and detergents; residue profits; pollution problems.
Library facilities: Open to all users; no loan services; reprographic services.
Information services: Bibliographic services; translation services; literature searches.
Consultancy: Yes.
Publications: *Grasas y Aceites*, bimonthly research reports.

Instituto del Frío   13.46
[Refrigeration Institute]
Address: Ciudad Universitaria, 28040 Madrid, Spain
Telephone: (91) 449 61 62; 449 61 66
Affiliation: Consejo Superior de Investigaciones Científicas
Enquiries to: Director, Fernando Beltrán Cortés
Founded: 1951
Subject coverage: Refrigeration engineering; food conservation by freezing.
Library facilities: Open to all users; no loan services; no reprographic services.
Information services: Bibliographic services.
Consultancy: Yes.

SWEDEN

Djupfrysningsbyrån 13.47
[Frozen Food Institute]
Address: Box 1542, S-111 85 Stockholm, Sweden
Telephone: (08) 10 82 18
Enquiries to: Manager, Per-Olof Carlbaum
Founded: 1953
Subject coverage: Information research, education, and reports on frozen foods.
Library facilities: No library facilities.
Information services: No bibliographic services; no translation services; no literature searches; no access to on-line information retrieval systems.
Consultancy: No telephone answers to technical queries; no writing of technical reports; compilation of statistical trade data; no market surveys in technical areas.
Publications: Research reports; statistics; irregular publications on production and consumption of frozen foods in Sweden.

Köttforskningsinstitutet 13.48
[Swedish Meat Research Institute]
Address: PO Box 504, S-244 00 Kävlinge, Sweden
Telephone: (046) 73 22 30
Telex: 32206 Meatres S
Facsimile: (046) 73 61 37
Affiliation: Swedish Meat Marketing Association; Swedish Farmers' Meat Marketing Organization
Enquiries to: Information officer, Olle Holmqvist
Founded: 1967
Subject coverage: Meat and meat products; biochemistry; microbiology; technology.
Library holdings: About 5000 bound volumes; 200 current periodicals; about 3000 reports.
Information services: Services are provided subject to availability of staff.
Consultancy: Consultancy (limited · to affiliated cooperative organizations).
Publications: List on request.

SIK - Svenska 13.49
Livsmedelsinstitutet *
[SIK - the Swedish Food Institute]
Address: Box 5401, S-402 29 Göteborg, Sweden
Telephone: (031) 400120
Telex: 21651 SIK S
Parent: Foundation SIK - The Swedish Food Institute; Swedish Board for Technical Development (STU)
Enquiries to: Librarian, Birgitta Berg
Founded: 1946
Subject coverage: Processing, preservation and storage of food and food products, and related matters concerning food raw materials, packaging, distribution and consumption.

Library facilities: Open to all users; loan services (nc); reprographic services (c).
Information services: Bibliographic services; translation services; literature searches; access to on-line information retrieval systems.
Member services only.
Access is available to SDC, Dialog, and IRS.
Consultancy: Yes.
Publications: Catalogue on request.

SWITZERLAND

Eidggenössische 13.50
Forschungsanstalt für Obst-,
Wein- und Gartenbau *
– FAW
[Swiss Federal Research Station for Fruit Growing, Viticulture and Horticulture]
Address: Schloss, CH-8820 Wädenswil, Switzerland
Telephone: (01) 780 13 33
Affiliation: Department of Public Economy, Division of Agriculture
Enquiries to: Librarian, G. Schwarz
Founded: 1890
Subject coverage: Fruit growing, viticulture, horticulture; biochemistry, agricultural chemistry; entomology, nematology; phytopathology; weeds; biology and chemistry of wines and fruit juices; storage and processing of fruits and vegetables; mushrooms.
Library facilities: Not open to all users; loan services (nc); reprographic services (c).
The library has microfiche reader/printer facilities.
Information services: Bibliographic services; no translation services; no literature searches; no access to on-line information retrieval systems (but access is planned).
Consultancy: Yes.
Publications: Catalogue on request.

Schweizerischer Verband der 13.51
Ingenieur-Agronomen und der
Lebensmittel-Ingenieure
[Swiss Association of Agricultural and Foodstuff Engineers]
Address: 3052 Zollikofen, Switzerland
Telephone: (031) 57 06 68

UNITED KINGDOM

AFRC Institute of Food Research, Bristol Laboratory 13.52

Address: Langford, Bristol BS18 7DY, UK
Telephone: (0924) 852661
Telex: 449095
Affiliation: Agricultural and Food Research Council
Enquiries to: Information officer, Dr M.A. Winstanley
Founded: 1964 (as Meat Research Institute; renamed in 1985)
Subject coverage: Red meat and poultry meat research - carcass composition and meat quality; refrigeration and processing technology; microbiology, muscle biology, instrumental and sensory evaluation; analysis.
Library facilities: Open to all users (at director's discretion - nc); loan services (to libraries only); reprographic services.
Information services: Bibliographic services (nc); literature searches (nc); access to on-line information retrieval systems (c).
Industrial Development Group for technical enquiries.
Consultancy: No consultancy.
Publications: *Biennial Report and Appendix*; quarterly newsletter; quarterly updated publications lists.

AFRC Institute of Food Research, Norwich Laboratory 13.53

Address: Colney Lane, Norwich NR4 7UA, UK
Telephone: (0603) 56122
Telex: 975453
Affiliation: Agricultural and Food Research Council
Enquiries to: Liaison officer
Founded: 1964
Subject coverage: Chemistry, physics, biochemistry and microbiology of foods and food components including the physical and engineering principles of food processing.
Library facilities: Open to all users (for reference purposes only, and subject to librarian's discretion - nc); loan services (via accredited libraries only - nc); reprographic services (to accredited libraries only - nc).
Library holdings: 15 000 bound volumes; 350 current periodicals.
Information services: The National Collection of Yeast Cultures (NCYC), housed at the institute, forms a resource and information centre. Services include computer searches of the NCYC strain data base and of the NCYC literature data base. On-line computer identification of unknown yeasts using computer networking system (PSS via AGRINET) is planned.

Consultancy: Telephone answers to technical queries (limited to topics related to research programme - nc); no writing of technical reports; no compilation of statistical trade data; no market surveys in technical areas.
Advice can be obtained from the NCYC on matters relating to yeasts.
Courses: NCYC courses.
Publications: Biennial report; quarterly newsletters; occasional reports on specialist topics; quarterly publications lists; NCYC leaflet.

Biotechnology Centre, Wales 13.54
– BTCW
Address: Singleton Park, Swansea SA2 8PP, UK
Telephone: (0792) 296396
Telex: 48358 ULSWAN-G
Enquiries to: Director, Dr R.N. Greenshields
Founded: 1983
Subject coverage: Biotechnology, with application in the food industry.
Library facilities: No library facilities.
Information services: No information services.
Consultancy: Telephone answers to technical queries (c); writing of technical reports (c); no compilation of statistical trade data; market surveys in technical areas (c).
Biotechnological services (c).
Publications: *International Industrial Biotechnology*, bimonthly bulletin.

Brewing Research Foundation * 13.55
Address: Lyttel Hall, Nutfield, Redhill, Surrey RH1 4HY, UK
Telephone: (073) 782 2272
Affiliation: Brewers' Society
Enquiries to: Information officer, G. Jackson
Founded: 1951
Subject coverage: Brewing, including genetics, microbiology, biochemistry, chemistry and brewing technology.
Library facilities: Open to all users; no loan services; reprographic services (c).
The library is open to non-staff users at the director's discretion; loan services are also occasionally provided on discretion.
Information services: Translation services; literature searches (at the director's discretion); no access to on-line information retrieval systems (but this service is planned).
Consultancy: Yes.
Publications: *Bulletin of Current Literature*, monthly review of brewing and allied journals.

British Food Manufacturing Industries Research Association * 13.56

– Leatherhead Food RA

Address: Randalls Road, Leatherhead, Surrey KT22 6RY, UK

Telephone: (0372) 376761

Telex: 929846

Enquiries to: Sales manager, J.R. Swift

Founded: 1919

Subject coverage: Food manufacturing science and technology: analytical chemistry; microbiology; nutrition; hygiene; oils and fats; biochemistry; waste treatment.

Library facilities: Not open to all users; loan services (nc); reprographic services (nc).

Full library service is available to members only.

Information services: Bibliographic services; no translation services; literature searches; access to on-line information retrieval systems (c).

The library maintains data bases on: food science and technology, food legislation, commercial data, food commodities, and food processing plant.

Consultancy: Yes (c).

Campden Food Preservation Research Association * 13.57

– CFPRA

Address: Chipping Campden, Gloucestershire GL55 6LD, UK

Telephone: (0386) 840319

Telex: 337017 CFPRA G

Enquiries to: Librarian, C.J. Willcox

Founded: 1919

Subject coverage: Food science and technology; agriculture (crops suitable for processing); chemistry; microbiology.

Library facilities: Open to all users (non-members by appointment only); loan services; reprographic services.

Information services: Literature searches (c).

Consultancy: Yes (c).

Flour Milling and Baking Research Association 13.58

– FMBRA

Address: Chorleywood, Rickmansworth, Hertfordshire WD3 5SH, UK

Telephone: (09278) 4111

Telex: 8952883

Enquiries to: Senior information officer, D.A. Williams

Founded: 1967

Subject coverage: Botany and biochemistry of the wheat grain; milling processes and machinery; flour quality and improvement; biochemistry of raw materials for bread, biscuits, and flour-confectionery; extrusion cooking processing methods and machinery; food preservation; nutrition; food legislation.

Library facilities: Not open to all users; loan services; reprographic services (c).

Non-members of FMBRA are admitted to the library at the director's discretion, and are not allowed loan services.

Information services: Bibliographic services (nc); no translation services; literature searches (c); access to on-line information retrieval systems (c).

Information services are available for members only. The library maintains a computerized information data base, established 1978, covering areas listed above. The data base contains 32 000 entries to date in the form of bibliographic details and keywords, and is available on-line to members of FMBRA.

Consultancy: No consultancy.

Publications: Abstracts.

Torry Research Centre * 13.59

Address: PO Box 31, Aberdeen AB9 8DG, UK

Telephone: (0224) 877071

Affiliation: Ministry of Agriculture, Fisheries and Food

Enquiries to: Head of Information Services, J.J. Waterman

Founded: 1929

Subject coverage: Fish handling and processing.

Library facilities: Not open to all users; loan services (at director's discretion); reprographic services (c).

Consultancy: Yes (c).

14 HEALTH AND SAFETY

BELGIUM

Archives Belges de Médecine Sociale, Hygiène, Médecine du Travail et Médecine Légale
14.1

[Belgian Archives for Social Medicine, Hygiene, Occupational Medicine and Forensic Medicine]
Address: Cité administrative de l'État, Quartier Esplanade 6, B-1010 Bruxelles, Belgium
Affiliation: Ministère de la Santé Publique et de la Famille
Enquiries to: Secretary-General/Administrator, Dr M. Luyckx
Founded: 1937
Library facilities: Open to all users; no loan services; no reprographic services.
Information services: Bibliographic services; no translation services; literature searches; no access to on-line information retrieval systems.
Consultancy: No consultancy.

Association Nationale pour la Protection contre l'Incendie
14.2

[National Association for Protection against Fire]
Address: BP 1A, Parc Scientifique, B-1348 Ottignies, Louvain-la-Neuve, Belgium
Telephone: (010) 41 87 12
Library facilities: Library.

BULGARIA

Centre for Scientific Information in Medicine and Public Health *
14.3

Address: Georgi Sofijski I, Sofia 1431, Bulgaria
Telephone: 51 87 42
Affiliation: Medical Academy
Enquiries to: Deputy Director, Dr Borjana Stantcheva
Founded: 1965
Subject coverage: All branches of medicine and public health organization and management.
Library facilities: Open to all users; loan services (c); reprographic services (c).
Information services: Bibliographic services (c); translation services (c); literature searches (c); access to on-line information retrieval systems (c).
Biosis; Inis; Inspec; AGRIS; MSIS-NIR; MEDIC.
Consultancy: Yes (c).

CYPRUS

Cyprus Ministry of Health, Nicosia General Hospital Library *
14.4

Address: Medical and Public Health Services, Nicosia, Cyprus
Telephone: 403165
Enquiries to: Librarian
Subject coverage: Provides substantial information and reading facilities for the hospital staff.
Information services: No information services.
Consultancy: Yes (nc).

Cyprus Ministry of Health, Public Health Inspectors' School Library

14.5

Address: Medical and Public Health Services, Nicosia, Cyprus
Telephone: (02) 402425
Enquiries to: Chief Health Inspector
Founded: 1963
Subject coverage: Anything related to the training of health inspectors.
Library facilities: Not open to all users; loan services (nc); no reprographic services.
Library holdings: 185 bound volumes; 40 current periodicals; 4 reports.
Information services: No information services.
Consultancy: Telephone answers to technical queries (nc); writing of technical reports (nc); no compilation of statistical trade data; no market surveys in technical areas.

Ministry of Health, School of Nursing Library *

14.6

Address: Medical and Public Health Services, Nicosia, Cyprus
Telephone: 403165
Enquiries to: Librarian
Founded: 1953
Subject coverage: Anything related to the training of nurses.
Library facilities: Open to all users (nc).
The library provides loan services and reading facilities to student nurses, teachers and lecturers.
Information services: No information services.

CZECHOSLOVAKIA

Ústav Vědeckých Lékařských Informací

14.7

[Medical Information Institute]
Address: Vítězného února 31, 121 32 Praha 2, Czechoslovakia
Telephone: 29 99 56-9
Enquiries to: Head, Jiří Drbálek. Bibliographic and Information Department
Founded: 1947
Subject coverage: Biomedical sciences; public health; health legislation.
Library facilities: Open to all users (nc); loan services (nc); reprographic services (c).
Library holdings: Over 230 000 bound volumes; 1290 current periodicals; about 8000 reports.
Information services: Bibliographic services (nc); no translation services; literature searches (nc); access to on-line information retrieval systems (nc).
On-line access through national information centre ÚVTEI-ÚTZ to retrieval systems stored in: MCNTI (Moscow) and CINTI (Bulgaria); Datastar, SDC, INKA, Prestel, etc (limited access). Machine-readable data files; Czechoslovak medical literature data base; *Excerpta Medica* data base and SDI searches.
Consultancy: No consultancy.

DENMARK

Dansk Brandvaerns-Komité

14.8

[Danish Fire Protection Association]
Address: Datavej 48, DK-3460 Birkeroed, Denmark
Telephone: (02) 82 00 99
Enquiries to: Librarian, Dorrit Hansen
Founded: 1920
Subject coverage: Fire - prevention, protection, extinguishing.
Library facilities: Open to all users (nc); loan services (nc); reprographic services (c).
Library holdings: 2000 bound volumes; 60 current periodicals; 4000 reports.
Information services: Bibliographic services (c); no translation services; literature searches (c); no access to on-line information retrieval systems.
Consultancy: Telephone answers to technical queries (nc); writing of technical reports (c); no compilation of statistical trade data.
Occasional market surveys in technical areas.
Courses: About fifteen courses per annum, and two theme days - topics include: natural gas in industry; protection against fire in industrial buildings.
Publications: Quarterly survey of literature about fire; *Katalog*, annual.

Sundhedsstyrelsen, Biblioteket

14.9

[National Board of Health, Library]
Address: St Kongensgade 1, DK-1264 København K, Denmark
Telephone: (01) 14 10 11
Affiliation: Ministry of the Interior
Enquiries to: Librarian, Ilse Løye
Founded: 1909
Subject coverage: Danish health services.
Library facilities: Not open to all users; loan services (nc); reprographic services (nc).
Library holdings: 27 500 bound volumes; 300 current periodicals.
Information services: Bibliographic services (nc); no translation services; literature searches (nc); no access to on-line information retrieval systems.
Consultancy: No consultancy.

Publications: List on request.

World Health Organization 14.10
Regional Office of Europe *
– WHO/EURO
Address: 8 Scherfigsvej, DK-2100 København 0, Denmark
Telephone: (01) 29 01 11
Telex: 15348
Enquiries to: Scientist, Health and Biomedical Documentation, Dr M.C. Thuriaux
Founded: 1946
Subject coverage: Health.
Library facilities: Open to all users; loan services (interlibrary - nc); no reprographic services.
Library holdings: 400 journals; approximately 4000 volumes with emphasis on health services, health policy, health programmes; geographical coverage - Europe, Algeria, and Morocco.
Information services: Bibliographic services (on limited basis to developing countries - nc); no translation services; literature searches (limited to countries without easy access - nc); no access to on-line information retrieval systems..
Data bases available on health legislation and training in public health.
Consultancy: Yes (to member states only - nc).

FINLAND

Kansanterveyslaitos 14.11
[National Public Health Institute]
Address: Mannerheimintie 166, SF-00280 Helsinki 28, Finland
Telephone: (90) 47441
Telex: 121394 SF
Enquiries to: Head librarian, Marita Antila
Founded: 1910
Subject coverage: Bacteriology; biochemistry; epidemiology; immunology; nutrition; toxicology; vaccine production; virology.
Library facilities: Not open to all users; loan services (nc); reprographic services (c).
Information services: Bibliographic services; no translation services; literature searches; access to on-line information retrieval systems.
Consultancy: Yes (nc).
Publications: List on request.

Liikunnan ja Kansanterveyden 14.12
Tutkimuslaitos, Tietopalvelu
[Research Institute of Physical Culture and Health, Information Service]
Address: Seminaarinkatu 15, SF-40100 Jyväskylä, Finland
Telephone: (41) 291 562
Affiliation: Foundation for Promotion of Physical Culture and Health
Enquiries to: Information specialist, Anitta Pälvimäki
Founded: 1971
Subject coverage: Physical education, sports sciences, and public health.
Library facilities: The Information Service functions in cooperation with the Jyväskylä University Library, which is the national centre for publications in sports sciences. It oversees the national and international exchange of information in sports and related areas, using the contents of the university library.
Information services: Bibliographic services (nc); no translation services; literature searches (nc); access to on-line information retrieval systems (c).
Access to national and international on-line data bases, notably: automation unit of Finnish Research Libraries - KDOK/MINTTU (data bases: the Finnish national bibliography; the Finnish index of periodicals); Lockheed Information Systems, LMSC - Dialog Information Retrieval System; System Development Corporation - SDC Search Service.
Consultancy: Telephone answers to technical queries (nc); no writing of technical reports; no compilation of statistical trade data; no market surveys in technical areas.
Publications: Reports; special bibliographies; yearbook; annual bibliography *Finnish literature on Physical Education and Sport* (in Finnish; also available on-line in the KDOK/MINTTU system as *Finsport* data base).

Säteilyturvakeskus 14.13
– STUK
[Finnish Centre for Radiation and Nuclear Safety]
Address: PO Box 268, SF-00101 Helsinki 10, Finland
Telephone: (90) 61671
Telex: 122691 STUK-SF
Enquiries to: Librarian, Armi Länkelin
Founded: 1958
Subject coverage: Radiation protection; nuclear safety.
Library facilities: Open to all users; loan services; reprographic services.
Library holdings: 20 000 bound volumes; 290 current periodicals; 15 000 reports.
Information services: Bibliographic services (nc); no translation services; literature searches (nc); no access to on-line information retrieval systems.

Consultancy: Telephone answers to technical queries (nc); no writing of technical reports; no compilation of statistical trade data; no market surveys in technical areas.
Publications: List on request.

Suomen Palontorjuntaliitto ry 14.14
[Finnish Fire Protection Association]
Address: Iso Roobertinkatu 7A4, SF-00120 Helsinki 12, Finland
Telephone: (90) 649 233
Library facilities: Library.

Työterveyslaitos * 14.15
[Occupational Health Institute, Library]
Address: Hartmaninkatu 1, SF-00290 Helsinki 29, Finland
Telephone: (90) 474 7383
Telex: 125070 TYO SF
Enquiries to: Chief Librarian, A. Larmo
Founded: 1945
Subject coverage: Occupational safety and health.
Library facilities: Open to all users; loan services; reprographic services.
Information services: Bibliographic services; literature searches (limited); access to on-line information retrieval systems (for staff only).
Consultancy: Yes.

FRANCE

Association Interprofessionnelle 14.16
de France pour la Prévention des
Accidents et de l'Incendie
[Interprofessional Association of France for the Prevention of Accidents and Fire]
Address: BP 259, rue de l'Orangerie, Zone Industrielle, 59472 Seclin Cedex, France
Telephone: 20 97 93 26
Telex: 131230

Centre de Recherche d'Étude et 14.17
de Documentation en Économie
de la Santé
– CREDES
[Research Centre for the Study and Documentation of Health Economics]
Address: 1 rue Paul Cézanne, 75008 Paris, France
Telephone: 42 25 63 00
Affiliation: CNAMTS
Enquiries to: Research director, M. Mizrahi. Information service
Founded: 1985

Library facilities: No library facilities.
Library holdings: 160 current periodicals; 2200 reports.
Information services: No information services.
Consultancy: Telephone answers to technical queries; writing of technical reports; compilation of statistical trade data; market surveys in technical areas.

École Nationale de la Santé 14.18
Publique *
– ENSP
[National College of Public Health]
Address: avenue du Professeur Léon Bernard, 35043 Rennes Cedex, France
Telephone: 99 59 29 36
Affiliation: Ministère des Affaires Sociales et de la Solidarité Nationale
Enquiries to: Director, Dr Jean-Paul Picard
Founded: 1945
Subject coverage: Public health.
Library facilities: Not open to all users (generally only open to the college students and those who have a special permit); reprographic services.
Information services: Bibliographic services.
Consultancy: Yes.

Institut National de la Santé et 14.19
de la Recherche Médicale
– INSERM
[Health and Medical Research National Institute]
Address: 101 rue de Tolbiac, 75654 Paris Cedex 13, France
Telephone: (1) 45 84 14 41
Telex: 270 532
Enquiries to: Director general, Philippe Lazar
Subject coverage: Promotes, conducts and develops biomedical research, from molecular biology to public health, with emphasis on pathology.
Publications: Série Santé Publique; Série Statistiques, Nomenclature; Collection Colloques INSERM; research reports.

Institut National de Recherche et 14.20
de Sécurité pour la Prévention
des Accidents du Travail et des
Maladies Professionnelles
– INRS
[National Research and Safety Institute for the Prevention of Occupational Accidents and Diseases]
Address: 30 rue Olivier-Noyer, 75680 Paris Cedex 14, France
Telephone: (1) 45 45 67 67
Telex: 203 594 F
Enquiries to: Head of documentation service
Founded: 1968

Subject coverage: Prevention of occupational accidents and diseases; occupational risk prevention.
Library facilities: Open to all users; no loan services; reprographic services.
Library holdings: 12 000 bound volumes; 800 current periodicals.
Information services: Bibliographic services; translation services; literature searches; access to on-line information retrieval systems.
Consultancy: Telephone answers to technical queries; writing of technical reports; no compilation of statistical trade data; no market surveys in technical areas.
Courses: Courses available include correspondence courses on occupational risk prevention, and scientific and technical refresher courses. Further details of these and other courses are given in the INRS publication *Stages de Formation.*
Publications: Periodicals: *Travail et Sécurité* (Work and Safety); *Cahiers de Notes Documentaires* (information data sheets); *Les Risques du Métier* (Occupational Hazards); *Documents pour le Médecin du Travail* (fact sheets for the occupational physician); *Bulletin de documentation* (abstract bulletin). Brochures. Publications list on request.
INRS is the secretariat of the International Section for Research on Prevention of Occupational Risks set up by the International Social Security Association. In this capacity, INRS publishes a newsletter, *Safety Research News*, and the *Bulletin of Applied Research for the Protection of Man at Work.*

Laboratoire National de la Santé 14.21
[National Public Health Laboratory]
Address: 25 boulevard Saint-Jacques, 75680 Paris Cedex 14, France
Telephone: (1) 47 07 45 69
Affiliation: Ministère de la Santé et de la Sécurité Sociale
Enquiries to: Director, Dr R. Netter
Founded: 1958
Subject coverage: Medical control, including drugs, sera and blood derivatives; vaccines; virus epidemiology; hydrology; toxicology.
Library facilities: No library facilities.
Consultancy: Yes.
Publications: List on request.

GERMAN DEMOCRATIC REPUBLIC

Zentralinstitut für Arbeitsmedizin der DDR * 14.22
– ZAM
[Central Institute for Occupational Medicine of the GDR]
Address: Nöldnerstrasse 40/42, DDR-1134 Berlin, German Democratic Republic
Telephone: (02) 55 099 01
Affiliation: Ministry of Health
Founded: 1948
Subject coverage: Occupational medicine; industrial hygiene; occupational health service.
Library facilities: Library facilities are provided by the library of the Academy for Postgraduate Medical Training of the GDR.
Information services: Bibliographic services; literature searches.
Consultancy: Yes.

GERMAN FEDERAL REPUBLIC

Bundesanstalt für Arbeitsschutz 14.23
– BAU
[Federal Institute for Occupational Safety]
Address: Postfach 17 02 02, Vogelpothsweg 50-52, D-4600 Dortmund 1, German Federal Republic
Telephone: (0231) 17631
Telex: 822 153
Enquiries to: Chief Librarian
Founded: 1972
Subject coverage: Ergonomics; occupational medicine; occupational safety.
Library facilities: Open to all users (nc); loan services (nc); reprographic services (paper copies - nc).
Library holdings: 35 630 bound volumes; about 510 current periodicals; about 5000 reports.
Information services: Bibliographic services (nc); no translation services; literature searches (nc); no access to on-line information retrieval systems.
DI (nc).
Consultancy: Telephone answers to technical queries (answers by scientific staff only on special request - nc); no writing of technical reports; compilation of statistical trade data (by scientific staff only on special request - nc); no market surveys in technical areas.
Courses: Courses for safety-related personnel.

Bundesgesundheitsamt 14.24

– BGA

[Federal Health Office]

Address: Thielallee 88-92, D-1000 Berlin 33, German Federal Republic

Telephone: (030) 8308-1

Telex: 0184 016

Affiliation: Ministry for Youth, Family Affairs and Health

Enquiries to: Head of Information Office, Klaus J. Henning

Founded: 1952

Subject coverage: Consumers' health protection; reduction of environmental hazards; control of diseases; scientific consultant functions; approval and monitoring, particularly in the fields of narcotics and drugs legislation.

Information services: No information services.

Consultancy: Yes.

Publications: *Bundesgesundheitsblatt; Tätigkeitsberichte des BGA; Schriftenreihe des Bundesgesundheitsamtes.*

Bundesgesundheitsamt - Robert Koch-Institut 14.25

[Federal Health Office - Robert Koch Institute]

Address: Nordufer 20, D-1000 Berlin 65, German Federal Republic

Telephone: (030) 45031

Enquiries to: Head, Dr Wilhelm Weise

Subject coverage: Virology; bacteriology; immunology; biochemistry; cytology. The World Health Organization has named the institute a reference centre for the following areas: influenza; salmonella; blood group research; vibrio research; yellow fever.

Publications: *RKI-Berichte.*

Deutsches Krankenhausinstitut 14.26

– DKI

[German Hospital Institute]

Address: Tersteegenstrasse 9, D-4000 Düsseldorf 30, German Federal Republic

Telephone: (0211) 43 44 22

Affiliation: Universität Düsseldorf

Enquiries to: Head of Information and Documentation, Dr Hans-Jürgen Seelos

Founded: 1953

Subject coverage: Improvement of the medical, nursing, social, and economic efficiency of hospitals; interdisciplinary research, documentation, training and education, and dissemination of information in the field of health and hospital services.

Library facilities: Open to all users; no loan services; reprographic services (c).

Information services: Bibliographic services (c); no translation services; literature searches (c); access to on-line information retrieval systems (c).

HECLINET (Health Care Literature Information Network), a bibliographic data base on hospital administration and health care, is available on-line through the host DIMDI, Köln. HECLINET is a joint venture of the national hospital institutions of Denmark, Sweden, Austria, Switzerland, and West Germany. The total file contains 68 000 references from 1969 to the present with about 4500 references being added per year. 30 per cent of the data base is English in origin. However, all descriptors and keywords are in English and German. HECLINET includes all areas of hospital administration and the non-clinical aspects of health care. The data base contains information on hospital design and construction, hospital maintenance, hospital hygiene, and administration. Relevant information on the economic, political and legal aspects of the field is also included.

Consultancy: Yes (c).

Publications: *Health Care Information Service,* bimonthly.

Universität Karlsruhe, Forschungsstelle für Brandschutztechnik * 14.27

[Karlsruhe University, Fire Protection Research Station]

Address: Hertzstrasse 16, D-7500 Karlsruhe 21, German Federal Republic

Telephone: (0721) 608 4473

Enquiries to: P.G. Seeger

Founded: 1950

Library facilities: Open to all users; no loan services; reprographic services (c).

Information services: Bibliographic services; literature searches (c).

Consultancy: Yes.

Publications: List on request.

Vereinigung zur Förderung des Deutschen Brandschutzes eV 14.28

[German Fire Protection Union]

Address: Buchenallee 18, D-4417 Altenberge, German Federal Republic

Telephone: (02505) 26 17

Library facilities: Library.

GREECE

Occupational Health and Safety Centre 14.29

Address: 6 Dodekanissou Street, 174 56 Alimos, Athinai, Greece
Telephone: (01) 9919566
Affiliation: Ministry of Labour
Enquiries to: Director, Ch. Vassilopoulos
Founded: 1978
Subject coverage: Investigation of physical and chemical nuisance factors of the workplace environment; technical and theoretical assistance to public agencies dealing with the promotion of industrial safety and hygiene.
Library facilities: No library facilities.
Library holdings: 250 bound volumes.
Information services: Bibliographic services will be available in the near future.
Open exhibition of personal protective equipment.
Consultancy: Telephone answers to technical queries (nc); writing of technical reports; no compilation of statistical trade data; no market surveys in technical areas.
Courses: Training of safety technicians and occupational courses for physicians held every three months.

HUNGARY

Országos Közegészségügyi Intézet * 14.30
– OKI
[National Institute of Hygiene]
Address: Gyáli útca 2-6, H-1097 Budapest, Hungary
Telephone: 142-250
Telex: 22-5349 oki
Affiliation: Ministry of Health
Enquiries to: Section head, Dr László Kóti
Founded: 1925
Subject coverage: Environmental hygiene; epidemiology; virology; hygiene microbiology.
Library facilities: Open to all users; loan services; reprographic services.
Information services: Bibliographic services (selected information on articles in the field of environmental hygiene); translation services (from and into Russian, English, German, and French); no literature searches; no access to on-line information retrieval systems.
Consultancy: No consultancy.

Országos Munkavédelmi Tudományos Kutató Intézet 14.31
– OMTKI
[National Scientific Research Institute for Occupational Safety]
Address: Postafiók 7, H-1281 Budapest 27, Hungary
Telephone: 36-1 164440
Telex: 22 70 79 mtki h
Affiliation: National Board of Occupational Safety Inspection
Enquiries to: Director, Nagy Gyula
Founded: 1954
Subject coverage: Elaboration and development of procedures and techniques for improving working conditions, protection of workers; research, planning, legislation, codification, and realization in the same field.
Library facilities: Not open to all users (nc); no loan services; reprographic services (c).
Library holdings: 7500 bound volumes; 102 current periodicals.
Information services: Bibliographic services (c); translation services (c); literature searches (c); no access to on-line information retrieval systems.
Consultancy: No telephone answers to technical queries; writing of technical reports (c); no compilation of statistical trade data; no market surveys in technical areas.
Publications: *Occupational Health and Safety*, quarterly; *Up-to-date Working Conditions*, *Occupational Safety*, fortnightly; *Occupational Accident Analysis*, annually; *Occupational Safety Information*; *Studies on Occupational Safety*; *Lectures, Research, Publications*, periodical.

ICELAND

Vinnueftirlit Rikisins * 14.32
[Occupational Safety and Health Administration Institute]
Address: PO Box 5295, Sidumuli 13, 125 Reykjavík, Iceland
Telephone: 82970
Affiliation: Ministry of Health and Social Affairs
Enquiries to: Information officer, Daniel Benediktsson
Founded: 1981
Subject coverage: Occupational safety; occupational health; occupational diseases.
Library facilities: Open to all users; no loan services; reprographic services (nc).
Information services: Bibliographic services (nc); no translation services; literature searches (nc); no access to on-line information retrieval systems.

There are central facilities for on-line searching provided by the National Science Council and the institute has access to this capacity. For any search a charge is made in agreement with internationally used procedures and standards in this area, determined by local regulations for unit costs.
Consultancy: Yes (nc).

IRELAND

National Industrial Safety Organisation   14.33
– NISO
Address: Davitt House, Mespil Road, Dublin 4, Ireland
Telephone: (01) 765861
Enquiries to: Secretary
Founded: 1963
Subject coverage: Industrial safety, health, and welfare.
Library facilities: Open to all users (mainly to member firms and organizations); loan services; no reprographic services.
Information services: Bibliographic services; literature searches.
NISO runs an advisory service to member firms and organizations and an industrial safety film library service for its members. It is the national centre for the Republic of Ireland of the International Occupational Safety and Health Information Centre (CIS).

ITALY

Associazione Italiana per l'Igiene   14.34
e la Sanita Pubblica
[Italian Association for Hygiene and Public Health]
Address: Via Principe Amedeo 126/B, 00185 Roma, Italy
Telephone: (06) 7314350
Telex: AIISP
Enquiries to: President
Founded: 1922
Library facilities: Not open to all users.
Some publications on hygiene and public health are distributed.
Information services: Information on hygiene and public health in Italy can be obtained on request from the association.
Consultancy: Yes.

LUXEMBOURG

Direction de la Santé   14.35
[Health Directorate]
Address: 57 boulevard de la Pétrusse, Luxembourg
Telephone: 4 08 01
Telex: 2546 sante lu
Parent: Ministry of Health
Enquiries to: Director, Dr Joseph Kohl
Library facilities: No library facilities.
Information services: No information services.
Consultancy: Telephone answers to technical queries (nc).
Publications: Various information booklets.

MALTA

Department of Health *   14.36
Address: 15 Merchants Street, Valletta, Malta
Telephone: 24071
Telex: 1100 MODMLT MT
Enquiries to: Chief Government Medical Officer
Subject coverage: Health.
Information services: Information on health matters provided to organizations and individuals, but there is no regular information service or library facilities.

NETHERLANDS

Bureau Industriële Veiligheid   14.37
TNO
– BIV-TNO
[Industrial Safety Bureau TNO]
Address: Postbus 45, 2280 AA Rijswijk (ZH), Netherlands
Telephone: (15) 138777
Telex: 38034 pmtno
Parent: Nederlandse Organisatie voor Toegepast-Natuurwetenschappelijk Onderzoek (TNO)
Enquiries to: Head, Ir A.C. van Mameren
Founded: 1971
Subject coverage: Industrial safety in the following industries: food, beverages, and tobacco; chemicals, plastics, and rubber; pharmaceuticals; mechanical and electrical engineering; petroleum extraction and processing; transport.
Library facilities: Open to all users (c); loan services (limited services - c); reprographic services (limited

services - c).

The main TNO libraries are open to all users.

Information services: Bibliographic services (articles, brochures, and documentation sheets are available - c); translation services (limited services - c); literature searches (limited services - c); access to on-line information retrieval systems (c).

TNO data bank *FACTS* concerning accidents during processing, handling, transportation, and storage of dangerous materials.

Consultancy: Telephone answers to technical queries (nc); writing of technical reports (c); compilation of statistical trade data (c); market surveys in technical areas (c).

Further details of consultancy can be obtained from the bureau.

Courses: An intensive course, lasting two days, related to processing, storage, handling and transport of inflammable, explosive and toxic materials.

Publications: Bibliographic list; booklets.

Ministerie van Welzÿn, Volksgezondheid en Cultuur, Bibliotheck Volksgezondheid *

14.38

– WVC

[Ministry of Welfare, Health and Cultural Affairs, Health Library]

Address: Postbus 439, 2260 AK Leidschendam, Netherlands

Telephone: (070) 209260

Telex: 32347WYCL

Enquiries to: Librarian, Jac J. Brakel

Founded: 1971

Subject coverage: Public health.

Library facilities: Open to all users; loan services (nc); reprographic services (c).

Information services: Bibliographic services (nc); no translation services; literature searches (c); access to on-line information retrieval systems (c).

Consultancy: Yes (nc).

Radiologische Dienst TNO

14.39

– RD-TNO

[Radiological Service TNO]

Address: Postbus 9034, Utrechtseweg 310, 6800 ES Arnhem, Netherlands

Telephone: (085) 56 93 33

Telex: 45016 KEMA

Facsimile: (085) 51 56 06

Affiliation: Nederlandse Organisatie voor Toegepast-Natuurwetenschappelijk Onderzoek

Enquiries to: Director, Dr H.W. Julius

Founded: 1946

Subject coverage: Health physics and radiation protection.

Library facilities: Open to all users; loan services; reprographic services.

Information services: No information services.

Consultancy: Telephone answers to technical queries (nc); writing of technical reports (c); no compilation of statistical trade data; no market surveys in technical areas.

Veiligheidsinstituut *

14.40

– VI

[Safety Institute]

Address: Postbus 5665, De Boelelaan 32, 1007 AR Amsterdam, Netherlands

Telephone: (020) 445655

Enquiries to: Library manager, C. Oosthuizen

Founded: 1891

Subject coverage: Occupational safety, health and well-being.

Library facilities: Open to all users; loan services (nc); reprographic services (c).

Information services: Bibliographic services (nc); no translation services; literature searches (c); access to on-line information retrieval systems (c).

HSELINE, Safety Science Abstracts, CIS (ILO), Labordoc.

Consultancy: Yes (nc).

Publications: New books and periodicals list, (monthly); bibliographies on specific safety subjects; data bank on hazardous chemicals and preparations in the working place.

NORWAY

Arbeidsforskningsinstituttene

14.41

[Work Research Institutes]

Address: Postboks 8149 Dep., Gydas vei 8, N-0033 Oslo 1, Norway

Telephone: (02) 466850

Enquiries to: Chief librarian, Sindre Varran Biblioteket

Founded: 1963

Subject coverage: Occupational health, work physiology, work psychology, muscle physiology.

Library facilities: Open to all users (nc); loan services (nc); reprographic services (nc at the moment).

Library holdings: 20 000 bound volumes; 300 current periodicals.

Information services: Bibliographic services (nc); no translation services; literature searches (nc); access to on-line information retrieval systems (nc at the moment).

Consultancy: Telephone answers to technical queries (nc); writing of technical reports (research reports); no compilation of statistical trade data; no market surveys in technical areas.

Norsk Brannvern Forening 14.42
– NBF
[Norwegian Fire Protection Association]
Address: PO Box 6703, Wesselsgaten 8, St Olavs plass, Oslo 1, Norway
Telephone: (020) 20 01 54
Founded: 1923
Subject coverage: Fire safety; fire prevention.
Library facilities: Open to all users; loan services; reprographic services.
Information services: Bibliographic services; translation services (occasionally); literature searches; access to on-line information retrieval systems.
Consultancy: Yes.

Norsk Institutt for 14.43
Sykehusforskning
– NIS
[Norwegian Institute for Hospital Research]
Address: Strindveien 2, N-7034 Trondheim - NTH, Norway
Telephone: (07) 592571
Telex: 55620 sintf n
Affiliation: SINTEF
Enquiries to: Librarian, Grete Sletten
Founded: 1972
Subject coverage: Research and development in administration, organization, and construction of hospitals and other health care institutions.
Library facilities: Open to all users (nc); loan services (nc); reprographic services (nc).
Library holdings: 4500 bound volumes; 115 current periodicals.
Information services: Bibliographic services (nc); no translation services; literature searches (c); access to on-line information retrieval systems (c).
Publications: List on request.

Statens Institutt for Folkehelse * 14.44
– SIFF
[National Institute of Public Health]
Address: Geitmyrsveien 75, N-0462 Oslo 4, Norway
Telephone: (02) 356020
Affiliation: Norwegian Health Services
Enquiries to: Librarian, Torleif Bertelsen
Founded: 1929
Subject coverage: Medical microbiology; virology; immunohaematology; immunology; public health; human toxicology; medical entomology; sanitary engineering; pharmacy; biophysics; epidemiology; informatics; chemistry; biochemistry; veterinary science.
Library facilities: Open to all users; loan services; reprographic services (c).
Information services: No information services.
Access to information retrieval systems from own terminal. Searches are done for staff users.

Publications: *NIPH Annals*, semi-annually; *MSIS Ukerapport*, weekly; annual report; *Årsrapport* (annual report, Norwegian version); *SIFF-SK Rapport* (report series in Norwegian).

POLAND

Biblioteka Państwowego Zakładu 14.45
Higieny
[National Institute of Hygiene, Library]
Address: ˙ Chocimska 24, Warszawa, Poland
Telephone: 49-40-51
Affiliation: National Institute of Hygiene
Enquiries to: Head of library, Barbara Wysocka
Founded: 1918
Subject coverage: Epidemiology; microbiology; vaccines and sera control; radiobiology; immunopathology; hygiene; foodstuffs; toxicology; education; medical statistics.
Library facilities: Open to all users (nc); loan services (nc); reprographic services (c).
Library holdings: 42 000 bound volumes; 448 current periodicals.
Information services: Bibliographic services (nc); no translation services; literature searches (nc); no access to on-line information retrieval systems.
Consultancy: Telephone answers to technical queries (nc); writing of technical reports (nc); no compilation of statistical trade data; no market surveys in technical areas.

Centralny Instytut Ochrony 14.46
Pracy
– CIOP
[Central Institute for Labour Protection]
Address: Ulica Tamka 1, 00-349 Warszawa, Poland
Telephone: 26-34-21
Affiliation: Ministry of Labour, Wages and Social Affairs
Enquiries to: Director, Dr Stanisław Dabrowski.
Library and Information Centre
Founded: 1950
Subject coverage: Ergonomics; chemical and physical hazards; acoustics; noise, vibration; personal protection.
Library facilities: Open to all users (nc); loan services (nc); reprographic services (c).
Library holdings: 2045 bound volumes; 259 current periodicals.
Information services: Bibliographic services (nc); translation services (c); literature searches (c); no access to on-line information retrieval systems.

Consultancy: Telephone answers to technical queries (nc); writing of technical reports (nc); no compilation of statistical trade data; no market surveys in technical areas.
Publications: Publications list on request.

Główna Biblioteka Pracy i Zabezpieczenia Społecznego   14.47
– GBP
[Central Library of Labour and Social Security]
Address: Ulica Mysia 2, 00-496 Warszawa, Poland
Telephone: 29-96-33
Affiliation: Ministry of Labour, Wages and Social Affairs
Enquiries to: Director, Małgorzata Kłossowska
Founded: 1974
Subject coverage: Labour, wages, social affairs and related problems; work organization, management, employment, productivity, health and labour protection, safety at work, labour law, social security, living conditions of working people.
Library facilities: Open to all users (nc); loan services (nc); reprographic services (nc).
Library holdings: 43 938 bound volumes (books); 12 942 bound volumes of periodicals; 475 current periodicals.
Information services: Bibliographic services (nc); no translation services; literature searches (nc); no access to on-line information retrieval systems.
Publications: Bibliography of economic and social problems, annually; documentation review, monthly; library communiqué, monthly; *Labour-Wages-Social Affairs*, weekly; occasional bibliographies.

PORTUGAL

Instituto de Higiene e Medicina Tropical *   14.48
[Hygiene and Tropical Medicine Institute]
Address: Rua da Junqueira 96, 1300 Lisboa, Portugal
Telephone: (019) 632141
Affiliation: Ministério da Educaçăo e Ciência
Enquiries to: Director of the Library
Founded: 1902
Subject coverage: Tropical medicine; parasitology; microbiology; public health.
Library facilities: Open to all users; no loan services; reprographic services (c).
Information services: No information services.

ROMANIA

Societatea de Ingienă şi Sănătate Publică   14.49
[Hygiene and Public Health Society]
Address: c/o USSM, Strada Progresului 8-10, 70754 Bucureşti, Romania
Telephone: 14 10 71

SPAIN

Instituto Nacional de Medicina y Seguridad del Trabajo   14.50
[National Medical and Occupational Safety Institute]
Address: Pabellón No 8, Facultad de Medicina, Madrid 3, Spain
Enquiries to: Director, Professor D. Manuel Dominguez Carmona

SWEDEN

Arbetarskyddsstyrelsen *   14.51
[National Swedish Board of Occupational Safety and Health]
Address: S-171 84 Solna, Sweden
Telephone: 08-730 90 00
Founded: 1949
Library facilities: Open to all users; loan services; reprographic services.
Information services: Literature searches (on data bases); access to on-line information retrieval systems. CIS-ILO data base; NIOSHTIC data base; AMILIT data base. In-house services which are also open to the public via the Nordic data communication network.
Publications: Newsletter; *Abetarskydd; Arbete och Hälsa* (Work and Health); catalogue on request.

Statens Strålskyddsinstitut *   14.52
– SSI
[National Institute of Radiation Protection]
Address: Box 60 204, S-104 01 Stockholm, Sweden
Telephone: 08-24 40 80
Telex: 117 71
Enquiries to: Librarian, Ann-Marie Lindholm
Founded: 1965
Subject coverage: Radiation protection.

Library facilities: Open to all users; loan services; reprographic services.
Consultancy: Yes.
Publications: The information department provides SSI reports, free of charge, mostly in Swedish; complete list on request.

Svenska Brandförsvarsföreningen 14.53
– SBF
[Swedish Fire Protection Association]
Address: Tegeluddsvägen 100, S-115 87 Stockholm, Sweden
Telephone: 08-783 70 00
Telex: 118 09
Enquiries to: Librarian, Ingrid Roberts
Founded: 1919
Subject coverage: Fire protection and prevention.
Library facilities: Open to all users; loan services; reprographic services.
Information services: Bibliographic services (c); literature searches (c); no access to on-line information retrieval systems.
Subscription can be made to library cards of registered and classified articles from journals on the subject of fire and similar topics.
Consultancy: Yes (c).
Publications: *News from the library*, six issues per year.

SWITZERLAND

International Occupational 14.54
Safety and Health Information
Centre
– CIS
Address: International Labour Office, CH-1211 Genève 22, Switzerland
Telephone: (022) 99 67 40
Telex: 22.271
Affiliation: International Labour Office
Founded: 1959
Subject coverage: Occupational safety; occupational medicine and physiology; industrial hygiene and toxicology; accident prevention and safety engineering; safety management and training.
Library facilities: Open to all users; no loan services; reprographic services (c).
Copies of original documents available on microfiche.
Information services: Bibliographic services (c); no translation services; literature searches (c); access to on-line information retrieval systems (c).
On-line data base in English and French: CIS; CIS-ILO; CISDOC. CISDOC contains 27 000 records,

each in English and French covering the worldwide occupational safety and health literature from 1974 to the present. Available from Télésystèmes-Questel and ESA-IRS.
Consultancy: Yes (limited - nc).
Publications: *CIS Abstracts* (English); *Bulletin CIS* (French); bibliographies.

Schweizerische Gesellschaft für 14.55
Gesundheitspolitik
[Swiss Society for Health Policy]
Address: Brunnenwiesli 7, CH-8810 Horgen, Switzerland
Telephone: (01) 725 78 10

Schweizerische Gesellschaft für 14.56
Sozial- und Präventivmedizin
[Swiss Society for Social and Preventive Medicine]
Address: c/o Institut de Médecine Sociale et Préventive, Université de Genève, Quai Charles-Page 27, CH-1211 Genève 4, Switzerland

World Health Organization 14.57
Address: Avenue Appia, CH-1211 Genève 27, Switzerland
Subject coverage: In the fields of medicine and public health the World Health Organization carries out a variety of programmes including: research promotion and development through a network of collaborating national laboratories; development of national health services family health; mental health; prevention and control of communicable diseases; promotion of environmental health; development of health manpower suited to particular needs; quality control of prophylactic, diagnostic and therapeutic substances and development of policies to enable third world countries to meet their needs for pharmaceuticals.
Publications: *Bulletin* (WHO scientific papers) bimonthly; *Chronicle*, monthly; *International Digest of Health Legislation*; *World Health Statistics Report*, quarterly; *World Health Forum*, quarterly; *World Health Statistics Annual*; *Weekly Epidemiological Record*; *Monograph Series*; *Public Health Papers*; *Technical Report Series*; *Official Records*; *WHO Offset Publications*, irregular; *World Health*, monthly magazine.

UNITED KINGDOM

BIOS (Consultancy and Contract Research) Limited   14.58

Address: Pinewood, College Ride, Bagshot, Surrey GU19 5ER, UK
Telephone: (0276) 73363
Telex: 265871 MON REF G REF SJJ137
Enquiries to: Manging director, R.K. Greenwood
Founded: 1975
Subject coverage: Health care research and development: analytical chemistry, pharmaceutical formulation, microbiology, biology, bioavailability, pharmacokinetics, clinical trials in volunteers and patients, regulatory affairs.
Library facilities: Open to all users (c); no loan services; no reprographic services.
Library holdings: About 70 bound volumes; 16 current periodicals.
Information services: Bibliographic services (c); translation services (c); literature searches (c); access to on-line information retrieval systems (c).
Literature reviews and assessments (c).
Consultancy: Telephone answers to technical queries (c); writing of technical reports (c); compilation of statistical trade data (only in close conjunction with clients - c); market surveys in technical areas (restricted to preliminary attitudinal research in technical areas, and searching for potential licensees - c).
Regulatory affairs (c).
Courses: Biannual residential course in industrial expertise for the pharmaceutical industry.

British Occupational Hygiene Society   14.59

Address: 1 St Andrew's Place, Regent's Park, London NW1 4LB, UK
Telephone: (01) 486 4860
Enquiries to: Honorary secretary, G.W. Crockford
Founded: 1954
Library facilities: Not open to all users; no loan services; reprographic services (c).
Information services: Bibliographic services (limited); no translation services; no literature searches; no access to on-line information retrieval systems.
Consultancy: No telephone answers to technical queries; writing of technical reports (c); no compilation of statistical trade data; market surveys in technical areas (c).
Courses: Annual conference held each year in April.

Central Public Health Laboratory   14.60

Address: 61 Colindale Avenue, London NW9 5HT, UK
Telephone: (01) 200 4400
Telex: 8953942 DEFEND G
Affiliation: Public Health Laboratory Service
Enquiries to: Chief Librarian, Susan Bloomfield
Founded: 1946
Subject coverage: Medical microbiology; communicable diseases; epidemiology.
Library facilities: Open to all users (nc); loan services (c); reprographic services (c).
The library is open to bona fide enquirers by prior arrangement only. Loans and photocopies are provided to other libraries only of items not easily obtainable elsewhere in the UK.
Library holdings: 30 000 bound volumes; 450 current periodicals; 5000 reports.
Information services: No information services.
Consultancy: Telephone answers to technical queries (nc); no writing of technical reports; no compilation of statistical trade data; no market surveys in technical areas.
Publications: *PHLS Library Bulletin*, a weekly current awareness service in diagnostic medical microbiology; *PHLS Microbiology Digest*, a quarterly journal of brief up-to-date reviews; list of publications.

Centre for Applied Microbiology and Research *   14.61

Address: Porton Down, Salisbury, Wiltshire SP4 0JG, UK
Telephone: (0980) 610391
Telex: 47683 PHCAMR
Affiliation: Public Health Laboratory Service
Publications: Annual report.

Fire Research Station   14.62
– FRS
Address: Melrose Avenue, Boreham Wood, Hertfordshire WD6 2BL, UK
Telephone: (01) 953 6177
Telex: 8951648
Affiliation: Building Research Establishment, Department of the Environment
Enquiries to: Head of Fire Research
Founded: 1947
Subject coverage: Fire statistics; ignition and growth of fire; structural aspects of fire in buildings; detection, extinction, and suppression of fire; special fire hazards in industries and materials.
Library facilities: Not open to all users (open only by prior arrangement with the librarian - charge for some services); loan services (BLLD form); reprographic services (c).
Books may be borrowed only via inter-library loans; journals may not be borrowed. Photocopies can be

made, but only for non-BLLD material, and subject to staff availability.

Library holdings: 67 000 reports and books; about 300 current periodicals.

Information services: Bibliographic services (occasional charge); no translation services (translations made for staff may be available); literature searches (if staff are available - c); access to on-line information retrieval systems (charge for lengthy searches).

In-house data base - Fire Research Library Automated Information Retrieval system (FLAIR). This has been running for over 5 years and currently has over 22 000 items on line - including references back to around 1977, and most FRS publications. It is planned to have FLAIR available via Dialtech/ESA.

Consultancy: Telephone answers to technical queries (occasional charge); no writing of technical reports; no compilation of statistical trade data; no market surveys in technical areas.

Publications: *Fire Science Abstracts*, quarterly; bibliographic list.

Health and Safety Executive *   14.63
– HSE

Address: Red Hill, off Broad Lane, Sheffield S3 7HQ, UK

Telephone: (0742) 78141

Telex: 54556

Affiliation: Health and Safety Commission

Enquiries to: Head of Library and Information Services, S. Pantry

Founded: 1974

Subject coverage: All aspects of health and safety in all workplaces in the United Kingdom.

Library facilities: Open to all users; loan services (c); reprographic services (c).

The Library and Information Service also acts as a referral point for any enquiries from the public.

Information services: No bibliographic services; translation services (c); no literature searches; access to on-line information retrieval systems (c).

HSELINE is HSE's publicly available data base and contains 50 000 references. It dates from 1977 up to date, but does contain much information prior to 1977.

Consultancy: No consultancy.

Health Education Council   14.64
– HEC

Address: 78 New Oxford Street, London WC1A 1AH, UK

Telephone: (01) 631 0930

Enquiries to: Information Officer

Founded: 1968

Subject coverage: Health education and promotion.

Library facilities: Open to all users (nc); loan services (UK only, and audio-visual services reference only -

nc); reprographic services (photocopy service only - c).

Library holdings: About 6000 bound volumes; 350 current periodicals; 200 reports.

Information services: Bibliographic services (nc); no translation services; literature searches (manual - no charge within the UK); no access to on-line information retrieval systems.

Consultancy: No consultancy.

International Commission on Radiological Protection   14.65
– ICRP

Address: Clifton Avenue, Sutton, Surrey SM2 5PU, UK

Telephone: (01) 642 4680

Telex: 895 1244 ICRPG

Affiliation: International Society of Radiology

Enquiries to: Scientific secretary, Dr M.C. Thorne

Founded: 1928

Subject coverage: Basic standards in radiation protection.

Library facilities: Open to all users (nc); no loan services; no reprographic services.

Library holdings: 500 volumes of reports; 2000 reprints.

Information services: No information services.

Consultancy: Telephone answers to technical queries (nc); no writing of technical reports; no compilation of statistical trade data; no market surveys in technical areas.

Publications: *Annals of the ICRP*, quarterly bibliographic list.

National Radiological Protection Board   14.66
– NRPB

Address: Chilton, Didcot, Oxfordshire OX11 0RQ, UK

Telephone: (0235) 831600

Telex: 837124

Facsimile: (0235) 833891

Enquiries to: Information officer

Founded: 1970

Subject coverage: Radiological protection, including health and safety aspects of nuclear power; biological/ medical effects of ionizing and non-ionizing radiations; radioactivity in consumer products; environmental radioactivity including natural radiation; dosimetry.

Library facilities: Open to all users (by arrangement - nc); loan services (to other libraries - nc); no reprographic services.

The library answers queries on publications excluding those of the board. The information office answers queries of a general nature and on the publications of the board.

Library holdings: 10 000 bound volumes; 100 current periodicals; 5000 reports.
Information services: Bibliographic services (nc); no translation services; no literature searches; no access to on-line information retrieval systems.
Consultancy: No consultancy.
Courses: Regular courses on various aspects of radiological protection.

Paint Research Association 14.67

Address: Waldegrave Road, Teddington, Middlesex TW11 8LD, UK
Telephone: (01) 977 4427
Telex: 928720
Enquiries to: Librarian, S.C. Haworth
Founded: 1926
Subject coverage: Surface coatings science and technology, including paints, pigments, oils, resins, polymers, additives; painting; health hazards; safety and environmental regulations; paint defects; corrosion and fouling; chemical and physical tests and analysis; microbiology; industrial commercial information.
Library facilities: Open to all users (by appointment only - c); loan services (members only); reprographic services (c).
Information services: Bibliographic services (c); translation services (c); literature searches (c); access to on-line information retrieval systems (c).

YUGOSLAVIA

Institut za medicinska 14.68
istraživanja i medicinu rada *
[Institute for Medical Research and Occupational Health]
Address: PO Box 291, Moše Pijade 158, 41001 Zagreb, Yugoslavia
Telephone: 041-434-188
Enquiries to: Head librarian, Nada Vajdička
Founded: 1947
Subject coverage: Toxicology of metals and mineral metabolism; toxicology of pesticides; atmospheric pollution; radiological protection; psychophysiology; occupational diseases and intoxications; epidemiology of chronic diseases; anthropology.
Library facilities: Open to all users; loan services (nc); reprographic services (c).

15 MANAGEMENT, BUSINESS AND PRODUCTIVITY

AUSTRIA

Österreichisches Produktivitäts- und Wirtschaftlichkeits- Zentrum
15.1
– ÖPWZ
[Austrian Centre for Productivity and Efficiency]
Address: Rockhgasse 6, A-1014 Wien, Austria
Telephone: (0222) 63 86 36
Telex: 11-5718
Enquiries to: Head, Dr René Pusch.
Department for Information and Documentation
Founded: 1950
Subject coverage: Promotion and support of productivity, efficiency and development of better skills in Austria.
Library facilities: Open to all users; loan services (c); reprographic services (c).
Library holdings: 5200 bound volumes; 330 current periodicals.
Information services: Bibliographic services (c); no translation services; literature searches (c); no access to on-line information retrieval systems.
Information services are mainly based on two documentation services, which are published as *Technik Aktuell* and *Betriebswirtschaftliche Dokumentation*.
Consultancy: No telephone answers to technical queries; no compilation of statistical trade data; no market surveys in technical areas.
Courses: Training courses; seminars; workshops.

BELGIUM

Office Belge pour l'Accroissement de la Productivité
15.2
– OBAP
[Belgian Productivity Centre]
Address: 60 Rue de la Concorde, Bruxelles 5, Belgium
Telephone: (02) 11 81 55
Affiliation: Ministry of Economic Affairs
Subject coverage: Economics; applied social sciences; building, distribution and textiles; regional development; staff and management training.
Publications: *Synopsis*, six issues a year; *Activity Report*, annual.

CZECHOSLOVAKIA

Institut Řízení
15.3
– IŘ
[Management Institute]
Address: Jungmannova 29, 115 49 Praha 1, Czechoslovakia
Telephone: 2119
Affiliation: Office of the Federal Government Praesidium
Enquiries to: Director, Dr L. Schánĕl.
Information Centre
Founded: 1965
Subject coverage: Training of top-level managers; economic management; information services in management.
Library facilities: Open to all users; loan services; reprographic services (c).
A considerable portion of the library's holdings comprises foreign publications.
Library holdings: 33 000 bound volumes; 400 current periodicals; 11 000 reports.

Information services: Bibliographic services; no translation services; literature searches; no access to on-line information retrieval systems.

The centre is involved in publication of Czechoslovak studies and translations from other languages.

Consultancy: No consultancy.

Courses: Four week courses for top managers; two week courses and special short seminars for managers.

Publications: *New Publications; Moderni Řizeni* (Modern Management), monthly; *Organizace a Řizeni* (Organization and Management), bimonthly.

DENMARK

Dansk Management Center 15.4
– DMC
[Danish Management Centre]
Address: 7 Kristianiagade, DK-2100 København Ø, Denmark
Telephone: (01) 38 97 77
Founded: 1971
Subject coverage: Management education.
Library facilities: Not open to all users; loan services; reprographic services.
Information services: Access to on-line information retrieval systems (Management Course Information).

FINLAND

Helsingin Kauppakorkeakoulun 15.5
Kirjasto *
[Helsinki School of Economics Library]
Address: Runeberginkatu 22-24, SF-00100 Helsinki 10, Finland
Telephone: (90) 0 43131
Telex: 122220
Enquiries to: Chief librarian, Dr Henri Broms
Founded: 1911
Subject coverage: Economics and business sciences.
Library facilities: Open to all users (nc); loan services (nc); reprographic services (c).
Library holdings: 220 000 bound volumes; 1600 current periodicals.
Information services: Bibliographic services (c); no translation services; literature searches (c); access to on-line information retrieval systems (c).
The library maintains the following data bases, accessible on-line: *Scimp, SCANP, Bild, FINP* and *Thes.*
Consultancy: Telephone answers to technical queries.
Courses: Courses for online-users, twice a year.
Publications: Printed versions of the data bases listed above.

Keskuskauppakamari 15.6
– KKK
[Central Chamber of Commerce]
Address: Fabianinkatu 14, PO Box 1000, SF-00101 Helsinki, Finland
Telephone: (90) 650 133
Telex: centralchamber 121394
Facsimile: (90) 650133
Enquiries to: Information officer, Lasse Lehman.
Library and Information Services
Founded: 1918
Subject coverage: Business information.
Library facilities: Not open to all users; loan services (nc); reprographic services (nc).
Library holdings: 390 current periodicals.
Information services: Bibliographic services (nc); no translation services; literature searches (nc); access to on-line information retrieval systems (c).
The KKK maintains a register of approved translators.
Consultancy: No telephone answers to technical queries; no writing of technical reports; compilation of statistical trade data (nc); no market surveys in technical areas.

FRANCE

Organization for Economic 15.7
Cooperation and Development/
Organisation de Coopération et
de Développement Économiques
– OECD/OCDE
Address: 2 rue André-Pascal, 75775 Paris Cedex 16, France
Telephone: (4) 524 8200
Telex: 62 160 OCDE PARIS
Enquiries to: Chief librarian
Founded: 1961
Subject coverage: The organization promotes economic and social welfare among the 24 OECD member countries.
The library's collection covers international and national economics, social affairs, research and development, education, environment.
Library facilities: Open to all users; no loan services; reprographic services.
Publications: *Catalogue of Periodicals* , annual; *New Books and Periodicals*, monthly; *Special Annotated Bibliography*, semiannual.

GERMAN FEDERAL REPUBLIC

Institut für Wirtschaftsforschung- Hamburg
15.8

– HWWA

[Institute for Economic Research, Hamburg]

Address: Neuer Jungfernstieg 21, D-2000 Hamburg 36, German Federal Republic
Telephone: (40) 3562-1
Telex: 211 458
Enquiries to: Information officers, C. Bacmeister, D. Passarge
Informationszentrum des HWWA (HWWA Information Centre)
Founded: 1908
Subject coverage: The information services of the centre cover all fields of national and international economics and related disciplines. These services extend to data on products, institutions and personnel.
Library facilities: Open to all users; no loan services; reprographic services (c).
The library is one of Europe's largest libraries specializing in economic literature; the press documentation and archives department compiles press cuttings from 150 newspapers and periodicals from 40 countries.
The information centre has a particularly large stock of national and international statistics, branch and product reviews, company and industrial directories, occasional company papers, and literature on former German colonies.
The library, in its function as Depository Library of the UN, FAO, GATT, OECD and the EC, contains nearly all publications of these organizations since their foundation.
Information services: Bibliographic services (c); no translation services; literature searches (c); no access to on-line information retrieval systems.
The information service department offers the following services: immediate supply of data on economic and political facts; systematic compilation of information on particular subjects on request and compilation of specific bibliographies; regular documentation - continuous supply of information on particular subjects on request. Documentation is supplied in the following different forms: bibliographical documentation (titles with informative summaries); full-text documentation (giving the unabridged text, classified by subject headings); statistical documentation (extracts of statistics); photocopying service (subject to copyright regulations).
Consultancy: Yes (nc).

Landesgewerbeamt Baden-Wuerttemberg
15.9

[Baden-Wuerttemberg Trade and Industry Office]

Address: Kienestrasse 18, PO Box 831, D-7000 Stuttgart 1, German Federal Republic
Telephone: (0711) 2020-1
Telex: 07 23 931 lga
Affiliation: Baden-Wuerttemberg Ministry of Economics
Enquiries to: Eberhard Albrecht.
Professional Information Library
Subject coverage: Economic and industrial development of Baden-Wuerttemberg.
Library facilities: Open to all users (nc - special library for professional information); loan services (only in Baden-Wuerttemberg - nc); reprographic services (c).
Information services: Bibliographic services (only for libraries in Baden-Wuerttemberg - nc); no translation services; literature searches (c); access to on-line retrieval systems (c).

GREECE

Greek Productivity Centre
15.10

Address: Kapodistriou 28, Athinai 147, Greece
Telephone: 600 411-19
Enquiries to: Information Centre, Technical Applications Service

ICELAND

Stjórnunarfélag Íslands
15.11

– SFI

[Icelandic Management Association]

Address: Sidmuli 23, PO Box 155, 105 Reykjavík, Iceland
Telephone: (91) 82930
Founded: 1961

IRELAND

Irish Management Institute *
15.12

– IMI

Address: Sandyford Road, Dublin 16, Ireland
Telephone: (01) 983911
Telex: 30325
Enquiries to: Librarian, M. Ainscough
Founded: 1952
Subject coverage: Management education, training and development.
Library facilities: Not open to all users; loan services (nc); reprographic services (c).

Library services are provided principally for members; open to non-members for reference only.

Information services: Bibliographic services (nc); no translation services; literature searches (nc); access to on-line information retrieval systems (c - Dialog, Blaise).

Consultancy: No consultancy.

ITALY

Centro Italiano di Ricerche e d'Informazione sull'Economia delle Imprese Pubbliche e di Pubblico Interesse *

15.13

– CIRIEC

[Italian Centre for Research and Information on Public Enterprise]

Address: Via Fratelli Gabba 6, 20121 Milano, Italy

Telephone: (02) 872565

Affiliation: Centre Internationale de Recherches et d'Information sur l'Économie Publique, Sociale et Cooperative (Liège, Belgium)

Enquiries to: General secretary, Dr Alberto Mortara

Subject coverage: Economics, organization, legal and financial aspects of public enterprise, both regionally and in the developing countries.

Library facilities: Open to all users; no loan services; reprographic services (c).

Information services: Bibliographic services (nc); no translation services; literature searches (nc).

Consultancy: Yes (nc).

Publications: *Economica Publica*, monthly.

NETHERLANDS

Exportbevorderings- en'Voorlichtings- dienst *

15.14

– EVD

[Netherlands Foreign Trade Agency]

Address: Bezuidenhoutseweg 151, 2594 AG 's-Gravenhage, Netherlands

Telephone: (070) 798933

Telex: 31099 ecozanl

Affiliation: Ministry of Economic Affairs

Enquiries to: Head, J.H. Ypma.

Hoofdafdeling Documentaire Research en Informatieverstrekking

Founded: 1936

Subject coverage: Export promotion; worldwide economic information; foreign markets; trade; investment climates; import regulations; commerce and business.

Library facilities: Open to all users; loan services (nc); reprographic services (c).

Library holdings: Over 100 000 volumes; 1800 periodicals; 3500 directories.

Information services: Bibliographic services (nc); no translation services; literature searches (nc); access to on-line information retrieval systems (c - Dialog).

The library offers computerized retrieval and SDI services from its data base *Foreign Trade Abstracts* which holds approximately 145 000 items from 1974 onwards, updated semimonthly, and is available through Dialog and through Euronet Diane hosts Datastar and Belindis. It also provides a market intelligence service investigating foreign markets.

Consultancy: Yes (nc).

Publications: *Economic Titles/Abstracts*, bimonthly with annual cumulative subject index: each issue contains around 600 original language abstracts with complete bibliographical data, and is also available on COM Microfiche; *Key to Economic Science*, bimonthly with annual subject and author indexes.

Nederlandse Vereniging voor Management *

15.15

– NIVE

[Dutch Management Association]

Address: Postbus 90730, 2509 LS 's-Gravenhage, Netherlands

Telephone: (070) 26 43 41

Telex: 32626

Enquiries to: Librarian, A.L.i G. Regtop.

NIVE Library

Founded: 1925

Subject coverage: Promoting effective functioning of management at all levels (directors, production managers, department heads, foremen) in all sectors: industry, government, education, health care, welfare, sports, culture.

Library facilities: Open to all users; loan services (c - within the Netherlands only); reprographic services (c).

Stichting Kwaliteitsdienst *

15.16

– KDI

[Dutch Foundation for Quality Control]

Address: 734 Weena, 3014 DA Rotterdam, Netherlands

Telephone: (010) 147466

Enquiries to: R. Knaapen

Founded: 1953

Library facilities: Not open to all users; no loan services; reprographic services (c).

Information services: No bibliographic services; no translation services.

Consultancy: Yes (c).

Publications: Two papers on quality control: *Sigma*, 6 issues a year; *Specifiek*, 8 issues a year.

NORWAY

Norges Handelshøgskole Biblioteket *   15.17
[Norwegian School of Economics and Business Administration, Library]
Address: Helleveien 30, N-5035 Bergen, Sandviken, Norway
Telephone: (05) 256500
Telex: 40642 NHHN
Enquiries to: Chief librarian, John E. Steen
Founded: 1936
Library facilities: Open to all users; loan services; reprographic services (c).
Information services: Bibliographic services; no translation services; literature searches; access to on-line information retrieval systems (c).
Consultancy: Yes.

Norsk Produktivitetsinstitutt   15.18
– NPI
[Norwegian Productivity Institute]
Address: Akersgaten 64 VII, PO Box 8401, Hammersborg, Oslo 1, Norway
Telephone: (02) 20 94 75
Subject coverage: The institute has the following departments: mercantile; technical; leadership and cooperation; international and local productivity; information.
Publications: *NPI-nytt*, 10 issues a year.

POLAND

Główna Biblioteka Pracy i Zabezpieczenia Społecznego *   15.19
– GBP
[Central Library of Labour and Social Security]
Address: Ulica Mysia 2, 00-496 Warszawa, Poland
Telephone: 29 96 33
Affiliation: Ministry of Labour, Wages and Social Affairs
Enquiries to: Director, Małgorzata Kłossowska
Scientific Information Centre
Founded: 1974
Subject coverage: Labour, wages, social affairs and related problems: work organization, management, employment, productivity, health and labour protection, safety at work, labour law, social security, living conditions.

Library facilities: Open to all users; loan services (nc); reprographic services (nc).
Information services: Bibliographic services (nc); no translation services; literature searches (nc); no access to on-line information retrieval systems.
The library also edits reference works, and provides retrospective and current awareness services; it has a card index of articles in periodicals, a card index of books, and subject catalogues.
Consultancy: Yes (nc).
Publications: *Bibliography of economic and social problems of labour*, annual; *Documentation Review*, monthly; *Library's Communique*, monthly; *Labour-Wages-Social Affairs*, weekly.

Instytut Organizacji Przemysłu Maszynowego *   15.20
– IOPM
[Institute for Organization of the Machine Industry]
Address: 00-987 Warszawa, Poland
Telephone: 20 48 31
Telex: 812720 CO MPM PL
Enquiries to: Manager, Alicja Kawecka
Centre for Organizational and Economic Information
Founded: 1953
Subject coverage: Organization and management principles, as well as computer science. The information centre prepares data for the institute's own scientists as well as for staff across the whole metallurgy and mechanical engineering industries.
Library facilities: Open to all users.
Library holdings: The library holds 15 000 books, over 200 national and foreign periodicals, Polish standards, and legal regulations.
Publications: *Literature Abstracts Review*, 10 issues a year, containing about 240 abstracts each; *Bulletin of IOPM - Information for the Enterprises of Metallurgy and Machine Industry*, monthly; *Analytical-Synthetic Information*, 20 issues a year.

Instytut Organizacji Zarzadzania i Doskonalenia Kadr   15.21
– IOZiDK
[Management Organization and Development Institute]
Address: Ulica Wawelska 56, skr poczt 69, 02-067 Warszawa, Poland
Telephone: (022) 25 12 81
Telex: 81-5527 ODKA
Affiliation: Urząd Rady Ministrów
Enquiries to: Head, Bogdan Radomski
Branzowy Ośrodek Informacji Naukowej, Technicznej i Ekonomicznej (Scientific Information Centre)
Founded: 1959
Subject coverage: Management training and development; management organization of public administration and industry; personnel management.

Library facilities: Open to all users (nc); loan services (for institute's users and other libraries - nc); reprographic services (for institute's users only - nc).
Library holdings: 21 127 bound volumes; 313 current periodicals.
Information services: Bibliographic services (nc); translation services (for institute's users only - c); literature searches (for institute's users only - nc); access to on-line information retrieval systems (for training only - nc).
The information centre provides selective, addressed, retrospective and current information profiles, for national and foreign users. Services are based on ISIS/Integrated Set of Information Systems CDS/ SDI.
Consultancy: Yes (nc).
Courses: Information Centre Services, twice a year.
Publications: *Doskonalenie Kadr Kierowniczych,* monthly; *Organizacja - Metody - Technika,* monthly.

Instytut Wzornictwa Przemysłowego 15.22
– IWP
[Institute of Industrial Design]
Address: Ulica Áwiętojerska 5/7, 00-236 Warszawa, Poland
Telephone: 31 22 21
Affiliation: Ministry of Science and Education
Enquiries to: Director, R. Terlikowski
Founded: 1950
Library facilities: Open to all users; loan servicesprographic services (c).
Information services: Bibliographic services; translation services; literature searches; no access to on-line information retrieval systems.
Consultancy: Yes.
Publications: *IWP News; Documental Review of Industrial Design; Works and Materials;* annual report.

PORTUGAL

Associação Industrial Portuguesa 15.23
– AIP
[Portuguese Industrial Association]
Address: Praça das Indústrias, 1399 Lisboa Codex, Portugal
Telephone: 63 90 44/8
Telex: 12282 FIPORT P
Enquiries to: Head.
Serviço de Documentação e Publicações
Subject coverage: Development and promotion of Portuguese industry, including manufacturing, raw materials, distribution and marketing sectors.
Library facilities: Open to all users; loan services; reprographic services.

Information services: Bibliographic services.
Consultancy: Yes.

SWITZERLAND

Eidgenössische Technische Hochschule Zürich 15.24
[Swiss Federal Institute of Technology, Department of Industrial Management]
Address: Zürichbergstrasse 18, CH-8032 Zürich, Switzerland
Telephone: (01) 47 08 00
Enquiries to: Head, Professor Ernst Brem
Subject coverage: Industrial management and engineering: layout, scheduling and control of production; management of development and research; marketing; product evaluation, manufacturing equipment and market relations of small to medium size enterprises; applications of computers for industrial control.
Consultancy: Yes.

Schweizerische Arbeitsgemeinschaft für Qualitätsförderung * 15.25
– SAQ
[Swiss Association for the Promotion of Quality]
Address: Postfach 2613, CH-3001 Bern, Switzerland
Telephone: (031) 22 03 82
Telex: 33 528 atag ch
Founded: 1965
Library facilities: Not open to all users; loan services (infrequently); reprographic services (infrequently).
Consultancy: Yes.

TURKEY

Milli Prodüktivite Merkezi 15.26
– MPM/NPC
[National Productivity Centre]
Address: Gelibolu Sokak 5, Güvenevler, Kavaklidere, Ankara, Turkey
Telephone: (41) 286700
Telex: tr mipm 46041
Affiliation: European Association of National Productivity Centres (EANPC)
Enquiries to: Director, Zühal Kutes.
Training and Publications Department
Founded: 1965
Subject coverage: Productivity promotion; research and development, including socio-economic aspects.
Library facilities: Open to all users (nc); no loan services; no reprographic services.

An exchange system has been established between the library and all others; internal and non-staff requests are processed through the centre (charge made).

Library holdings: 10 500 bound volumes; 105 current periodicals.

Information services: Bibliographic services (nc); no translation services; no literature searches; no access to on-line information retrieval systems.

Consultancy: Telephone answers to technical queries (nc); writing of technical reports (c); no compilation of statistical trade data; no market surveys in technical areas.

Courses: The centre offers three or five day courses for training and planning supervisors and managers, on the following subjects: working conditions, job satisfaction, personnel management, job evaluation, public relations, auditing, production planning and control, accounting, productivity measurement, quality control circles, inventory planning and control, budget control, cost accounting, etc.

UNITED KINGDOM

British Institute of Management 15.27
– BIM

Address: Cottingham Road, Corby, Northamptonshire NN17 1TT, UK
Telephone: (0536) 204222
Facsimile: (0536) 201651
Enquiries to: Head, information services, Richard Withey.
Management Information Centre
Founded: 1947
Subject coverage: Management science.
Library facilities: Open to all users (nc but only reference facilities for non-members of the institute); loan services (nc but UK BIM members only); reprographic services (c).
Library holdings: 70 000 bound volumes; 4000 current periodicals.
Information services: Bibliographic services (c - 170 reading lists available: £30 full set to members, £60 to non-members); no translation services; literature searches (nc); access to on-line information retrieval systems (c).
Consultancy: Telephone answers to technical queries (charged service for in-depth answers available to non-members); writing of technical reports (nc); no compilation of statistical trade data; no market surveys in technical areas.
Publications: List of publications available on request.

British Library Business 15.28
Information Service

Address: 25 Southampton Buildings, London WC2A 1AW, UK
Telephone: (01) 404 0406
Telex: 266959
Facsimile: (01) 430 1906
Affiliation: Science Reference and Information Service, British Library
Enquiries to: Head, David King
Founded: 1981
Subject coverage: All aspects of business information.
Library facilities: Open to all users (nc); loan services (very restricted - nc); reprographic services (c).
Library holdings: 2500 directories; product literature from 15 000 companies; about 4000 trade and business journals; about 2000 market research reports.
Information services: No bibliographic services; no translation services; literature searches (very limited - nc); access to on-line information retrieval systems (business only - c).
Consultancy: Telephone answers to technical queries (brief answers to business queries only - nc); no writing of technical reports; no compilation of statistical trade data.
Market surveys in business areas (nc); referral services (nc).
Courses: Four courses per annum on business information; two courses per annum on business information for public librarians.

Japan Business Services Unit 15.29
Address: University of Sheffield, Sheffield S10 2TN, UK
Telephone: (0742) 768555
Telex: 547216 UGSHEF G
Facsimile: (0742) 739826
Affiliation: University of Sheffield
Enquiries to: Business development manager, R.J. Yates
Founded: 1983
Subject coverage: Language and business information related to Japan and Korea. The unit provides a range of services including interpreting, information on companies and markets, and briefing sessions.
Library facilities: Not open to all users (access by appointment - c); no loan services; reprographic services (c).
Information services: Bibliographic services (c); translation services (c); literature searches (c); access to on-line information retrieval systems (c).
Consultancy: Telephone answers to technical queries (c); writing of technical reports (c); compilation of statistical trade data (c); market surveys in technical areas (c).
Courses: All courses are tailor-made to meet the requirements of the unit's clients. They cover language

at any level required, spoken and written, business and social etiquette and practical orientation aimed at those doing business with Japan and Korea.

National Materials Handling Centre *   15.30

Address: Cranfield Institute of Technology, Cranfield, Bedford MK43 0AL, UK
Telephone: (0234) 750323
Telex: 825072 CITECH G
Enquiries to: Head, Information Services, H.P. Keeble
Founded: 1970
Subject coverage: Warehouse design; plant layout; equipment testing; market research; distribution systems planning.
Library facilities: Open to all users (free for members); loan services (members only); reprographic services (c).
Library holdings: The centre has a specialized collection on materials handling, including reports and over 100 periodicals, a trade literature collection of UK mechanical handling equipment, and a slide collection.
Information services: Bibliographic services (c); literature searches (c) services are free to members.
Publications: *International Distribution and Handling Review*, bimonthly with annual index; abstracts journal, issued free to members; publications list on request.

YUGOSLAVIA

Inštitut za Organizaciju i Razvoj   15.31
– ORGANOMATIK
[Organization and Development Institute]
Address: Milana Rakića 35, 11000 Beograd, Yugoslavia
Telephone: (011) 417 433
Telex: 12956 YU ORG BGD
Affiliation: Executive Committee, Assembly of the Socialist Republic of Serbia
Enquiries to: Librarian
Founded: 1956
Subject coverage: Management science, industrial engineering, industrial economics, strategic planning and related activities such as training and publishing; software build-up and implementation.
Library facilities: Not open to all users; no loan services; reprographic services (c).
The library is also a software package storing and issuing service.
Library holdings: 4000 titles; 25 periodicals.
Information services: Translation services.

16 MATERIALS TESTING

AUSTRIA

Technische Versuchs- und Forschungsanstalt
16.1

[Technical Testing and Research Institute]
Address: Karlsplatz 13, A-1040 Wien, Austria
Telephone: (0222) 5880 13429
Telex: 131 000
Affiliation: Technische Universität Wien
Enquiries to: Head, Professor T. Varga
Founded: 1981
Subject coverage: Materials technology.
Library facilities: Open to all users (nc); loan services (nc); reprographic services (c).
Library holdings: 3180 bound volumes; 30 current periodicals.
Information services: No information services.
Consultancy: Telephone answers to technical queries (nc); writing of technical reports (c); no compilation of statistical trade data; no market surveys in technical areas.
Courses: Course on structural failure, product liability, and technical insurance.
Publications: List on request.

BELGIUM

Association Belge pour l'Étude, l'Essai et l'Emploi des Matériaux
16.2

– ABÉM
[Belgian Association for the Study, Testing and Use of Materials]
Address: c/o H.Matteu, Rue d'Arlon 53, B-1040 Bruxelles, Belgium
Telephone: 230 30 10

CZECHOSLOVAKIA

Technický a skúšobný Ústav Stavabný
16.3

[Construction Engineering and Testing Institute]
Address: Studena 9, 826 34 Bratislava, Czechoslovakia
Telephone: (07) 225 429; 224 268
Enquiries to: Director, Ján Kováč
Subject coverage: Building materials, elements and buildings.
Publications: Institute proceedings.

DENMARK

Dansk Selskab for Materialprøvning og-Forskning
16.4

[Danish Society for Materials Research and Testing]
Address: c/o Dansk Ingeniørforening, Vester Farimagsgade 31, DK-1606 København V, Denmark
Telephone: (01) 15 65 65

Korrosionscentralen
16.5

– ATV
[Danish Corrosion Centre]
Address: Park Allé 345, DK-2605 Brøndby, Denmark
Telephone: (2) 631100
Telex: 33 888
Facsimile: (2) 96 26 36
Affiliation: Academy of Technical Sciences
Enquiries to: Managing director, Erik Nielsen
Founded: 1965
Subject coverage: Consultancy, failure analysis, testing, research, inspection and control in the field of corrosion, protection against corrosion, and metallurgy.
Library facilities: Not open to all users; no loan services; reprographic services (c).

Information services: Bibliographic services (c); literature searches (c); no access to on-line information retrieval systems.
Consultancy: Yes (c).

FINLAND

Valtion Teknillinen Tutkimuskeskus, Metallilaboratorio * 16.6
– VTT
[Technical Research Centre of Finland, Metals Laboratory]
Address: Metallimiehenkuja 6, SF-02150 Espoo 15, Finland
Telephone: (90) 4561
Telex: 122972 vttha sf
Enquiries to: Information officer, Marjatta Kimonen.
Technical Information Service, Technical Research Centre of Finland, Vuorimiehentie 5, SF-02150 Espoo, Finland.
Founded: 1942
Subject coverage: Metallic materials; destructive and non-destructive testing; quality assurance; welding; and machine and machine shop technology.
Library facilities: Not open to all users; loan services (c); no reprographic services.
Consultancy: Yes (c).

FRANCE

Association Française de Recherches et d'Essais sur les Matériaux et les Constructions 16.7
[French Association for Reseach and Testing of Materials and Buildings]
Address: 12 rue Brancíon, 75737 Paris Cedex 15, France
Telephone: (1) 45 39 22 33

GERMAN DEMOCRATIC REPUBLIC

Zentralinstitut für Festkörperphysik und Werkstofforschung 16.8
[Central Institute for Solid-State Physics and Materials Research]
Address: Helmholtzstrasse 20, DDR-8027 Dresden, German Democratic Republic
Telephone: (051) 46590
Parent: Akademie der Wissenschaften der DDR
Enquiries to: Director, Professor Johannes Barthel
Subject coverage: Topics include fundamental research in physics - materials with high purity; mechanical, corrosive and wear properties; ceramics; non-destructive determination of materials.
Publications: Biannual reports; conference proceedings.

GERMAN FEDERAL REPUBLIC

Bundesanstalt für Materialprüfung * 16.9
– BAM
[Federal Institute for Materials Testing]
Address: Unter den Eichen 87, D-1000 Berlin 45, German Federal Republic
Telephone: (030) 81 04-1
Telex: 0183261 bamb d
Affiliation: Bundesministerium für Wirtschaft
Enquiries to: Head, Information and Public Relations, Dr H.-U. Mittmann
Founded: 1871
Subject coverage: Research in problems of materials science and materials testing for metals, construction materials and organics. Physical, chemical and biological testing; chemical safety engineering; development of new materials and of methods for non-destructive testing; special topics such as tribology, colour metrics, welding, nuclear power.
Library facilities: Open to all users; loan services (nc); reprographic services (c).
Information services: Bibliographic services (nc); no translation services; literature searches (c); access to on-line information retrieval systems (c).
BAM has seven documentation centres: Measurement of Mechanical Quantities (MmG), Colour, Rheology, Tribology, Welding Technology (DS), Non-destructive Testing (ZfP), Biological Materials Testing (BIO). Each documentation centre issues title bibliographies with abstracts and search literature according to individual profiles. For consultancy services there is access to many data bases (including in-house data

bases on materials, technology, and chemistry). Information is supported by scientists from all BAM departments.

Consultancy: Yes (c).

Publications: Amts- und Mitteilungsblatt; bibliographies, abstracts and author indices in the fields listed above; reports; summary of published reference works; publications list.

Fachinformationszentrum Werkstoffe eV   16.10

– FIZ-W

[Materials Information Centre]

Address: Unter den Eichen 87, D-1000 Berlin 45, German Federal Republic

Telephone: (030) 830001-0

Telex: 186387 fizw d

Affiliation: Bundesanstalt für Materialprüfung

Enquiries to: Managing director, Dr P. Büttner

Founded: 1983

Subject coverage: Documentation and information services in materials sciences and applications.

Library facilities: Open to all users (c); no loan services; reprographic services (c).

Library facilities are supported by the resources of the Bundesanstalt für Materialprüfung.

Information services: Bibliographic services (c); no translation services; literature searches (c); access to on-line information retrieval systems (c).

FIZ-W maintains and works with four bibliographic data bases (trilingual, 400 000 records with abstracts, accessible through Euronet and DATEX P): SDIM1 (finished 1979) and SDIM2 (from 1979) - materials science of metals, metallurgy, processing and production techniques, testing and measuring techniques, environmental aspects, corrosion, economics; TRIBO - tribology, friction, wear and lubrication; RHEO - rheology.

Consultancy: No consultancy.

GKSS-Forschungszentrum Geesthacht GmbH   16.11

[GKSS Research Centre Geesthacht GmbH]

Address: Postfach 11 60, Max-Planck-Strasse, D-2054 Geesthacht, German Federal Republic

Telephone: (04152) 12-1

Telex: 0218712 gkssg

Enquiries to: Siegfried Otto

Founded: 1956

Subject coverage: Materials technology; underwater technology; environmental research, meteorological research, environmental techniques.

Library facilities: Open to all users; no loan services; no reprographic services.

Information services: Bibliographic services (nc); no translation services; no literature searches.

On-line information retrieval systems for staff users only.

Consultancy: Yes (nc).

Publications: Wissenschaftlich-technischeBerichte der GKSS, annually.

ITALY

Gruppo Nazionale di Struttura della Materia   16.12

[National Research Group for the Structure of Matter]

Address: ·Viale dell'Università 11, 00185 Roma, Italy

Telephone: (06) 492489; 4952258

Parent Consiglio Nazionale delle Ricerche

Enquiries to: Director, Professor Giovanni Signorelli

Founded: 1962

Subject coverage: The group consists of 34 research units and five laboratories. Basic research: topics include atomic and molecular physics, collision physics and non-linear optics; applied research: topics include the physics and technology of materials, failure physics, and advanced technology with the use of laser.

NETHERLANDS

Bond voor Materialenkennis   16.13

[Netherlands Society for Materials Science]

Address: Postbus 390, 3330 AJ Zwijndrecht, Netherlands

Telephone: (078) 19 22 30

Gemeenschappelijk Centrum voor Onderzoek GCO/JRC *   16.14

[Joint Research Centre]

Address: Postbus 2, 1755 ZG Petten, Netherlands

Telephone: (02246) 5656

Telex: 57211

Affiliation: Commission of the European Communities

Enquiries to: Head, B. Seysener

Library and Documentation

Founded: 1959

Subject coverage: Nuclear science; materials science.

Library facilities: Open to all users; loan services (c); reprographic services (c).

Information services: Bibliographic services (c); translation services (c); literature searches (c); access to on-line information retrieval systems (c).

The centre has access to the following on-line retrieval systems: LIRS, ESA, INKA, CAS, Echo.

Consultancy: Yes (c).

Nederlandse Vereniging voor Niet Destructief Onderzoek 16.15

[Netherlands Society for Non-Destructive Testing]
Address: c/o Bond voor Materialenkennis, Postbus 390, 3330 AJ Zwijndrecht, Netherlands
Telephone: (078) 19 22 30

NORWAY

Norges Branntekniske Laboratorium 16.16

[Norwegian Fire Research Laboratory]
Address: N-7034 Trondheim NTH, Norway
Telephone: (07) 595190
Telex: 55620 sintf n
Affiliation: SINTEF
Enquiries to: Secretary, Dorid Helland
Founded: 1934
Subject coverage: Fire testing, research and information.
Library facilities: Open to all users; no loan services; no reprographic services.
Consultancy: Consultancy (c).
Publications: Reports summarizing the laboratory's research work.

SPAIN

Instituto de Estructura de la Materia 16.17

[Structure of Matter Research Institute]
Address: Serrano 119, Madrid 28006, Spain
Telephone: (91) 261 94 00
Telex: 42182 csie e
Affiliation: Consejo Superior de Investigaciones Científicas
Enquiries to: Director, Dr J.M. Orza
Subject coverage: Theoretical and high-energy physics; quantum chemistry; molecular and atomic collisions; molecular physics; molecular spectroscopy; macromolecular physics.

SWEDEN

Statens Provningsanstalt 16.18

– SP
[National Testing Institute]
Address: Brinellgatan 4, Box 857, S-501 15 Borås, Sweden
Telephone: (033) 16 50 00
Telex: 362 52 testing A
Facsimile: 135502

Affiliation: Ministry of Industry
Enquiries to: Librarian, Elisabeth Sandström
Founded: 1920
Subject coverage: Materials testing, metrology, building research, mechanics, heating and ventilation, fire research, acoustics, electrical engineering, electronics, chemical analysis.
SP is responsible for developing new standardization and metrological techniques.
Library facilities: Not open to all users; loan services (nc); reprographic services (c).
Library holdings: 16 500 bound volumes; 680 current periodicals.
Information services: No information services.
Consultancy: No consultancy.
Publications: Technical reports - list on request.

Stenindustrins Forskningsinstitut AB 16.19

– SFI
[Swedish Stone Industry Research Institute]
Address: Gullmarsvägen 13, S-121 41 Johanneshov, Stockholm, Sweden
Telephone: (08) 81 86 00
Founded: 1946
Subject coverage: Technical problems regarding quarrying, uses and machining of natural stone.
Library facilities: No library facilities.
Information services: No information services.
Consultancy: Yes (c).

SWITZERLAND

Eidgenössische Materialprüfungs- und Versuchsanstalt für Industrie, Bauwesen und Gewerbe * 16.20

– EMPA
[Swiss Federal Laboratory for Materials Testing and Research]
Address: Unterstrasse 11, Postfach 977, CH-9001 St Gallen, Switzerland
Telephone: (071) 20 91 41
Telex: 71278
Enquiries to: Librarian, Gertrud Luterbach EMPA-Bibliothek
Founded: 1880
Subject coverage: Textiles, leather; printing; paper, packaging; oils, fats, detergents; wood preservation.
Library facilities: Open to all users; loan services (c); reprographic services (c).
Information services: Literature searches (c); access to on-line information retrieval systems (c).

The library has on-line access to the following: Dialog; Chemical Information System; Pergamon InfoLine; DataStar; Télésystèmes Questel; DIMDI.

UNITED KINGDOM

Chatfield Applied Research Laboratories Limited   16.21

Address: 13 Stafford Road, Croydon, Surrey CR0 4NG, UK
Telephone: (01) 688 5689
Telex: 296500 INTEGR
Enquiries to: Managing director, Dr C.J. Chatfield
Founded: 1946
Subject coverage: Surface coatings.
Library facilities: Not open to all users; no loan services; reprographic services.
Library holdings: About 250 bound volumes; 2000 current periodicals; 2000 reports.
Information services: No bibliographic services; no translation services; literature searches (c); no access to on-line information retrieval systems.
Consultancy: Telephone answers to technical queries (occasional charge); writing of technical reports (c); no compilation of statistical trade data; no market surveys in technical areas.

Fulmer Research Institute Library   16.22

Address: Hollybush Hill, Stoke Poges, Slough, Buckinghamshire SL2 4QD, UK
Telephone: (02816) 2181
Telex: 849374
Enquiries to: Librarian, R.F. Flint
Founded: 1946
Subject coverage: Materials.
Library facilities: Not open to all users (library open to institute staff, others can be admitted by special arrangement); loan services.
Library holdings: 10 000 bound volumes; 200 current periodicals.
Information services: Bibliographic services (c); literature searches (c); access to on-line information retrieval systems (c).
Consultancy: Consultancy (c).

Industrial Unit of Tribology, Leeds University   16.23

Address: Woodhouse Lane, Leeds LS2 9JT, UK
Telephone: (0532) 431751
Telex: 51311 Relays G
Enquiries to: Manager, Dr C.N. March
Founded: 1969
Subject coverage: Tribology; lubrication; bearings; wear; friction.

Library facilities: No library facilities.
Information services: No bibliographic services; no translation services; literature searches (c); access to on-line information retrieval systems (c).
Consultancy: Telephone answers to technical queries (free to membership scheme subscribers only); writing of technical reports (c); no compilation of statistical trade data; market surveys in technical areas (c). Design appraisal - c.

National Nondestructive Testing Centre   16.24

Address: AERE Harwell, Building 149, Didcot, Oxfordshire OX11 0RA, UK
Telephone: (0235) 24141
Telex: 83135
Affiliation: United Kingdom Atomic Energy Authority
Enquiries to: Manager, R.S. Sharpe
Founded: 1967
Subject coverage: Non-destructive testing.
Library facilities: No loan services; no reprographic services.
Open to users by arrangement only.
Library holdings: 1000 bound volumes; 20 current periodicals; 1000 reports.
Information services: Bibliographic services; no translation services; literature searches (nc); access to on-line information retrieval systems (nc).
The computerized NDT data base holds approximately 33 000 literature references.
Consultancy: Telephone answers to technical queries (no charge if brief); writing of technical reports (c); no compilation of statistical trade data; no market surveys in technical areas.
Publications: *NDT Info*, published in *NDT International*, contains information on published NDT research *QT Handbook; QT News* (three per annum).

Spring Research and Manufacturers' Association   16.25
– SRAMA

Address: Henry Street, Sheffield S3 7EQ, UK
Telephone: (0742) 760771
Telex: 547676
Enquiries to: Director, J.A. Bennett
Founded: 1946
Subject coverage: All aspects of springs, spring materials, and spring manufacturing technology.
Library facilities: Open to all users; no loan services (but volumes may be loaned informally to members); reprographic services (c - but no charge to members). Library charges vary according to the extent of use, but most services are free to SRAMA members.

Library holdings: About 550 bound volumes; 40 current periodicals; 5000 reports.

Information services: Bibliographic services (c); no translation services; literature searches (c); access to on-line information retrieval systems (c).

Consultancy: Telephone answers to technical queries (c); writing of technical reports (as cooperative research or contract - c); compilation of statistical trade data (limited by availability of non-confidential statistics - c); market surveys in technical areas (as part of research programme - may be undertaken as contract work).

Other technical and professional consultancies: quality assurance, materials selection etc.

A charge may be made for complex technical queries answered by phone; non-members may be required to write in and be billed.

Courses: Courses on technical subjects - spring design, material selection, quality assurance, SPC, heat treatment and inspection, etc - about three times per annum.

Publications: Annual report; list on request.

Yarsley Technical Centre Limited
16.26

Address: Trowers Way, Redhill, Surrey RH1 2JN, UK

Telephone: (0737) 65070

Telex: 8951511

Affiliation: Fulmer Research Institute Limited

Enquiries to: Marketing services manager, R.C. Watkins

Founded: 1931

Subject coverage: Consultancy on the development, testing, design and evaluation of all non-metallic materials and the products and components manufactured from them; specialization: chemistry technology and applications of polymers and plastics materials.

Library facilities: No library facilities.

Information services: No information services.

Consultancy: Telephone answers to technical queries (occasional charge); writing of technical reports (c); no compilation of statistical trade data; market surveys in technical areas (c).

Polymer design; contract research, design, development and testing of non-metallic materials - c.

YUGOSLAVIA

Inštitut za Ispitivanje Materijala SR Srbije
16.27

– IMS

[Materials Testing Institute, Serbia]

Address: Bulevar Vojvade Mišića 43, 11000 Beograd, Yugoslavia

Telephone: (011) 650322

Telex: 12403 yu ims

Enquiries to: Director, Djordje Mešecković

Subject coverage: Building materials. The institute holds patented systems for prestressing, for prefabricated housing and other buildings, and for precast construction system for halls and bridges.

Publications: Informacije-Center za Stanovanje (housing Centre Information); bulletin.

Inštitut za Tehnologiju Nuklearnih i Drugih Mineralnih Sirovina
16.28

[Institute for Technology of Nuclear and other Mineral Raw Materials]

Address: PO Box 390, Franche D'Epere 86, Beograd, Yugoslavia

Telephone: (011) 685571

Telex: 11581 yu intems

Enquiries to: Director general, Dr Rade Ćosović

17 MATHEMATICS AND PHYSICS

AUSTRIA

Zentralbibliothek für Physik in Wien 17.1
– ZBPH
[Central Library for Physics in Vienna]
Address: Boltzmanngasse 5, A-1090 Wien, Austria
Telephone: (0222) 34 11 68
Telex: 76222 Physia
Affiliation: Bundesministerium für Wissenschaft und Forschung
Enquiries to: Director, Dr W. Kerber
Founded: 1946
Subject coverage: Physics and related fields.
Library facilities: Open to all users; loan services; reprographic services (c).
Library holdings: 107 023 volumes; 2477 periodicals; 470 370 atomic energy reports; 180 000 reprints.
As a university library and a depository library for Reports of the USAEC (now DOE-TIC), the library has a large collection of non-conventional literature; it also holds 90 various abstracts and index journals (printed versions).
Information services: Bibliographic services; literature searches; access to on-line information retrieval systems.
The library has access to the following on-line information retrieval systems: Blaise, Inis, INKA, Dialog.
Consultancy: Consultancy.

BELGIUM

Institut von Karman de Dynamique des Fluides * 17.2
– IVK
[Von Karman Institute for Fluid Dynamics]
Address: Chaussée de Waterloo 72, B-1640 Rhode Saint Genèse, Belgium
Telephone: (02) 358 19 01

Enquiries to: Librarian, N.V. Toubeau
Founded: 1956
Subject coverage: Aerodynamics; physics; wind effects on buildings and structures, laser-doppler velocimetry, computational fluid dynamics, pollution; turbomachinery; axial and centrifugal compressors.
Library facilities: Open to all users; loan services ; reprographic services (c).
The institute is the Belgian deposit library for NASA publications.
Information services: Not available.

BULGARIA

Centre for Research in Physics 17.3
Address: V.I. Lenin 72, 1184 Sofia, Bulgaria
Telephone: 73 41
Affiliation: Bulgarian Academy of Sciences
Enquiries to: Director, Milko Borisov
Founded: 1972
Subject coverage: Fundamental physics, including structure of matter, semiconductors, acoustics, optics, liquid crystals, superconductors, solar energy, plasma physics; quantum electronics, lasers, holography, optical storage in computers; nuclear physics, high-energy physics, neutron physics, reactor technology; astrophysics, extragalactic astronomy.
Publications: Bolgarskij fizičeskij žurnal; Jodrena Energija; Astronomičeski Kalendar na Observatorijata v Sofia; Astrofizičeskie issledovanija.

Mathematics Institute * 17.4
Address: PO Box 373, 1090 Sofia, Bulgaria
Affiliation: Bulgarian Academy of Sciences
Enquiries to: Head, D. Văcov, Documentation Office Information Centre for Mathematics and Natural Sciences of the Bulgarian Academy of Sciences, Ul 7-mi Noemvri, 1000 Sofia, Bulgaria.
Founded: 1954
Library facilities: Open to all users; loan services; reprographic services.

Loan and reprographic services are available through the Central Library of the Bulgarian Academy of Sciences.
Information services: Not available.

CZECHOSLOVAKIA

Jednota Československých Matematiků a Fysiků *   17.5
[Czechoslovak Mathematicians' and Physicists' Association]
Address: Husova 5, 110 00 Praha 1, Czechoslovakia

Ústřední Informační Středisko pro Jaderný Program *   17.6
– ÚISJP
[Czechoslovak Atomic Energy Commission, Centre for Scientific and Technical Information]
Address: Zbraslav nad Vltavou Czechoslovakia
Telephone: 59 15 83

DENMARK

Danmarks Tekniske Højskole, Laboratoriet for Køleteknik   17.7
[Technical University of Denmark, Refrigeration Laboratory]
Address: Building 402, DK-2800 Lyngby, Denmark
Telephone: (02) 88 46 22
Telex: 37529 DTHDIA DK
Enquiries to: Librarian, Birger Rosendahl
Founded: 1939
Subject coverage: Refrigeration techniques; heat pump systems.
Library facilities: Open to all users; loan services (nc); reprographic services (c).
Loan services are available only through the National Technological Library of Denmark.
Information services: Not available.

Isotopcentralen   17.8
[Danish Isotope Centre]
Address: 2 Skelbaekgade, DK-1717 København V, Denmark
Telephone: (01) 21 41 31
Telex: 16600 fotex DK attn isotopcent
Affiliation: Danish Academy of Technical Sciences
Enquiries to: Head, Torben Sevel
Founded: 1957
Subject coverage: Application of isotopes in industry, environment and science.
Library facilities: Not open to all users; no loan services; reprographic services.

Consultancy: Consultancy (c).

Jydsk Selskab for Fysik og Kemi *   17.9
[Society for Physics and Chemistry in Jutland]
Address: c/o Århus Universitets Kemisk Institut, Langelandsgade 140, DK-8000 Århus C, Denmark
Telephone: (06) 12 46 33

Lydteknisk Institut   17.10
– LI
[Danish Acoustical Institute]
Address: Building 356, Akademivej, DK-2800 Lyngby, Denmark
Telephone: (02) 93 12 11
Telex: 37529 dth/dia dk
Affiliation: Danish Academy of Technical Sciences
Enquiries to: Head, Knud Skovgård
Founded: 1941
Subject coverage: Acoustics, noise control, vibration control.
Library facilities: Not open to all users; no loan services; reprographic services (c).
Information services: No information services.
Consultancy: Telephone answers to technical queries (nc); writing of technical reports (c); no compilation of statistical trade data; no market surveys in technical areas.

FINLAND

Teknillinen Korkeakoulu   17.11
[Helsinki University of Technology, Department of Technical Physics]
Address: Rakentajanaukio 2 C, SF-02150 Espoo, Finland
Telephone: (90) 4512474
Telex: 125161
Enquiries to: Librarian, Silja Rummukainen
Subject coverage: Material physics; nuclear engineering; information technology; energy.
Library facilities: Loan and reprographic services are available through the main university library.
Library holdings: 500 000 reports in the field of nuclear science and energy.
Publications: List on request.

FRANCE

Institut Laue-Langevin   17.12
– ILL
[Laue-Langevin Institute]
Address: 156 X, 38042 Grenoble Cedex, France
Telephone: (76) 48 70 20

Telex: ILL A320-621F
Affiliation: Centre National de la Recherche Scientifique (France); Kernforschungszentrum Karlsruhe (German FR); Science and Engineering Research Council (UK)
Enquiries to: Librarian, C. Castets
Founded: 1969
Subject coverage: Solid-state physics; nuclear physics; physical chemistry; materials science; biochemistry. Central facility: thermal neutron research reactor.
Library facilities: Open to all users (nc); no loan services; reprographic services (c - articles less than 20 pages).
Library holdings: 20 000 bound volumes; 200 current periodicals; 9000 reports, including theses.
Information services: Bibliographic services (in-house bibliographic information on thermal neutron scattering experiments); no translation services; no literature searches; no access to on-line information retrieval systems.
Consultancy: Telephone answers to technical queries (nc - restricted to in-house research); writing of technical reports (30 in-house technical reports per annum); no compilation of statistical trade data; no market surveys in technical areas.
Publications: Annual report; *ILL Experimental Reports and Theory College Activities; News for Reactor Users*, newsletter; *Neutron Research Facilities at the ILL High Flux Reactor*; bibliography: publications related to ILL neutron scattering measurements.

GERMAN DEMOCRATIC REPUBLIC

Physikalische Gesellschaft der DDR * 17.13
[Physics Society of the GDR]
Address: Am Kupfergraben 7, DDR-1080 Berlin, German Democratic Republic
Telephone: 200 06 91
Library facilities: Library.

Zentralinstitut für Festkörperphysik und Werkstofforschung * 17.14
[Central Institute for Solid-State Physics and Materials Research]
Address: Helmholtzstrasse 20, DDR-8027 Dresden, German Democratic Republic
Telephone: (051) 46590
Parent: Akademie der Wissenschaften der DDR
Enquiries to: Director, Professor Johannes Barthel

Subject coverage: Topics include fundamental research in physics - materials with high purity; mechanical, corrosive and wear properties; ceramics; non-destructive determination of materials.
Publications: Biannual reports; conference proceedings.

GERMAN FEDERAL REPUBLIC

Fachinformationszentrum Energie, Physik, Mathematik GmbH 17.15
– FIZ Karlsruhe
[Energy, Physics, and Mathematics Information Centre]
Address: Eggenstein-Leopoldshafen 2, D-7514, German Federal Republic
Telephone: (07247) 82 4600/4601
Telex: 17724710
Enquiries to: Online and Marketing Manager, Dr B. Jenschke
Founded: 1977
Subject coverage: Production of data bases and data compilations in energy and technology, aeronautics and astronautics, space research, physics and astronomy, mathematics and computer science.
Library facilities: Open to all users (nc); loan services (c); reprographic services (c).
Library holdings: 1.9m documents of non-conventional literature (reports, conference papers, theses, standards, patents, publications of firms).
Information services: Bibliographic services; no translation services; literature searches (c); access to on-line information retrieval systems (c).
INKA; STN International (on-line service offered jointly by FIZ Karlsruhe and the American Chemical Society). Magnetic tape services, information supply (retrospective searches, SDI profiles, technical information). Editing, distribution of printed information; document delivery.
Consultancy: Telephone answers to technical queries; no writing of technical reports; no compilation of statistical trade data; no market surveys in technical areas.
Publications: *Facts and Figures*, information leaflet.

Gesellschaft für Mathematik und Datenverarbeitung mbH 17.16
– GMD
[Mathematics and Data Processing Research Society]
Address: Schloss Birlinghoven, Postfach 1240, D-5205 St Augustin 1, German Federal Republic
Telephone: (02241) 141
Telex: 889 469 gmd d
Affiliation: Governments of German FR and the state of Nordrhein-Westfalen

Enquiries to: Abteilung für Informationswesen
Founded: 1968
Subject coverage: Data processing; mathematics; social sciences.
Library facilities: Generally not available to non-staff users.
Publications: List on request.

GREECE

Greek Mathematical Society 17.17
Address: Odos Panepistimiou 34, Athinai 143, Greece
Enquiries to: General secretary, D. Chassapis
Library holdings: 2000 volumes.

HUNGARY

Központi Fizikai Kutató Intézete * 17.18
– KFKI
[Central Research Institute for Physics]
Address: PO Box 49, Konkoly Thege ut, H-1525 Budapest XII, Hungary
Telephone: 698 566
Telex: KFKI H 224722
Affiliation: Magyar Tudományos Akadémia (Hungarian Academy of Sciences)
Enquiries to: Director, Dr Károly Szegő
Founded: 1950
Subject coverage: Physics, chemistry, computer science, electronics, mathematics.
Library facilities: Not open to all users; loan services (nc); reprographic services (c).
Information services: Not available.
Publications: Lists of current periodicals (annual), and of books and reports recently acquired (weekly).

Műszaki Fizikai Kutató Intézete, 17.19
Könyvtár *
– MFKI
[Technical Physics Research Institute, Library]
Address: PO Box 76, Fóti Útca 56, H-1325 Budapest IV, Hungary
Telephone: 692 100
Affiliation: Magyar Tudományos Akadémia (Hungarian Academy of Sciences)
Enquiries to: Librarian, Katalin Horváth
Subject coverage: Technical physics.
Library facilities: Not open to all users; loan services (nc); reprographic services (nc).
Loan and reprographic services are available to libraries only.
Information services: Bibliographic services; literature searches.

Consultancy: Yes.

IRELAND

Dublin Institute for Advanced 17.20
Studies, School of Theoretical
Physics
– DIAS-STP
Address: 10 Burlington Road, Dublin 4, Ireland
Telephone: (01) 68 07 48
Telex: 31687 DIAS
Enquiries to: Librarian, E.R. Wills
Founded: 1940
Subject coverage: Theoretical physics; applied mathematics.
Library facilities: Not open to all users (nc); no loan services; reprographic services (subject to staff availability - nc).
The library is primarily intended for the school's research workers, but visitors working in the field of theoretical physics are welcome.
Library holdings: 9000 bound volumes; 200 current periodicals.
Information services: Bibliographic services (nc); no translation services (apart from occasional very brief translations); literature searches (nc); access to on-line information retrieval systems (occasional change). Services are limited by staff availability, and on-line retrieval services are limited to the school's area of work.
Consultancy: Telephone answers to technical queries (limited - nc); no writing of technical reports; no compilation of statistical trade data; no market surveys in technical areas.

ITALY

Fondazione Giorgio Ronchi * 17.21
– FGR
[Giorgio Ronchi Foundation]
Address: Largo Enrico Fermi 1, 50125 Arcetri, Firenze, Italy
Telephone: (055) 22 11 63
Affiliation: Italian government
Enquiries to: President, Professor Vasco Ronchi
Founded: 1944
Subject coverage: Research in the field of optics from scientific, philosophical and technical points of view.
Library facilities: Not open to all users; no loan services; reprographic services.

Gruppo Nazionale di Struttura della Materia
17.22

[National Research Group for the Structure of Matter]
Address: Viale dell'Università 11, 00185 Roma, Italy
Telephone: (06) 492489; 4952258
Parent Consiglio Nazionale delle Ricerche
Enquiries to: Director, Professor Giovanni Signorelli
Founded: 1962
Subject coverage: The group consists of 34 research units and five laboratories. Basic research: topics include atomic and molecular physics, collision physics and non-linear optics; applied research: topics include the physics and technology of materials, failure physics, and advanced technology with the use of laser.

Istituto Nazionale d'Ottica *
17.23

– INO
[National Institute of Optics]
Address: Largo Enrico Fermi 6, 50125 Arcetri, Firenze, Italy
Telephone: (055) 22 11 79; 22 17 35; 22 61 29
Enquiries to: Librarian
Founded: 1927
Subject coverage: Optics, in all branches.
Library facilities: Open to all users; no loan services; reprographic services (c).
Information services: Bibliographic services (nc); no translation services; literature searches (nc); access to on-line information retrieval systems.
Consultancy: Yes (nc).

Laboratori Nazionali di Frascati *
17.24

– LNF
[Frascati National Laboratories]
Address: PO Box 13, 00044 Frascati (Roma), Italy
Telephone: (06) 94031
Telex: 614122 LNFI
Affiliation: Istituto Nazionale di Fisica Nucleare
Enquiries to: Librarian
Founded: 1955
Subject coverage: Nuclear physics.
Library facilities: Open to all users; no loan services; no reprographic services.
Consultancy: Yes.
Publications: Bollettino d'Informazione , 3 issues a year.

Società Italiana Fisica *
17.25

[Physics Society of Italy]
Address: Via L. Degli Andalò 2, 40124 Bologna, Italy
Telephone: (051) 33 15 54; 58 15 69
Library facilities: Library.

NETHERLANDS

Centrum voor Wiskunde en Informatica *
17.26

– CWI
[Mathematics and Computer Science Centre]
Address: Postbus 4079, Kruislaan 413, 1009 AB Amsterdam, Netherlands
Telephone: (020) 592 9333
Telex: 12571
Affiliation: Stichting Mathematisch Centrum
Enquiries to: Chief librarian
Founded: 1946
Subject coverage: Mathematics and its applications; operations research and system theory; computer science.
Library facilities: Open to all users; loan services (nc); reprographic services (c).
Loan services are available for books and scientific reports only.
Information services: Bibliographic services (c); no translation services; no literature searches; access to on-line information retrieval systems (c).
The centre has on-line access to the following: Lis, ESA/IRS, ISI.
Consultancy: Yes (c).
Publications: List on request.

Instituut TNO voor Wiskunde, Informatieverwerking en Statistiek *
17.27

[Mathematics, Information Processing, and Statistics Institute - TNO]
Address: Postbus 297, 2501 BD 's-Gravenhage, Netherlands
Telephone: (070) 82 41 61
Telex: 31707 wstno nl
Affiliation: Nederlandse Organisatie voor Toegepast-Natuurwetenschappelijk Onderzoek
Enquiries to: Director, J. Remmelts
Founded: 1945
Subject coverage: Statistics; traffic research; numerical mathematics; operations research; data processing in business; computation and information processing.

Nationaal Instituut voor Kernfysica en Hoge-Energie Fysica, sectie-Kernfysica
17.28

– NIKHEF-K
[National Institute for Nuclear and High-Energy Physics, Nuclear Physics Division]
Address: Postbus 4093, Ooster Ringdijk 18, 1009 AJ Amsterdam, Noord-Holland, Netherlands
Telephone: (020) 5922080

Telex: IKO 11538 NL
Enquiries to: Librarian, N. Kuijl
Founded: 1946
Subject coverage: Nuclear physics; radiochemistry.
Library facilities: Open to all users (nc); loan services (nc); reprographic services (c).
Library holdings: 12 000 bound volumes; 147 current periodicals; 1000 reports.
Information services: No bibliographic services; no translation services; no literature searches; access to on-line information retrieval systems (c).
Consultancy: No consultancy.
Publications: Annual report; list on request.

Technische Hogeschool Delft 17.29
[Delft University of Technology, Department of Technical Physics]
Address: Lorentzweg 1, 2628 CJ Delft, Netherlands
Library facilities: Library.

NORWAY

Institutt for Energiteknikk 17.30
– IFE
[Energy Technology Institute]
Address: PO Box 40, N-2007 Kjeller, Norway
Telephone: (02) 71 25 60
Telex: 16361 energ n
Enquiries to: Librarian, Solveig Opheim
Founded: 1953 (as Institutt for Atomenergi; name and mandate changed in 1980)
Subject coverage: Energy modelling and system studies; energy conservation; new energy technologies; reactor safety, fuelling and control; isotope production and applications; irradiation technology; process simulation and control; materials technology; petroleum reservoir studies; multi-phase flow studies; condensed matter (physics).
Library facilities: Open to all users; loan services (c); reprographic services (c).
The institute's library is a depositary library for publications issued by the International Atomic Energy Agency; it has a special collection of scientific and technical reports. The main part of the collection is on microfiche.
Information services: No bibliographic services; no translation services; literature searches (c); access to on-line information retrieval systems (c).
The library has on-line access to the following: Nordic Energy Index (NEI); INKA On-line Service; Dialog.
Consultancy: Yes (informal).

POLAND

Instytut Matematyczny PAN 17.31
[Mathematics Institute of the Polish Academy of Sciences (PAN)]
Address: Ulica Sniadeckich 8, skr poczt 137, 00-950 Warszawa, Poland
Telephone: (022) 28 24 71
Telex: 816112
Parent: Polska Akademia Nauk
Enquiries to: Chief librarian, Maria Mostowska Central Mathematics Library
Founded: 1949
Subject coverage: Mathematics and its applications.
Library facilities: Open to all users (nc); loan services (international interlibrary only - nc); reprographic services (c).
Library holdings: 100 000 bound volumes; 484 current periodicals.
Information services: Bibliographic services (nc); no translation services; literature searches (nc); no access to on-line information retrieval systems.
Consultancy: No consultancy.
Publications: Publications list on request.

Instytut Podstawowych 17.32
Problemów Techniki
– IPPT
[Fundamental Technological Research Institute]
Address: Świetokrzyska 21, 00-049 Warszawa, Poland
Telephone: (022) 26 12 81
Telex: 815638 Ippt pl
Affiliation: Polska Akademia Nauk
Enquiries to: Head librarian
Founded: 1953
Subject coverage: Mathematics and physics; theory of continuous media; theory of structures; mechanical deformation; acoustics and ultrasonic and electromagnetic waves; design and calculation of machine and building structures; medical ultrasonic diagnosis; structure and synthesis of Polish speech; atomic frequency standards.
Library facilities: Open to all users; loan services; reprographic services.
Information services: Bibliographic services; translation services.
Consultancy: Yes.
Publications: *Polish Analytical Bibliography of Mechanics* ; publications list on request.

ROMANIA

Centrul de Fizica Tehnica Iaşi *   17.33
[Technical Physics Centre]
Address: Splai Bahlui 47, Iaşi, Romania
Enquiries to: Director, Dr N. Rezlescu
Library holdings: 45 000 volumes.

Institutul Central de Fizica *   17.34
[Central Physics Institute]
Address: PO Box MG-6, Bucureşti-Măgurele 76900, Sector 5, Romania
Telephone: (90) 807040
Telex: 11397 csen
Enquiries to: Head, Dr M. Ibvaşcu
Information and Documentation Office
Subject coverage: Atomic and nuclear physics; solid-state physics; theoretical physics; nuclear engineering; laser and plasma physics; accelerators; physics of materials; seismology; astronomy and astrophysics.
Library facilities: Open to all users; loan services (nc); reprographic services (c).
Library holdings: 363 000 volumes.
Information services: Bibliographic services (c); translation services (c); literature searches (c); no access to on-line information retrieval systems.
Consultancy: Yes.

SPAIN

Instituto de Estructura de la Materia *   17.35
[Structure of Matter Research Institute]
Address: Serrano 119, Madrid 28006, Spain
Telephone: (91) 261 94 00
Telex: 42182 csie e
Affiliation: Consejo Superior de Investigaciones Científicas
Enquiries to: Director, Dr J.M. Orza
Subject coverage: Theoretical and high-energy physics; quantum chemistry; molecular and atomic collisions; molecular physics; molecular spectroscopy; macromolecular physics.

Instituto de Optica 'Daza de Valdés'   17.36
['Daza de Valdés' Optics Institute]
Address: Serrano 121, 28006 Madrid, Spain
Telephone: 261 68 00
Affiliation: Consejo Superior de Investigaciones Científicas
Enquiries to: M. Morales
Library and Publications Service
Founded: 1946

Library facilities: Open to all users; loan services; reprographic services.
Reprographic services are available through CSIC's Information Institute, ICYT, at the following address: Joaquin Costa 22, 28006 Madrid.
Library holdings: 5000 bound volumes; 70 current periodicals.
Information services: Bibliographic services (through the ICYT - c); no translation services; no literature searches; access to on-line information retrieval systems (through the ICYT).
Consultancy: No consultancy.

Real Sociedad Española de Física *   17.37
[Royal Spanish Society of Physics]
Address: Facultad de Ciencias Químicas, Ciudad Universitaria, Madrid 3, Spain
Enquiries to: Secretary general, S. Martínez Carrera

Real Sociedad Matemática Española *   17.38
[Royal Spanish Mathematical Society]
Address: Serrano 123, Madrid 6, Spain
Telephone: 2619800
Enquiries to: Secretary, Juan Llovet Verdugo

SWITZERLAND

Organisation Européenne pour la Recherche Nucléaire *   17.39
– CERN
[European Organization for Nuclear Research]
Address: CH-1211 Genève 23, Switzerland
Telephone: (022) 83 61 11
Telex: 419 000 CER CH
Enquiries to: Head, Dr Alfred Günther
Scientific Information Service
Founded: 1954
Subject coverage: High-energy physics.
Library facilities: Open to all users (with prior agreement); loan services (interlibrary only); no reprographic services.
Information services: Bibliographic services (by special agreement); literature searches (by special agreement).
Publications: List on request.

UNITED KINGDOM

Atomic Energy Research Establishment
17.40

Address: Harwell, Didcot, Oxfordshire OX11 0RB, UK
Telephone: (0235) 24141
Telex: 83135
Affiliation: United Kingdom Atomic Energy Authority
Founded: 1946
Subject coverage: Nuclear science and technology; metallurgy; ceramics and materials in general; engineering; inorganic and analytical chemistry; chemical engineering; physics; health physics; computer science; hazardous materials; energy technology; environmental science.
Library facilities: The Harwell Library facilities are intended primarily for staff members.
Information services: Bibliographic services; no translation services; no literature searches; no access to on-line information retrieval systems.
RECAP - data base of UKAEA publications; BULLETIN - data base of publications relevant to AERE programme of work (also available as a weekly printed information bulletin); NDT - data base of information on non-destructive testing; LIBCAT - on-line library catalogue; Inis SDI - SDI service from Inis magnetic tapes.
Consultancy: No consultancy.
Publications: List of publications available.

Daresbury Laboratory
17.41

Address: Daresbury, Warrington, Cheshire WA4 4AD, UK
Telephone: (0925) 603000; 603519 (Information Officer)
Telex: 629609
Affiliation: Science and Engineering Research Council
Enquiries to: Information officer, J.C.C. Sharp. Technical Scientific and Information Services
Founded: 1963
Subject coverage: Nuclear structure; atomic, molecular and solid-state physics; biophysics and biochemistry; computing.
Library facilities: Not open to all users (available on application to the librarian only); loan services (interlibrary only); no reprographic services.
Library holdings: 4000 bound volumes; 250 current periodicals; 15 000 reports.
Information services: Bibliographic services (nc); no translation services; literature searches (nc); no access to on-line information retrieval systems.
The library has access to the following bibliographic bases: Synchrotron Radiation; Tandem Van de Graaff Accelerators; Daresbury Laboratory Publications.
Consultancy: No consultancy.

Publications: Publications are listed in the annual report.

Institute of Physics
17.42

Address: 47 Belgrave Square, London SW1X 8QX, UK
Telephone: (01) 235 6111
Telex: 918453
Enquiries to: Executive secretary, Dr L. Cohen
Founded: 1874
Subject coverage: Physics.
Library facilities: Open to all users (nc); no loan services; no reprographic services.
Library holdings: The library holds only the institute's own publications from 1874 onwards.
Information services: No information services.
Consultancy: No consultancy.
Publications: *Physics Bulletin; Physics Education; Physics in Technology; Journal of Physics; Reports on Progress in Physics; Physics in Medicine and Biology; Plasma Physics; Optica Acta;* list on request.

United Kingdom Atomic Energy Authority
17.43

– UKAEA
Address: 11 Charles II Street, London SW1Y 4QP, UK
Telephone: (01) 930 5454
Telex: 22565 ATOMLO
Facsimile: 01-930 5454 extension 274
Affiliation: Department of Energy
Enquiries to: Press Office
Founded: 1954
Subject coverage: Nuclear power.
Library facilities: Open to all users (by appointment only); no loan services; no reprographic services.
Information services: No information services.
Consultancy: Telephone answers to technical queries (if answer cannot be given immediately, information can usually be obtained - nc); no writing of technical reports; no compilation of statistical trade data; no market surveys in technical areas.
Publications: List on request.

YUGOSLAVIA

Savez Društava Matematičara, Fizičarai i Astronoma Jugoslavije *
17.44

[Union of Societies of Mathematicians, Physicists and Astronomers of Yugoslavia]
Address: c/o Komisija za Fisiku, Bijenicka C54 41000 Zagreb, Yugoslavia

18   MEDICAL SCIENCES

AUSTRIA

Bundesstaatliche Anstalt für Experimentellpharmakologische und Balneologische Untersuchungen *   18.1
[Federal State Establishment for Experimental Pharmacological and Balneological Research]
Address: Währinger Strasse 13A, A-1090 Wien, Austria
Enquiries to: Director, Dr I. Eichler

Ludwig Boltzmann Gesellschaft - Österreichische Vereinigung zur Förderung der Wissenschaftlichen Forschung *   18.2
[Ludwig Boltzmann Society - Austrian Association for the Promotion of Scientific Research]
Address: Postfach 33, Hofburg, A-1014 Wien, Austria
Telephone: (0222) 522777; 529902; 529582
Enquiries to: President, Dr Hertha Firuberg
Founded: 1960
Subject coverage: Addictive diseases; gerontology; public health; acupuncture; homoeopathy, and interrelated fields.

BELGIUM

Académie Royale de Médecine de Belgique *   18.3
[Belgian Royal Academy of Medicine]
Address: Palais des Académies, Rue Ducale 1, B-1000 Bruxelles, Belgium
Telephone: (02) 511 24 71
Founded: 1841
Library facilities: Open to all users; no loan services; reprographic services.

Archives Belges de Médecine Sociale, Hygiène, Médecine du Travail et Médecine Légale   18.4
[Belgian Archives for Social Medicine, Hygiene, Occupational Medicine and Forensic Medicine]
Address: Cité administrative de l'État, Quartier Esplanade 6, B-1010 Bruxelles, Belgium
Affiliation: Ministère de la Santé Publique et de la Famille
Enquiries to: Secretary-General/Administrator, Dr M. Luyckx
Founded: 1937
Library facilities: Open to all users; no loan services; no reprographic services.
Information services: Bibliographic services; no translation services; literature searches; no access to on-line information retrieval systems.
Consultancy: No consultancy.

Conseil National de l'Ordre des Médecins *   18.5
[Physicians' National Council]
Address: Place de Jamblinne de Meux 32, B-1040 Bruxelles, Belgium
Telephone: (02) 736 82 91
Library facilities: Library.

BULGARIA

Union of Scientific Medical Societies in Bulgaria *   18.6
Address: Serdika 2, 1000 Sofia, Bulgaria
Telephone: 88 31 11
Enquiries to: Secretary general, Professor S. Manolov

CZECHOSLOVAKIA

Československá Lékařská Společnost J E Purkyně * 18.7

[J E Purkyně Czechoslovak Medical Society]
Address: Vítězného února 31, 120 26 Praha 2, Czechoslovakia
Telephone: 29 09 00
Telex: 12 12 93
Subject coverage: Medicine; pharmacy.

DENMARK

Almindelige Danske Laegeforening * 18.8

[Danish Medical Association]
Address: Trondhjemsgade 9, DK-2100 København Ø, Denmark
Enquiries to: Director, N. Würtzen

Dansk Medicinsk Selskab * 18.9

[Danish Medical Society]
Address: 'Domus Medica', Trondhjemsgade 9 DK-2100 København Ø, Denmark
Enquiries to: Secretary, S. Walter

Dansk Toksikologi Center * 18.10

– DTC
[Danish Toxicology Centre]
Address: Bagsvaerd Hovedgade 141, DK-2880 Bagsvaerd, Denmark
Telephone: (02) 44 33 11
Affiliation: Danish Academy of Technical Sciences
Enquiries to: Lisbeth Valentin Hansen
Founded: 1982
Subject coverage: Toxicological consultants.
Library facilities: Not open to all users; no loan services; reprographic services (nc).
Information services: Bibliographic services (c); no translation services; literature searches (c); access to on-line information retrieval systems (c). Toxline; RTECS; TDB; Medline.
Consultancy: Yes (c).

FINLAND

Lääketieteellinen Keskuskirjasto 18.11

– LKK
[Central Medical Library]
Address: Haartmaninkatu 4, SF-00290 Helsinki 29, Finland
Telephone: 90-418 544
Telex: 121498 lkk sf

Affiliation: Helsingin Yliopisto (Helsinki University)
Enquiries to: Librarian
Founded: 1966
Library facilities: Open to all users (nc); loan services (charge only for interlibrary loans from abroad); reprographic services (c).
Library holdings: About 150 000 bound volumes; 2400 current periodicals.
Information services: Bibliographic services (c); no translation services; literature searches (c); access to on-line information retrieval systems (c).
Consultancy: Telephone answers to technical queries (nc); no writing of technical reports; no compilation of statistical trade data; no market surveys in technical areas.

Oulun Yliopisto, Lääketieteellisen tiedekunnan kirjasto * 18.12

[Oulu University Medical Library]
Address: Kajaanintie 52 A, SF-9022 Oulu 22, Finland
Telephone: 981-332 133
Telex: 32315 OYL SF
Enquiries to: Librarian, Päivi Kytömäki
Library facilities: Open to all users; loan services (nc); reprographic services (c).
Information services: Literature searches (c); access to on-line information retrieval systems (c).
Consultancy: Yes (c).

FRANCE

Centre International de Recherche sur le Cancer 18.13

– CIRC
[Cancer Research International Agency]
Address: 150 cours Albert Thomas, 69372 Lyon Cedex 08, France
Telephone: 78 75 81 81
Telex: 380 023
Affiliation: World Health Organization
Enquiries to: Librarian, Agnes Nagy-Tiborcz
Founded: 1966
Library facilities: Open to all users (nc); no loan services; reprographic services (photocopies can be made of journals which do not exist in any other library in Lyon - nc).
Library holdings: 15 700 bound volumes; 280 current periodicals; 2000 reports.
Information services: No information services.
Consultancy: No consultancy.

Information Médicale 18.14
Automatisée, Centre de
Documentation de l'INSERM
– IMA
[Automated Medical Information, INSERM Documentation Centre]
Address: Hôpital de Bicêtre, 78 rue du Général Leclerc, 94270 Le Kremlin-Bicêtre, France
Telephone: 46 71 86 87
Parent: Institut National de la Santé et de la Recherche Médicale
Enquiries to: Director, Dr Paulette Dostatni
Founded: 1968
Subject coverage: Biomedicine; pharmacology; toxicology; ethics; history of medicine; carcinology.
Library facilities: Reprographic services are provided by another INSERM organization: Service Signalements et Microfiches INSERM, Centre de Recherches de l'INSERM, 44 chemin de Ronde, 78110 Le Vesinet, France.
Information services: Bibliographic services (c); no translation services; access to on-line information retrieval systems (c).
Access in the biomedical field to NLM files (IMA is the French Medlars centre) and other bases available in France; help-desk about biomedical data bases available in France in the biomedical field.
Consultancy: No consultancy.
Courses: Training courses on NLM files access.

Institut National de la Santé et 18.15
de la Recherche Médicale *
– INSERM
[Health and Medical Research National Institute]
Address: 101 rue de Tolbiac, 75654 Paris Cedex 13, France
Telephone: (1) 45 84 14 41
Telex: 270 532
Enquiries to: Director general, Philippe Lazar
Subject coverage: Promotes, conducts and develops biomedical research, from molecular biology to public health, with emphasis on pathology.
Publications: Série Santé Publique; Série Statistiques, Nomenclature; Collection Colloques INSERM; research reports.

Institut Pasteur, Bibliothèque 18.16
[Pasteur Institute Library]
Address: 25 rue du Docteur Roux, 75724 Paris Cedex 15, France
Telephone: (1) 45 68 82 80
Telex: PASTEUR 250 609 F
Enquiries to: Head librarian, Nicole Dubois
Founded: 1888
Subject coverage: Biology; medical sciences.
Library facilities: Open to all users (c); loan services (nc); reprographic services (c).

Library holdings: 300 000 bound volumes; 600 current periodicals.
Information services: Access to on-line information retrieval systems (c).
Consultancy: No consultancy.

GERMAN DEMOCRATIC REPUBLIC

Gesellschaft für Klinische 18.17
Medizin der DDR *
[Clinical Medicine Society of the GDR]
Address: c/o Verlag Volk und Gesundheit, Neue Grünstrasse 18, DDR-1020 Berlin, German Democratic Republic

GERMAN FEDERAL REPUBLIC

Bundesgesundheitsamt 18.18
– BGA
[Federal Health Office]
Address: Thielallee 88-92, D-1000 Berlin 33, German Federal Republic
Telephone: (030) 8308-1
Telex: 0184 016
Affiliation: Ministry for Youth, Family Affairs and Health
Enquiries to: Head of Information Office, Klaus J. Henning
Founded: 1952
Subject coverage: Consumers' health protection; reduction of environmental hazards; control of diseases; scientific consultant functions; approval and monitoring, particularly in the fields of narcotics and drugs legislation.
Information services: No information services.
Consultancy: Yes.
Publications: Bundesgesundheitsblatt; Tätigkeitsberichte des BGA; Schriftenreihe des Bundesgesundheitsamtes.

Bundesgesundheitsamt - Robert 18.19
Koch-Institut *
[Federal Health Office - Robert Koch Institute]
Address: Nordufer 20, D-1000 Berlin 65, German Federal Republic
Telephone: (030) 45031
Enquiries to: Head, Dr Wilhelm Weise
Subject coverage: Virology; bacteriology; immunology; biochemistry; cytology. The World Health Organization has named the institute a reference centre for the following areas: influenza; salmonella; blood group research; vibrio research;

yellow fever.
Publications: *RKI-Berichte.*

Deutsche Forschungs- und Versuchsanstalt für Luft- und Raumfahrt eV, Institut für Flugmedizin *

18.20

[German Aerospace Research Institute, Aerospace Medicine Institute]
Address: Godesberger Allee 70, D-5300 Bonn 2, German Federal Republic
Telephone: (0228) 376970
Enquiries to: Librarian, Mrs Rodrian
Founded: 1934
Subject coverage: Aerospace physiology and psychology; space medicine and biology; environmental and underwater medicine; biophysics.

Deutsches Institut für Medizinische Dokumentation und Information

18.21

– DIMDI
[German Institute for Medical Documentation and Information]
Address: PO Box 42 05 80, Weisshausstrasse 27, D-5000 Köln 41, German Federal Republic
Telephone: (0221) 4724-1
Telex: 8881 364
Affiliation: Federal Ministry of Youth, Family and Health Affairs
Enquiries to: President, Dr Rolf Fritz
Founded: 1969
Subject coverage: Biosciences, including human, dental, and veterinary medicine; poisons; pharmaceutical products; agricultural sciences.
Library facilities: No library facilities.
Information services: No bibliographic services; no translation services; literature searches (c); access to on-line information retrieval systems (c).
Data bases include the following: ABDA-FAM, ABDA-INTER, ABDA-STOFF, ABDA-PHARMA, ADO, Agrep, AGRICOLA, AGRIS; ASFA; Biosis; CAB ABSTRACT; Cancerlit; CANCERPROJ; CHEMLINE; CLINPROT; EMbase; EMcancer; EMdrugs; EMforensic; EMhealth; EMtox; EMtrain; FSTA; GRIPSLEARN; HEALTH; HECLINET; IRCS; ISI/Biomed; ISI/ISTP&B; ISI/Multisci; Medlars 1 and 2; MEDITEC; PSTA; PSYCINFO; PSYNDEX; RTECS; TDB; TELEGENLINE; Toxline. DIMDI is a host and on-line vendor offering services on a national and international level. The institute is connected to Euronet, DATEX-P (German PTT), and other national packet switched networks like Transpac, Tymnet, and Telenet; it also operates its own network - DIMDINET. DIMDI has developed its own management and retrieval software - GRIPS;

Common Command Set is implemented.
Consultancy: No consultancy.
Courses: Comprehensive training programme in German or English - details on request.
Publications: *DIMDI Information* - includes a full list of data bases available.

Zentralbibliothek der Medizin

18.22

[National Medical Library]
Address: Joseph-Stelzmann-Strasse 9, D-5000 Köln 41, German Federal Republic
Telephone: (0221) 4785600
Telex: 8882214 zbmed
Enquiries to: Head of Information Department, U. Hoffmann
Founded: 1908
Subject coverage: Biomedical sciences.
Library facilities: Open to all users (nc); loan services (c); reprographic services (c - on-line ordering (DIMDI, DBI) is possible).
Library holdings: 640 000 bound volumes; 6600 current periodicals.
Information services: Bibliographic services (c); no translation services; literature searches (c); access to on-line information retrieval systems (c). DIMDI's data bases (Medlars, EMbase, Biosis, PSYCINFO).
Consultancy: No consultancy.

GREECE

Panhellenic Medical Association *

18.23

Address: Semitelou 2, Athinai 611, Greece
Telephone: 777 26 90

HUNGARY

Országos Orvostudományi Információs Intézet és Könyvtár

18.24

– OIK
[National Institute for Medical Information]
Address: Postafiók 452, Szentkirályi útca 21, H-1372 Budapest, Hungary
Telephone: (1) 343-789
Affiliation: Ministry of Health
Enquiries to: Director
Founded: 1949
Subject coverage: Medicine; pharmacy; human biology; biochemistry.
Library facilities: Not open to all users; loan services (nc); reprographic services (c).
Library holdings: 28 000 bound volumes; 583 current periodicals.

Information services: Bibliographic services (c); translation services (c); literature searches (c); access to on-line information retrieval systems (c). On-line access to Medline.
Consultancy: No consultancy.

IRELAND

Royal Academy of Medicine in Ireland   18.25

Address: 6 Kildare Street, Dublin 2, Ireland
Telephone: (01) 767650
Enquiries to: General secretary
Founded: 1832
Subject coverage: Medicine and allied sciences.
Library facilities: Open to all users (nc); no loan services; reprographic services.
Library holdings: All issues of *The Irish Journal of Medical Science* since 1832.
Information services: No information services.
Consultancy: No consultancy.
Publications: *The Irish Journal of Medical Science.*

ITALY

Biblioteca Medica Statale *   18.26
[State Medical Library]
Address: Viale del Policlinico 155, Roma, Italy
Telephone: (06) 490778
Affiliation: Ministero per i Beni Culturali e Ambientali
Enquiries to: Director, Dr Laura Posa
Founded: 1925
Subject coverage: Biomedical sciences.
Library facilities: Open to all users; loan services (nc); reprographic services (c).
Information services: Bibliographic services (nc); no translation services; literature searches (nc); access to on-line information retrieval systems (nc) .
Medlars system.
Consultancy: Yes (nc).

NETHERLANDS

Nederlandse Vereniging voor Medisch Orderwijs *   18.27
[Dutch Association for Medical Education]
Address: Bijlhouwerstraat 6, 3511 ZC Utrecht, Netherlands
Telephone: (030) 33 11 23

NORWAY

Den Norske Laegeforening *   18.28
[Norwegian Medical Association]
Address: Inkognitogata 26, Oslo 2, Norway
Telephone: (02) 56 62 90

Rikshospitalet *   18.29
[National Hospital, Medical Library]
Address: Pilestredet 32, Oslo 1, Norway
Telephone: (02) 20 10 50
Affiliation: Universitet i Oslo
Enquiries to: Director, Elisabeth Buntz
Founded: 1826
Subject coverage: Medicine.
Library facilities: Not open to all users (open only to staff, students, etc); loan services; reprographic services.
Information services: Generally available to staff only but other enquiries are considered.

POLAND

Głowna Biblioteka Lekarska *   18.30
[Central Medical Library]
Address: Chocimska 22,00-791 Warszawa, Poland
Telephone: 49 78 51
Enquiries to: Director, Dr J. Kapuścik
Library holdings: 523 000 volumes.

Instytut Chemii i Techniki Jadrowej *   18.31
[Nuclear Chemistry and Technology Institute]
Address: Ulica Dorodna 16, 03-195 Warszawa, Poland
Telephone: 11 06 56
Telex: 813 027
Enquiries to: Director, Dr J. Leciejewicz
Subject coverage: Radiobiology; nuclear fuel processing.
Library holdings: 30 000 volumes.

Instytut Farmakologii PAN   18.32
[Pharmacology Institute of the Polish Academy of Sciences]
Address: 12 Smętna Street, 31-343 Kraków, Poland
Telephone: (48) 12 374022
Telex: 0322696
Parent: Polska Akademia Nauk
Enquiries to: Barbara Morawska-Nowak Information Centre and Library
Founded: 1954
Subject coverage: Pharmacology and related sciences.
Library facilities: Open to all users (nc); loan services (nc); reprographic services (limited to cooperating establishments - nc).

Library holdings: 18 747 bound volumes; 200 current periodicals.
Information services: Bibliographic services (limited to cooperating establishments - nc); no translation services; literature searches (nc); no access to on-line information retrieval systems.
Consultancy: No consultancy.

PORTUGAL

Instituto de Higiene e Medicina Tropical * 18.33
[Hygiene and Tropical Medicine Institute]
Address: Rua da Junqueira 96, 1300 Lisboa, Portugal
Telephone: (019) 632141
Affiliation: Ministério da Educação e Ciência
Enquiries to: Director of the Library
Founded: 1902
Subject coverage: Tropical medicine; parasitology; microbiology; public health.
Library facilities: Open to all users; no loan services; reprographic services (c).
Information services: No information services.

Ordem dos Médicos * 18.34
[Physicians' Society]
Address: Avenida da Liberdade 65-1—DG, 1298 Lisboa Codex, Portugal
Telephone: 32 22 02
Library facilities: Library.

ROMANIA

Academia de Ştiinţe Medicale 18.35
– ASM
[Medical Sciences Academy]
Address: Boulevardul 1 Mai 11, 79173 Bucuresti, Romania
Telephone: (90) 502393
Affiliation: Ministry of Health
Enquiries to: Secretary of the Academy, Professor Emil Măgureanu
Founded: 1969
Library facilities: Open to all users (nc); no loan services; reprographic services.
Library holdings: 5870 bound volumes; 26 current periodicals.
Information services: No information services.
Consultancy: Telephone answers to technical queries; writing of technical reports; no compilation of statistical trade data; no market surveys in technical areas.

Biblioteca Centrala Medicala * 18.36
[Central Medical Library]
Address: Pitar Moş 15, Bucureşti, Romania
Telephone: 10 78 85
Enquiries to: Director, Sonia Punga
Library holdings: About 270 000 volumes.

Central de Documentare Medicala al Ministerului Sanatatii * 18.37
[Medical Documentation Centre of the Ministry of Health]
Address: Strada Polona 4, Bucureşti, Romania

Institutul de Medicină Si-Farmacie, Biblioteca Centrală 18.38
– IMFBC
[Medical and Pharmaceutical Institute, Central Library]
Address: Bd Dr Petru Groza 8, 76241 Bucureşti, Romania
Telephone: 49 30 30
Affiliation: Ministry of Education
Enquiries to: Director, Petre Silvică
Founded: 1857
Subject coverage: Human medicine, including pharmacy and stomatology.
Library facilities: Not open to all users; loan services; reprographic services.
The library is open only to students, professional staff, and specialist physicians.
Library holdings: 1 029 887 bound volumes; 9600 current periodicals.
Information services: Bibliographic services (c); no translation services.
Consultancy: Telephone answers to technical queries; no writing of technical reports; no compilation of statistical trade data; no market surveys in technical areas.

SPAIN

Real Academia Nacional de Medicina * 18.39
[Royal National Academy of Medicine]
Address: Arrieta 12, Madrid 13, Spain
Telephone: (91) 247 03 18
Library facilities: Library.

SWEDEN

Biomedicinska Centrum 18.40
– BIOMEDICUM
[Biomedical Centre]
Address: Box 570, S-751 23 Uppsala, Sweden
Telephone: 018-17 40 00
Telex: Sweden 76132
Facsimile: 018-15 17 59
Affiliation: Uppsala University
Enquiries to: Librarian, Per Syrén
Founded: 1968
Subject coverage: Biosciences; chemistry; pharmacy.
Library facilities: Open to all users; loan services; reprographic services (c).
Library holdings: over 2500 bound volumes; 650 current periodicals.
Information services: Bibliographic services (nc); no translation services; literature searches (c); access to on-line information retrieval systems (c).
Medlars, ESA-RECON, Dialog, Orbit.
Consultancy: On-the-phone answers to queries - nc.

Karolinska Institutets, Bibliotek 18.41
och Informationscentral
– KIBIC
[Karolinska Institute, Library and Information Centre]
Address: Box 60201, S-104 01 Stockholm, Sweden
Telephone: 08-34 05 60; Information Centre 08-23 22 70
Telex: 17179 KIBIC S
Enquiries to: Acting director, Göran Falkenberg Medical Information Centre (MIC-KIBIC)
Founded: 1810
Subject coverage: Human medicine including odontology, toxicology and allied topics.
Library facilities: Open to all users (nc); loan services (nc); reprographic services (c).
Library holdings: 522 000 bound volumes; 3200 current periodicals.
Information services: Bibliographic services (nc); no translation services; literature searches (c); access to on-line information retrieval systems (c).
The following data bases are available via Elhill at MIC/QZ: Medline, Cancerlit, Drugline, Nordser, Primline, RTECS, and Swemed.
Consultancy: No consultancy.
Courses: Annual courses for researchers; annual courses for medical students; Medlars on-line introductory course for intermediaries, about five per annum; Basic Medlars course for physicians and medical researchers, about ten per annum; continuing education Medlars seminars, about fifteen per annum; special courses for institutions, authorities and enterprises, about five per annum; courses in cooperation with other organizations, about five per annum.

Publications: List Bio-med: Biomedical Serials in Scandinavian Libraries, irregular; *KIBIC Rapport*, irregular; *MIC News*, irregular, published by MIC-KIBIC.

SWITZERLAND

Schweizerische Akademie der 18.42
Medizinischen Wissenschaften *
[Swiss Academy of Medical Sciences]
Address: Petersplatz 13, CH-4051 Basel, Switzerland
Telephone: (061) 25 49 77
Library facilities: Library.

TURKEY

Türk Rib Cemiyeti * 18.43
[Turkish Medical Society]
Address: Valikonagi Caddessi 10, Harbiye, Istanbul, Turkey
Enquiries to: Secretary, Dr A. Mukbil Atakam

UNITED KINGDOM

British Medical Association, 18.44
Nuffield Library *
– BMA
Address: Tavistock Square, London WC1H 9JP, UK
Telephone: (01) 387 4499
Enquiries to: Librarian, D.J. Wright
Founded: 1832
Subject coverage: Medicine and related subjects.
Library facilities: Not open to all users (members only; others at discretion of librarian); no loan services (except to members); reprographic services (c).
Microfilm and microfiche facilities available.
Information services: Bibliographic services (c); no translation services; literature searches (c); access to on-line information retrieval systems (c).
Computer access to Blaise and Datastar data bases.
Consultancy: Yes (c).

Brunel Institute for 18.45
Bioengineering, Information Unit
– BIB
Address: Brunel University, Uxbridge, Middlesex UB8 3PH, UK
Telephone: (0895) 71206
Telex: 261173
Affiliation: Brunel University

Enquiries to: Head of Information Services, A.C. Rickard
Founded: 1983
Subject coverage: Biomedical engineering; medical instrumentation; physiological monitoring devices; computers and automation in medicine; prosthetics and orthotics; aids for disabled people; laboratory instrumentation; microgravity techniques.
Library facilities: Open to all users (staff will assist personal visitors with their literature searches; charges made for special services); loan services (via the National Lending Library scheme only); reprographic services (within the terms of the Copyright Act).
Library holdings: The unit's data base contains approximately 40 000 indexed references to work published from 1961 to date and most of these are held in microform; a small stock of books and some specialist journals are held.
Information services: Bibliographic services (some 70 Standard Bibliographies are offered); literature searches (a contract service is in operation whereby organizations or individuals may commission specialized literature services); access to on-line information retrieval systems.
Blaise, Lockheed. An in-house system is being developed using a dedicated microcomputer; copies of the service's disks available to subscribers with compatible systems.
Consultancy: Yes (in the bio-medical engineering field; design and research services are also offered).
Publications: *EBRI - European Bioengineering Research Inventory* has been published on behalf of the EEC Commissioners; *Becan* current awareness bulletin, monthly.

Institute for the Study of Drug Dependence 18.46
– ISDD
Address: 1/4 Hatton Place, Hatton Garden, London EC1N 8ND, UK
Telephone: (01) 430 1991
Enquiries to: Information officer, Michael Ashton
Founded: 1968
Subject coverage: Drug misuse; drug dependence.
Library facilities: Open to all users (nc); no loan services; reprographic services (c).
Library holdings: 2000 bound volumes; about 100 current periodicals; 35 000 reports. English language collection only.
Information services: Bibliographic services (nc); no translation services; literature searches (nc); no access to on-line information retrieval systems.
Consultancy: Telephone answers to technical queries (nc); writing of technical reports (nc); no compilation of statistical trade data; no market surveys in technical areas.

Medical Research Council Library * 18.47
– MRC
Address: National Institute for Medical Research, The Ridgeway, Mill Hill, London NW7 1AA, UK
Telephone: (01) 959 3666
Telex: 922666 MRCNAT G
Enquiries to: Librarian
Founded: 1919
Subject coverage: Biomedical sciences.
Library facilities: Not open to all users (open to MRC staff only unless in exceptional circumstances); loan services (to other libraries); reprographic services (to other libraries).

Royal College of Surgeons of England 18.48
Address: 35-43 Lincoln's Inn Fields, London WC2A 3PN, UK
Telephone: (01) 405 3474
Enquiries to: Librarian, E.H. Cornelius
Founded: 1800
Subject coverage: Surgery, including its specialities such as plastic surgery, orthopaedics, ophthalmology, and otorhinolaryngology; anaesthesia; oral surgery; anatomy; physiology; pathology; history of medicine.
Library facilities: Not open to all users; loan services (to other libraries only - nc); reprographic services (c).
The library is for reference only, though books and periodicals are lent to other libraries. Items over 50 years old are not lent, nor is any manuscript item. The library is open to all diplomates of the college: others are admitted at the discretion of the librarian after application in writing, giving precise reasons for using the library, and supported by a letter of introduction from a Fellow of the College, the Dean of a Medical School, or the librarian of a recognized medical library.
Library holdings: 160 000 bound volumes; 590 current periodicals.
Information services: Bibliographic services (nc); no translation services; literature searches (nc); access to on-line information retrieval systems (c).
The library has its own Medline terminal.
Consultancy: No consultancy.

Royal Society of Medicine 18.49
Address: 1 Wimpole Street, London W1M 8AE, UK
Telephone: (01) 408 2119
Telex: 298902
Enquiries to: Librarian, David Stewart
Founded: 1805
Subject coverage: Biomedical sciences; clinical practice; clinical research.
Library facilities: Not open to all users; loan services (members only - nc); reprographic services (nc).

Library holdings: 500 000 bound volumes; 2000 current periodicals; 1000 reports.
Information services: Bibliographic services (nc); no translation services; literature searches (c); access to on-line information retrieval systems (c).
Data bases include Medline, Excerpta Medica, and Chemical, Biological, and Psychological Abstracts.
Consultancy: Telephone answers to technical queries (nc); no writing of technical reports; no compilation of statistical trade data; no market surveys in technical areas.

Sheffield University Biomedical Information Service 18.50

– SUBIS
Address: Sheffield S10 2TN UK
Telephone: (0742) 78555
Telex: 547216 UGSHEF G (Quote 'SUBIS')
Enquiries to: Head, J.K. Barkla
Founded: 1966
Subject coverage: Cell biology; physiology; immunobiology; neurobiology; biotechnology.
Library facilities: The service forms a part of the university library.
Information services: Bibliographic services (monthly current awareness service - c); literature searches (c); access to on-line information retrieval systems.
Own data base; Medline.
Consultancy: Yes (c).

YUGOSLAVIA

Centralna Medicinska Knjižnica * 18.51
[Central Medical Library]
Address: Vrazor trg 2, 61000 Ljubljana, Yugoslavia
Enquiries to: Librarian, S. Gorec
Library holdings: 140 00 books and periodicals; 1473 current periodicals.

19 METALLURGY

BELGIUM

Centre Belge d'Étude de la Corrosion 19.1
– CEBELCOR
[Belgian Centre for the Study of Corrosion]
Address: Avenue Paul Héger, Grille 2, B-1050 Bruxelles, Belgium
Telephone: (02) 649 63 96
Telex: 23069 unilib b
Enquiries to: Director, Antoine Pourbaix
Consultancy: Consultancy.
Publications: *Rapports Techniques Cebelcor*, irregular.

Centre Belgo-Luxembourgeois d'Information de l'Acier 19.2
– CBLIA
[Belgian-Luxembourg Steel Information Centre]
Address: rue Montoyer 47, B-1040 Bruxelles, Belgium
Telephone: (02) 513 38 20
Telex: 21287
Affiliation: Belgian and Luxembourg steel industry
Enquiries to: Librarian, Mrs Mertes
Founded: 1932
Subject coverage: Information and promotion in the field of steel applications.
Library facilities: Open to all users (nc); no loan services; no reprographic services.
Information services: No information services.
Consultancy: No consultancy.
Publications: List on request.

Centre de Recherches Scientifiques et Techniques de l'Industrie des Fabrications Métalliques 19.3
– CRIF
[Metalworking Industry Scientific and Technical Research Centre]
Address: Rue des Drapiers 21, B-1050 Bruxelles, Belgium
Telephone: (02) 511 23 70
Telex: (02) 21708
Affiliation: FABRIMETAL
Enquiries to: Philippe Jamin; L. Demol
Founded: 1949
Subject coverage: Metal manufacturing industry; mechanical engineering; electrotechnical engineering and electronics; transportation material; plastics conversion; rubber.
Library facilities: Open to all users; no loan services; reprographic services (c).
Information services: Bibliographic services; literature searches; access to on-line information retrieval systems (c).
ESA: Lockheed; Spidel. SDI service - charge made.
The centre is a member of ABD (Belgian Association for Documentation).
Consultancy: Yes.

Institut Belge de la Soudure/ Belgisch Instituut voor Lastechniek 19.4
[Belgian Institute of Welding]
Address: Rue des Drapiers 21, B-1050 Bruxelles, Belgium
Telephone: (02) 512 28 92
Enquiries to: Technical secretary, Mrs Ritzen
Founded: 1942
Subject coverage: Welding; brazing; soldering.
Library facilities: Not open to all users (c - membership); no loan services; reprographic services (c).

Library holdings: About 500 bound volumes; 30 current periodicals.
Information services: No information services.
Consultancy: Telephone answers to technical queries; no writing of technical reports; no compilation of statistical trade data; no market surveys in technical areas.

BULGARIA

Union of Mining Engineering, Geology and Metallurgy 19.5

Address: Rakovski 108, 1000 Sofia, Bulgaria
Telephone: 87 57 27
Enquiries to: Secretary, B. Petkov

CZECHOSLOVAKIA

Výskumný Ústav Zváračský, Odvetové Informačné Stredisko 19.6
– VÚZ
[Welding Research Institute, Branch Information Centre]
Address: 135 Februárového vitazstva, PO Box 436, 894 23 Bratislava, Czechoslovakia
Telephone: 07 728
Telex: 093384 zvar c
Subject coverage: The institute provides an advisory service to the industry, trains qualified welders and welding technologists, and manufactures special welding electrodes, fluxes, wires, and solders in small quantities (pilot production); it is also responsible for standards, patents, and information services for welding in Czechoslovakia.
Publications: *Zváračské správy* (Welding News); *Zváramie* (Welding); research reports; year book; seminar proceedings; catalogues of research results.

Výzkumný Ústav Hutnictví Železa * 19.7
– VUHŽ
[Iron and Steel Research Institute]
Address: 739 51 Dobrá, Czechoslovakia
Telephone: 4381-5
Telex: 52691
Enquiries to: Head of information centre, Boris Skandera
VÚHŽ - INFORMETAL
Founded: 1948
Subject coverage: Metallurgy.
Library facilities: Open to all users; loan services (nc); reprographic services (nc).
All services are intended for Czechoslovak users only.

Library holdings: The library of the national information system for metallurgy contains: 900 000 bound volumes; 400 000 special materials; and subscriptions to 11 500 periodicals.
Information services: Bibliographic services (nc); translation services (nc); literature searches (nc); access to on-line information retrieval systems (c).
INFORMETAL acts as a clearing house and coordinates the transfer of information within the Czechoslovakian national metallurgical information system. In conjunction with other collaborating information centres, it collects and stores bibliographic information on metallurgy for dissemination to Czechoslovak scientists. INFORMETAL maintains a computer-readable data base of this information, and provides abstracts journals, computerized literature searching, and SDI services. INFORMETAL maintains a data base of 100 000 references on magnetic tape. Computerized literature searching and SDI services are offered using the INFORMETAL data base. Introduction of on-line access within Czechoslovakia to this data base is planned.
Consultancy: Yes (nc).

DENMARK

Korrosionscentralen 19.8
– ATV
[Danish Corrosion Centre]
Address: Park Allé 345, DK-2605 Brøndby, Denmark
Telephone: (2) 631100
Telex: 33 888
Facsimile: (2) 96 26 36
Affiliation: Academy of Technical Sciences
Enquiries to: Managing director, Erik Nielsen
Founded: 1965
Subject coverage: Consultancy, failure analysis, testing, research, inspection and control in the field of corrosion, protection against corrosion, and metallurgy.
Library facilities: Not open to all users; no loan services; reprographic services (c).
Information services: Bibliographic services (c); literature searches (c); no access to on-line information retrieval systems.
Consultancy: Yes (c).

Svejsecentralen 19.9
– SVC
[Danish Welding Institute]
Address: Parke Alle 345, DK-2605 Brøndby, Denmark
Telephone: (02) 968800
Telex: 33388 SVC DK
Affiliation: Akademiet for de Tekniske Videnskaber

Enquiries to: Librarian, Birte Ziegler
Founded: 1940
Subject coverage: Welding; non-destructive testing; mechanical testing.
Library facilities: Open to all users; loan services; reprographic services (c).
There is an information sheet on every important book, report, standard or article, divided into profiles in 15 subjects; subscription rates on request.
Information services: Literature searches.
Consultancy: Yes (c).
Publications: List on request.

FINLAND

Teknillinen Korkeakoulu 19.10
[Helsinki University of Technology, Department of Mining and Metallurgy]
Address: Vuorimiehentie 2,SF-02150 Espoo, Finland
Telephone: (90) 455 4122
Telex: 125161
Enquiries to: Librarian, Marja Lampi-Dmitriev
Subject coverage: Metal physics; metal forming and heat treatment; theoretical process metallurgy; corrosion prevention; applied electrochemistry; applied process metallurgy; economic geology; mining engineering; mineral engineering.
Library facilities: Open to all users.
Loan and reprographic services are available through the main library of the university.
Publications: List on request.

Valtion Teknillinen 19.11
Tutkimuskeskus,
Metallilaboratorio *
– VTT
[Technical Research Centre of Finland, Metals Laboratory]
Address: Metallimiehenkuja 6, SF-02150 Espoo 15, Finland
Telephone: (90) 4561
Telex: 122972 vttha sf
Enquiries to: Information officer, Marjatta Kimonen.
Technical Information Service, Technical Research Centre of Finland, Vuorimiehentie 5, SF-02150 Espoo, Finland.
Founded: 1942
Subject coverage: Metallic materials; destructive and non-destructive testing; quality assurance; welding; and machine and machine shop technology.
Library facilities: Not open to all users; loan services (c); no reprographic services.
Consultancy: Yes (c).

Vuorimiesyhdistys ry 19.12
[Finnish Association of Mining and Metallurgical Engineers]
Address: c/o Rautaruukki Oy, PB 217, SF-90101 Oulu 10, Finland

FRANCE

Centre d'Information des Fontes 19.13
Moulées
– CIFOM
[Iron Castings Information Centre]
Address: 2 rue de Bassano, 75783 Paris Cedex 16, France
Telephone: (1) 47 23 55 50
Affiliation: Syndicat Général des Fondeurs de France; Chambre Syndicale des Producteurs de Fontes
Enquiries to: Director
Founded: 1960
Subject coverage: Technical information on iron castings.
Library facilities: Open to all users; no loan services; reprographic services.
Information services: Bibliographic services.
Consultancy: Yes.

Centre d'Information du Plomb 19.14
– CIP
[Lead Information Centre]
Address: 1 avenue Albert Einstein, BP 106, 78191 Trappes Cedex, France
Telephone: (1) 30 62 05 83
Telex: 69 67 45 F MITEC
Affiliation: IMETAL
Enquiries to: General secretary, F. Lechenet; F. Wilmotte
Founded: 1956
Library facilities: Open to all users; loan services (nc); reprographic services (c).
Information services: Bibliographic services (c); translation services (c); literature searches (c).
According to the statutes, the Lead Information Centre is commissioned to give information and advice to lead users; to help users in their research, studies and experiments in order to improve the techniques of production, transformation and utilization of lead; and to promote the uses of lead.
Consultancy: Yes (nc).
Publications: Lists; reviews; films.

Centre Technique des Industries 19.15
de la Fonderie *
– CTIF
[Foundry Industries Technical Centre]
Address: 12 avenue Raphaël, 75016 Paris, France
Telephone: (1) 45 04 72 50

Telex: CETIF-PARIS 611054
Enquiries to: Head of information service, Y. Rosenfeld
Founded: 1945
Subject coverage: All technical matters relative to foundry (no economics).
Library facilities: Open to all users; no loan services; reprographic services (c).
Information services: Bibliographic services (c); translation services (c - the centre translates about 100 items per year but does not offer an automatic service); literature searches (c).
literature searches (c).

Institut de Soudure *   19.16
– IS
[French Institute of Welding]
Address: 32 boulevard de la Chapelle, 75880 Paris Cedex 18, France
Telephone: (1) 42 03 94 05
Telex: 210 335 F
Enquiries to: Head of documentation department, Noëlle Fauriol
Founded: 1931
Subject coverage: Welding and allied techniques (brazing, soldering, thermal cutting, hot spraying).
Library facilities: Open to all users; no loan services; reprographic services (c).
Information services: Bibliographic services (c); translation services (c); literature searches (c); access to on-line information retrieval systems (c).
The centre produces, in cooperation with CNRS, section 245 (formerly 745) entitled 'Welding and allied techniques' of the data base Pascal. A paper edition is also available: *Bulletin Signalétique 245 (ex 745)*.
The data bases searched are those available on the following hosts: ESA/IRS; Télésystèmes-Questel; Lockheed.
Consultancy: Consultancy (c).

GERMAN DEMOCRATIC REPUBLIC

Montanwissenschaftliche   19.17
Gesellschaft der DDR
[Coal and Steel Society of the GDR]
Address: Wallstrasse 68, DDR-1020 Berlin, German Democratic Republic
Telephone: 279 27 81
Library facilities: Library.

GERMAN FEDERAL REPUBLIC

Beratungsstelle für   19.18
Stahlverwendung
[Steel Information Centre]
Address: Kasernenstrasse 36, Postfach 16 11, D-4000 Düsseldorf 1, German Federal Republic
Telephone: (0211) 829 1
Telex: 08582286
Founded: 1927
Library facilities: Open to all users; no loan services; no reprographic services.
Consultancy: Yes.
Publications: List on request.

Bundesanstalt für   19.19
Materialprüfung *
– BAM
[Federal Institute for Materials Testing]
Address: Unter den Eichen 87, D-1000 Berlin 45, German Federal Republic
Telephone: (030) 81 04-1
Telex: 0183261 bamb d
Affiliation: Bundesministerium für Wirtschaft
Enquiries to: Head, Information and Public Relations, Dr H.-U. Mittmann
Founded: 1871
Subject coverage: Research in problems of materials science and materials testing for metals, construction materials and organics. Physical, chemical and biological testing; chemical safety engineering; development of new materials and of methods for non-destructive testing; special topics such as tribology, colour metrics, welding, nuclear power.
Library facilities: Open to all users; loan services (nc); reprographic services (c).
Information services: Bibliographic services (nc); no translation services; literature searches (c); access to on-line information retrieval systems (c).
BAM has seven documentation centres: Measurement of Mechanical Quantities (MmG), Colour, Rheology, Tribology, Welding Technology (DS), Non-destructive Testing (ZfP), Biological Materials Testing (BIO). Each documentation centre issues title bibliographies with abstracts and search literature according to individual profiles. For consultancy services there is access to many data bases (including in-house data bases on materials, technology, and chemistry). Information is supported by scientists from all BAM departments.
Consultancy: Yes (c).
Publications: *Amts- und Mitteilungsblatt*; bibliographies, abstracts and author indices in the fields listed above; reports; summary of published reference works; publications list.

Fachinformationszentrum Werkstoffe eV

19.20

– FIZ-W

[Materials Information Centre]

Address: Unter den Eichen 87, D-1000 Berlin 45, German Federal Republic

Telephone: (030) 830001-0

Telex: 186387 fizw d

Affiliation: Bundesanstalt für Materialprüfung

Enquiries to: Managing director, Dr P. Büttner

Founded: 1983

Subject coverage: Documentation and information services in materials sciences and applications.

Library facilities: Open to all users (c); no loan services; reprographic services (c).

Library facilities are supported by the resources of the Bundesanstalt für Materialprüfung.

Information services: Bibliographic services (c); no translation services; literature searches (c); access to on-line information retrieval systems (c).

FIZ-W maintains and works with four bibliographic data bases (trilingual, 400 000 records with abstracts, accessible through Euronet and DATEX P): SDIM1 (finished 1979) and SDIM2 (from 1979) - materials science of metals, metallurgy, processing and production techniques, testing and measuring techniques, environmental aspects, corrosion, economics; TRIBO - tribology, friction, wear and lubrication; RHEO - rheology.

Consultancy: No consultancy.

Forschungsgemeinschaft ZINK eV

19.21

[Zinc Research Association]

Address: Friedrich-Ebert-Strasse 37/39, D-4000 Düsseldorf 1, German Federal Republic

Telephone: (0211) 350867

Affiliation: Zinkberatung eV

Enquiries to: Dr H. Johnen

Founded: 1939

Subject coverage: Zinc; zinc alloy; wrought zinc.

Library facilities: Open to all users; no loan services; reprographic services.

Consultancy: Yes.

Publications: *Zink-Mitteilungsblatt zer Zinkeratung*; leaflets.

Gesellschaft Deutscher Metallhütten- und Bergleute

19.22

– GDMB

[German Society of Metallurgical and Mining Engineers]

Address: POB 210, D-3392 Clausthal-Zellerfeld, German Federal Republic

Telephone: (05323) 3438

Telex: 953828 tu clz d (GDMB)

Enquiries to: Secretary

Founded: 1912

Subject coverage: Exploration; mining; ore dressing; metallurgy (non-ferrous); analysis of metals; industrial minerals.

Library facilities: Open to all users; loan services (postage charged); reprographic services (c).

Information services: Bibliographic services (c); literature searches (c).

Consultancy: Yes (c).

Publications: *Erzmetall*, monthly.

Informationsstelle Edelstahl Rostfrei *

19.23

[Stainless Steel Information Office]

Address: Kasernenstrasse 36, Postfach 2807, D-4000 Düsseldorf 1, German Federal Republic

Telephone: (0211) 829528

Telex: 8584 482

Enquiries to: Manager, Dr Gerd Montug

Founded: 1958

Subject coverage: Production of literature on stainless steel.

Consultancy: Consultancy (nc).

Publications: List on request.

Verein Deutscher Eisenhüttenleute, Informationszentrum Stahl und Bücherei

19.24

– VDEh

[Association of German Iron and Steel Engineers, Steel Information Centre and Library]

Address: Sohnstrasse 65, D-4000 Düsseldorf 1, German Federal Republic

Telephone: (0211) 6707-0

Telex: 8587086 vstd

Facsimile: (0211) 6707310

Enquiries to: Manfred Toncourt

Founded: 1860

Subject coverage: Iron and steel metallurgy; iron and steel production; shaping, treating and testing of steel; properties of iron and steel; constructional features of installations; environmental affairs; statistics.

Library facilities: Open to all users (nc); loan services (for members of the VDEh only or inter-library loan service); reprographic services (c).

Library holdings: 100 000 bound volumes; 430 current periodicals.

Information services: Bibliographic services (c); translation services (c); literature searches (c); access to on-line information retrieval systems (c).

Access to STN, FIZ-Technik, Datastar. The centre's own factual data banks are STEELFACTS and PLANTFACTS.

Consultancy: Telephone answers to technical queries (occasional charge); writing of technical reports

(occasionally); no compilation of statistical trade data; no market surveys in technical areas.
Compilation of constructional features; profile services (literature).
Publications: List available.

Verein Deutscher Giessereifachleute, Dokumentationsstelle und Bibliothek
19.25

– VDG-DOK
[German Foundrymen's Association, Documentation Centre and Library]
Address: Postfach 8225, Sohnstrasse 70, D-4000 Düsseldorf 1, German Federal Republic
Telephone: (0211) 6871344
Telex: 8586885
Enquiries to: Head of documentation centre and library, R. Brand
Founded: 1953
Subject coverage: Foundry technology.
Library facilities: Open to all users; loan services (nc); reprographic services (c).
Direct loan service only for VDG members.
Information services: Bibliographic services (c); translation services (c); literature searches (c); access to on-line information retrieval systems (c).
Access to hosts - FIZ-Technik; STN; Datastar; INKA; ESA/IRS; SDC. Data bases - SDIM, DOMA, WAA, Metadex.
Consultancy: Yes (c).
Publications: *VDG-Merkblätter; VDG-Fachberichte; VDG-Ubersetzungen; VDG-Fachbibliographien; Giesserei-Literaturschau.*

HUNGARY

Kohó- és Gépipari Tudományos Informatikai és Ipargazdasági Központ
19.26

– KG-Informatik
[Central Institute for Scientific Information and Industrial Economics of the Metallurgy and Machine Industry of Hungary]
Address: Arany János útca 24, H-1051 Budapest V, Hungary
Telephone: (01) 317-960
Telex: 22-5262 INFO-H
Enquiries to: Library manager
Founded: 1951
Subject coverage: Metallurgy; machine industry; industrial economics.
Library facilities: Not open to all users; loan services; reprographic services.

Information services: Bibliographic services; literature searches.
Consultancy: Yes.

Országos Magyar Bányászati és Kohászati Egyesület
19.27

[Hungarian Mining and Metallurgical Society]
Address: Anker köz 1, H-1061 Budapest, Hungary
Telephone: 423 943
Telex: 22 5369
Library facilities: Library.

PRODINFORM Müszaki Tanácsadó Vállalat
19.28

[PRODINFORM Technical Consulting Company]
Address: PO Box 453, Munkácsy M u 16, Budapest 1372, Hungary
Telephone: 323 770
Telex: prod-4 227750
Enquiries to: Library manager, József Bauer
Founded: 1951
Subject coverage: Metallurgy; machine industry; electrotechnology; electronics; mining.
Library facilities: Not open to all users (nc); loan services (nc); reprographic services (c).
Library holdings: 37 355 bound volumes; 503 current periodicals; 20 000 reports.
Information services: Bibliographic services (c); translation services (c); literature searches (c); no access to on-line information retrieval systems.
Consultancy: Telephone answers to technical queries (nc); no writing of technical reports; no compilation of statistical trade data; market surveys in technical areas (c).
Publications: *Iparjogvédelmi tájékoztató* (Information on Industrial Rights); *Ipari Szabványositás* (Industrial Standardization); *Korszerü technológiák* (Modern Technologies); *Minőség és megbizhatóság* (Quality and Reliability); *Kohászati uj külföldi könyvbeszerzések,* metallurgy bibliography; *Hiradástechnikai és müszeripari uj külföldi könyvbeszerzések,* communications, electronics and measurements bibliography; *Szerszámgépipari uj külföldi könyvbeszerzések,* machine tools industry bibliography.

ITALY

Alluminio Italia SpA/Istituto Sperimentale dei Metalli Leggeri 19.29
– ISML
[Light Metals Research Institute]
Address: Via G. Fauser 4, 28100 Novara, Italy
Telephone: (0321) 24701
Enquiries to: Chief librarian, Paolo Macciotta
Founded: 1937
Subject coverage: Research, testing, and assistance in aluminium, alloys metallurgy, technology, and applications.
Library facilities: Open to all users; no loan services; reprographic services (c).
Library holdings: 16 000 books; 300 journals.
Information services: Bibliographic services (nc); no translation services; no literature searches.
Consultancy: Yes.

Associazione Italiana di Metallurgia * 19.30
[Italian Metallurgical Association]
Address: Piazzale Rodolfo Morandi 2, 20100 Milano, Italy
Telephone: (02) 78 49 85
Library facilities: Library.

Centro Sperimentale Metallurgico SpA * 19.31
Address: Via di Castel Romano, 00129 Roma, Italy
Enquiries to: G. Odone
Library holdings: Over 25 500 volumes.

NETHERLANDS

Metaalinstituut TNO 19.32
[Metal Research Institute TNO]
Address: Postbus 541, 7300 AM Apeldoorn, Netherlands
Telephone: (055) 773344
Telex: 36395 TNOAP
Facsimile: (055) 419837
Affiliation: Bouw en Metaal, Nederlandse Organisatie voor Toegepast-Natuurwetenschappelijk Onderzoek
Enquiries to: Senior information officer, A.H.M. ten Have
Founded: 1949
Subject coverage: Production engineering (robots, computer-aided design and manufacture); welding; machining; forming; tribology; testing (destructive and non-destructive); energy and environmental studies.
Library facilities: Open to all users (nc); loan services (very limited - nc); reprographic services (c).
Microfiche and microfilm reader-printer - c.

Library holdings: 12 500 bound volumes; 527 current periodicals; 8300 reports.
Information services: Bibliographic services (c); translation services (c); literature searches (c); access to on-line information retrieval systems (c).
Access to STN, InfoLine, Inis, ESA, Derwent, SDC, Lockheed, Télésystèmes. Subscriptions to many abstract journals over a long period, so that it is possible to do manual literature searches.
Consultancy: Telephone answers to technical queries (nc); writing of technical reports (c); no compilation of statistical trade data; no market surveys in technical areas.
Courses: Courses on different aspects of production engineering, machining and welding.

Technische Hogeschool Delft 19.33
[Delft University of Technology, Sub-department of Metallurgy]
Address: Rotterdamseweg 137, 2628 AL Delft, Netherlands
Library facilities: Library.

POLAND

Biblioteka Glòwna Akademii Gòrniczo-Hutniczej im Stanisława Staszica w Krákowie * 19.34
– AGH
[Central Library of the Stanislaus Staszic University of Mining and Metallurgy]
Address: Al Mickiewicza 30, 39-059 Kraków, Poland
Telephone: 34 14 04
Telex: 0322311
Enquiries to: Director, Maria Świerczyńska
Founded: 1919
Subject coverage: Foundry, geology, metallurgy, mineral industry, mining, materials technology.
Library facilities: Open to all users; loan services; reprographic services.
National and international interlibrary loan service and reprographic service of scientific materials held by the library.
Information services: Bibliographic services.
Publications: List of Foreign Acquisitions, quarterly; List of Current Periodicals in the Central Library and the Institution Libraries, annual.

Instytut Metali Niezelaznych 19.35
– IMN
[Non-Ferrous Metals Institute]
Address: Ulica Sowińskiego 5, 44-101 Gliwice, Poland
Telephone: 31 72 21
Telex: 036345
Enquiries to: W. Młodozeniec
BIONTE
Founded: 1952
Subject coverage: Mineral processing; extractive metallurgy; metal working; analytical chemistry of non-ferrous metals.
Library facilities: Open to all users (c); loan services (c); reprographic services (c).
Library holdings: 40 000 bound volumes; 600 current periodicals; 8260 reports.
Information services: Bibliographic services (c); translation services (c); literature searches (c); no access to on-line information retrieval systems.
Consultancy: Telephone answers to technical queries (c); writing of technical reports (c); no compilation of statistical trade data; no market surveys in technical areas.
Courses: Courses may be provided on request.
Publications: *Transactions of the Institute of Non-Ferrous Metals*; information bulletin.

Instytut Metalurgii Zelaza im St Staszica 19.36
[Ferrous Metallurgy Institute]
Address: Ulica K. Miarki 12-14, 44-101 Gliwice, Poland
Telephone: 032-31 40 51
Telex: 036363
Affiliation: Ministry of Metallurgy and Machine Industry
Enquiries to: Stanislaw Pajak
Founded: 1945
Subject coverage: Ferrous metallurgy and related fields.
Library facilities: Open to all users; loan services; reprographic services.
Information services: Bibliographic services; translation services; access to on-line information retrieval systems.
Publications: *Przeglad Dokumentacyjny Hutnictwa; Informacje Ekspresowe Hutnictwa Zelaza; Biuletyn Informacynjy Postep w Hutnictwie Zelaza; Informacje Hutnicze; Bibliografie Problemów Wezllowych; Tllumaczenia; Przeglad Dokumentacyjny Opisów Patentowych; Prace Instytutu Metalurgii Zelaza im St Staszica*, quarterly.

Instytut Odlewnictwa 19.37
[Foundry Research Institute]
Address: Ulica Zakopiańska 73, 30-418 Kraków, Poland
Telephone: 094-646 40
Telex: PL-10 0322431
Affiliation: Ministry of Heavy and Agricultural Machines Industry
Enquiries to: Head of Information, Technical, and Economic Centre
Founded: 1946
Subject coverage: Research and development in the domain of foundry practice and implementation of the results into Polish industry.
Library facilities: Open to all users; loan services; reprographic services (c).
Information services: Bibliographic services (c); translation services (c); literature searches (c).
The institute has its own information system automatized by means of the Polish computer ODRA 1325. It utilizes foreign information systems such as Pascal and Ismec not directly but through the Engineering College of Wroclaw.
Consultancy: Yes (c).
Publications: *Prace Instytutu Odlewnictwa; Zeszyty Specjalne Instytutu Odlewnictwa; Informacja Ekspresowa Instytutu Odlewnictwa.*

PORTUGAL

Instituto de Soldadura e Qualidade 19.38
– ISQ
[Welding and Quality Institute]
Address: Rua Tomás de Figueiredo, 16 -A 1500 Lisboa, Portugal
Telephone: 707582
Telex: 13415 INSOLD P
Enquiries to: Fernando Carvalheiro
Centro de Documentação e Informação
Founded: 1965
Library facilities: Not open to all users (nc); loan services (nc); reprographic services (nc).
Library holdings: 10 000 bound volumes; 80 current periodicals.
Information services: Bibliographic services (nc); translation services (nc); literature searches (nc); access to on-line information retrieval systems (nc).
Consultancy: Telephone answers to technical queries (nc); no writing of technical reports; no compilation of statistical trade data; no market surveys in technical areas.

Siderurgia Nacional EP 19.39
– SN
[National Iron and Steel Works]
Address: Rua Braamcamp 7, 1297 Lisboa Codex, Portugal
Telephone: 533151
Telex: 12229
Enquiries to: Simone Raposeiro
Centro de Documentação
Founded: 1954
Subject coverage: Iron and steel; metallurgy.
Library facilities: Not open to all users; no loan services; reprographic services (nc).
Library holdings: 12 760 bound volumes; 233 current periodicals; 9785 reports.
Information services: Bibliographic services (nc); no translation services; no literature searches; no access to on-line information retrieval systems.
Consultancy: Telephone answers to technical queries (nc); no writing of technical reports; no compilation of statistical trade data; no market surveys in technical areas.

ROMANIA

Institutul de Cercetări Metalurgic 19.40
– ICEMET
[Metallurgical Research Institute]
Address: Strada Mehadia 39, 77769 Bucureşti Sector 6, Romania
Telephone: (90) 493380
Telex: ICEM 11369
Affiliation: Ministerul Industriei Metalurgice
Enquiries to: General manager, Dr Ianeu Drăgan
Subject coverage: All fields relating to the iron and steel industries.
Publications: Culegere de Cercetări Metalurgice, annual, with abstracts in English, French, German and Russian.

SPAIN

Asociación Técnica Española de 19.41
Estudios Metalúrgicos
[Spanish Technical Association for Metallurgical Studies]
Address: Avenida Diagonal 647, Barcelona, Spain
Telephone: (93) 240 81 23

Centro Nacional de 19.42
Investigaciones Metalúrgicas
– CENIM
[National Centre for Metallurgical Research]
Address: Avenida Gregorio del Amo 8, 28040 Madrid, Spain
Telephone: (91) 2538900
Telex: 42182
Affiliation: Consejo Superior de Investigaciones Cientificas
Enquiries to: Head of information, J. Fernández
Founded: 1964
Library facilities: Open to all users (nc); no loan services; reprographic services (c).
Library holdings: 12 000 books on metallurgy; 500 current periodicals.
Information services: Bibliographic services (c); no translation services; literature searches (c); access to on-line information retrieval systems (c).
Consultancy: Telephone answers to technical queries (nc); b (c); no compilation of statistical trade data; no market surveys in technical areas.
Courses: Topics for courses include the following: welding engineers; analytical techniques; metallography.

SWEDEN

Institutet för Verkstadsteknisk 19.43
Forskning
– IVF
[Swedish Institute of Production Engineering Research]
Address: Mölndalsvägen 85, S-412 85 Göteborg, Sweden
Telephone: 031-81 01 80
Telex: 27872 IVFGBG S
Enquiries to: Librarian, Göran Lindskog, Information officer, K.G. Fridman
Founded: 1964
Subject coverage: Welding; cutting; forging; heat treatment; surfacing; robotics; electronics; sheet metal forming; quality control; quality engineering; metals; plastics; composites; FMS; working environment; ergonomics; computer-aided design and manufacture.
Library facilities: Open to all users; loan services; reprographic services.
Information services: No information services.
Consultancy: Yes (c).
Publications: IVF-resultat, in Swedish with English summary; IVF-skrift, in Swedish; publications catalogue. Publications are sold by: Sveriges Mekanförbund, Box 5506, S-114 85 Stockholm, Sweden, telephone 08-783 80 00.

Jernkontoret 19.44
[Swedish Ironmasters' Association]
Address: Box 1721, S-111 87 Stockholm, Sweden
Telephone: 08-22 46 20
Telex: 10165 JERNGRC
Facsimile: 08-10 70 91
Subject coverage: Ore-based metallurgy, refractory materials, ladle metallurgy, non-destructive testing; powder metallurgy; scrap-based metallurgy, casting and solidification, metallurgical history; heating and furnace technology, environmental control, energy; properties - high-alloy steels, analytical chemistry, non-ferrous metals; processing - flat products, processing - long products, properties - commercial steels, properties - carbon and low-alloy steels.
Jernkontoret is the administrative centre for joint research activities in which steelworks in Sweden, Denmark, Finland, and Norway participate. Almost all Nordic iron, steel, and copper companies take part. Amongst the membership there are also suppliers to the metallurgical industries of raw materials, consumables, plant and equipment. The aim of the association is coordination of technical research, coupled with optimal utilization of resources, resulting in reduced cost to member companies.

Nordisk Förzinkningsförening 19.45
[Nordic Galvanizing Association]
Address: Kungsgatan 37 4 tr, S-111 56 Stockholm, Sweden
Telephone: 08-20 80 38
Affiliation: European General Galvanizers Association
Enquiries to: Managing director, Rune Thomas
Founded: 1950
Subject coverage: Use of zinc as corrosion protection on iron and steel.
Library facilities: Open to all users (nc - personal search or reading in the library cannot be allowed due to lack of space); loan services (nc); reprographic services (nc).
Information services: Bibliographic services (nc); no translation services; literature searches (nc); no access to on-line information retrieval systems.
Consultancy: Telephone answers to technical queries (nc); no writing of technical reports; no compilation of statistical trade data; no market surveys in technical areas.

Stålbyggnadsinstitutet 19.46
[Swedish Institute of Steel Construction]
Address: Drottning Kristinas väg 48, S-114 28 Stockholm, Sweden
Telephone: 08-24 29 80
Telex: Steelconstruc Fotex S 12442 or 12443
Enquiries to: Technical director, Lars Wallin
Founded: 1967
Subject coverage: Research, development, and information within the field of steel construction.
Library facilities: No library facilities.

Information services: No information services.
Consultancy: Consultancy (c).

Stiftelsen för Metallurgisk Forskning 19.47
[Metallurgical Research Foundation]
Address: PO Box 812, Arontorpsvägen 1, S-951 28 Luleå, Sweden
Telephone: 0920-556 40
Telex: 80482 MEFOS
Facsimile: 0920-558 32
Enquiries to: Managing director, Bertil Berg
Founded: 1963
Subject coverage: Process, metallurgical and metal working research in heavy pilot plant scale, consulting, engineering, process control.
Library facilities: No library facilities.
Information services: No bibliographic services; no translation services; no literature searches; access to on-line information retrieval systems (Steeldoc).
Consultancy: Telephone answers to technical queries (if enquiries are made by potential customers); writing of technical reports (as part of contract research for customers); no compilation of statistical trade data; no market surveys in technical areas.
Publications: Conference proceedings - details on request.

SWITZERLAND

Eisen-Bibliothek 19.48
[Iron Library]
Address: Klostergut Paradies, CH-8246 Langwiesen, Switzerland
Telephone: (053) 58063
Telex: 76222 gfsh CH For iron library
Affiliation: George Fischer Limited
Enquiries to: Scientific leader, Henry Lueling
Founded: 1948
Subject coverage: Technological history of iron and steel; metallurgy; metallography and testing of materials; mining arts and science; iron extraction and processing; basic science; industrial monographs.
Library facilities: Open to all users; no loan services; reprographic services (c).
Library holdings: The library contains over 30 000 volumes: catalogues according to international rules; catalogue of authors; catalogue of subjects (catchword-system); working rooms.
Information services: Bibliographic services (nc); no translation services; literature searches (c); no access to on-line information retrieval systems.
Consultancy: Yes (nc).

Schweizerische Aluminium AG 19.49
– ALUSUISSE
[Swiss Aluminium Limited]
Address: Research and Development, Bad Bahnhofstrasse 16, CH-8212 Neuhausen, Switzerland
Telephone: (053) 2 02 21
Telex: 896035 aluf ch
Enquiries to: Information officer, Dr H. Keller
Founded: 1888
Subject coverage: Aluminium (production, fabrication, end uses); chemical products and processes, plastics and composites.
Library facilities: Open to all users; loan services (nc); reprographic services (c).
Information services: Bibliographic services; no translation services; literature searches; access to on-line information retrieval systems.
Alusuisse belongs to the sponsor bodies of *World Aluminum Abstracts* a printed and on-line bibliographic information service, and is represented in the corresponding working panel. All services are restricted, used mainly as a customer service.
Consultancy: Yes.

UNITED KINGDOM

Aluminium Federation 19.50
– ALFED
Address: Broadway House, Calthorpe Road, Birmingham B15 1TN, UK
Telephone: (021) 455 0311
Telex: CHAMCOM B'HAM 338024 ALFED
Enquiries to: Librarian, D.J. Keevil
Founded: 1962
Subject coverage: Aluminium and its alloys.
Library facilities: Open to all users (by prior arrangement); loan services (c - postage); reprographic services (c).
The library is involved in the compilation and organization of *World Aluminum Abstracts* which is the key information source on the metal, with both printed and on-line versions.
Information services: Bibliographic services; no translation services; literature searches (members only); no access to on-line information retrieval systems.
Difficult technical enquiries are referred back to the industry. Enquiries by letter only (two copies required, one to R.E. Moult, the other to L.P. Matthews).
Consultancy: Consultancy (nc - limited).
Publications: List on request.

BCIRA 19.51
Address: Alvechurch, Birmingham B48 7QB, UK
Telephone: (0527) 66414
Telex: 337125
Enquiries to: Information manager, L.J. Stewardson
Founded: 1921

Subject coverage: All aspects of the technology of metal castings production including: environmental and working conditions; molten metal and mould and core production; improvement of foundry processes; casting design, application and choice of materials; quality control organization; operating economics and work simplification; design expansion and layout of foundries; production control and costing; application of computers to foundry operations. Advanced manufacturing technology and contract research and development.
Library facilities: Open to all users (charge to non-members); loan services (members only); reprographic services (c).
Library holdings: 10 000 bound volumes; 300 current periodicals; 1000 reports.
Information services: Bibliographic services (c); no translation services; literature searches (c); no access to on-line information retrieval systems.
Consultancy: Telephone answers to technical queries (nc); no writing of technical reports; no compilation of statistical trade data; market surveys in technical areas (c).

BNF Metals Technology Centre 19.52
Address: Grove Laboratories, Denchworth Road, Wantage, Oxfordshire OX12 9BJ, UK
Telephone: (02357) 2992
Telex: 837166
Enquiries to: Librarian, A. Greig; Information Officer
Founded: 1920
Subject coverage: Ferrous and non-ferrous metals.
Library facilities: Not open to all users; loan services (through BLLD); no reprographic services.
The library may be used by prior arrangement only.
Library holdings: 64 000 books and reports; 200 current periodicals.
Information services: Bibliographic services (c); no translation services; literature searches (c); access to on-line information retrieval systems (c).
Dialtech; Dialog.
Consultancy: Telephone answers to technical queries (c - but no charge to member companies); writing of technical reports (c); no compilation of statistical trade data; market surveys in technical areas.

Cadmium Association 19.53
– CA
Address: 34 Berkeley Square, London W1X 6AJ, UK
Telephone: (01) 499 8425
Telex: 261286
Enquiries to: Director of marketing and information services, Dr A.J. Wall
Subject coverage: All aspects of technology of cadmium and its applications - metal, alloys, and compounds.

Library facilities: Open to all users; loan services (nc); reprographic services (c).
Information services: Bibliographic services (c - depends on subject); translation services (c); literature searches (c - depends on subject); access to on-line information retrieval systems.
Information services operated jointly with those of Zinc Development Association and Lead Development Association.
All abstracted material from 1970 onwards (1955 onwards for materials on zinc coatings) is available on-line as the ZLC base through Pergamon InfoLine. ZDA library and abstracts service has on-line access to most other apppropriate data bases.
Consultancy: Yes (generally free).

Copper Development Association 19.54
– CDA
Address: Orchard House, Mutton Lane, Potters Bar, Hertfordshire EN6 3AP, UK
Telephone: (0707) 50711
Enquiries to: Information manager
Founded: 1933
Subject coverage: Information on copper and copper alloys and compounds.
Library facilities: No library facilities.
Information services: Commercial and technical information is available including sources of supply, property data, specifications, but not statistics or prices.
Consultancy: Telephone answers to technical queries (nc); writing of technical reports; no compilation of statistical trade data; no market surveys in technical areas.
Advice on availability and sources of supply of copper and copper alloys and compounds. Information on British and foreign specifications relating to copper and copper alloys.
Courses: Various seminars on copper and copper alloys relating to specific fields of application or alloy (irregular).
Publications: Range of technical publications includes *International Copper Information Bulletin*, giving details of recently published books, articles, papers, standards, etc, new products, forthcoming conferences (thrice yearly); *Focus* highlighting new developments in copper (biannual); full list on request.

Drop Forging Research 19.55
Association
– DFRA
Address: Shepherd Street, Sheffield S3 7BA, UK
Telephone: (0742) 27463
Enquiries to: Information officer
Founded: 1961
Subject coverage: Research on drop forging and related metal forming processes; reheating for forging;

environmental factors such as noise and ventilation.
Library facilities: Not open to all users (open mainly for use by members of the association); loan services (c); reprographic services (c).
Library facilities are limited to specialist works on metal forming, noise vibration energy furnaces, etc.
Information services: Bibliographic services (c); translation services (c); literature searches (c); no access to on-line information retrieval systems.
Services include: advice on plant, die materials and die design; factory energy surveys; advice on furnace plant; instrumentation; combustion analysis; furnace efficiency trials; noise service, including measurement and evaluation of noise, recommendation of control procedures; hearing protection; investigation of neighbourhood complaints of noise, etc.
Consultancy: Consultancy services offered on forging technology, plant, materials, design, dies, heat treatment, noise surveys, hearing protection, etc, and energy (c).

Institution of Mining and 19.56
Metallurgy
– IMM
Address: 44 Portland Place, London W1N 4BR, UK
Telephone: (01) 580 3802
Telex: 261410
Enquiries to: Librarian
Founded: 1892
Subject coverage: Economic geology, mining, mineral processing, and extractive metallurgy of non-ferrous and industrial minerals excluding coal.
Library facilities: Open to all users (non-members may be charged); loan services (nc); reprographic services (c).
Information services: Bibliographic services (c); no translation services; literature searches (c); access to on-line information retrieval systems (c).
Extensive and detail indexes to the international minerals industry literature available. On-line minerals industry bibliographic database, Immage, available.
Consultancy: No consultancy.
Publications: *IMM Abstracts*, 6 issues per year; microfiche version of 1892 - 1949 indexes available for sale in whole or part.

International Tin Research 19.57
Institute
– ITRI
Address: Kingston Lane, Uxbridge, Middlesex UB8 3PJ, UK
Telephone: (0895) 72406
Affiliation: International Tin Research Council
Enquiries to: Librarian, L. Hobbs
Founded: 1932
Subject coverage: Tin chemistry; metallurgy of tin based alloys and other alloys containing tin; tinplate

packaging; tin alloy coatings; tin containing solders.
Library facilities: Open to all users (notification of visit required - nc); loan services (nc); reprographic services (c).
Information services: Bibliographic services (nc); no translation services; literature searches (nc); no access to on-line information retrieval systems.
A nominal charge is made for subject bibliographies to cover the cost of reproduction.
Consultancy: Telephone answers to technical queries (nc); no writing of technical reports; no compilation of statistical trade data; no market surveys in technical areas.
Technical consultancy (nc). Consultancy services are free to industry and individuals who are using or considering using tin.

Lead Development Association 19.58
– LDA
Address: 34 Berkeley Square, London W1X 6AJ, UK
Telephone: (01) 499 8422
Telex: 261286
Enquiries to: Director of marketing and information services, Dr A.J. Wall
Founded: 1955
Subject coverage: All aspects of technology of lead and its applications - metals, alloys, and compound.
Library facilities: Open to all users; loan services (nc); reprographic services (c).
Information services: Bibliographic services (c - depends on subject); translation services (c); literature searches (c - depends on subject); access to on-line information retrieval systems.
Information services operated jointly with those of Zinc Development Association and Cadmium Association.
All abstracted material from 1970 onwards (1955 onwards for material on zinc coatings) is available on-line as the ZLC data base through Pergamon InfoLine. ZDA library and abstracts service has on-line access to most other appropriate data bases.
Consultancy: Consultancy (generally free).

Metals Information 19.59

Address: 1 Carlton House Terrace, London SW1 5DB, UK
Telephone: (01) 839 4071
Telex: 8814813
Affiliation: Metals Society; American Society for Metal
Enquiries to: Manager, W.G. Jackson
Founded: 1974
Subject coverage: Science, technology, and economics of metals production and use; physics and chemistry of metals and metallurgical processes; metallurgical information.

Information services: Bibliographic services (c); translation services (c); literature searches (c); access to on-line information retrieval systems (c).
Metals Information maintains 3 data bases: METADEX - the World of Metals literature, bibliographic; *World Aluminium Abstracts* - an international data base, bibliographic; Metals Datafile - the on-line source for alloy numerical data. International services - BISITS translations on microfiche.
Consultancy: Yes (c).
Publications: Data base user aids; specific processing and industry abstracts; metallurgical bibliographies; indexes, digests, and profiles; *The World Calendar*.

Metals Society 19.60
Address: 1 Carlton House Terrace, London SW1Y 5DB, UK
Telephone: (01) 839 4071
Telex: 8814813
Enquiries to: Librarian
Founded: 1974
Subject coverage: Metals, metallurgy, materials science and related subjects. The society was formed by the amalgamation of the Iron and Steel Institute and the Institute of Metals.
Library facilities: Open to all users (open to members; also non-members for a charge); loan services (nc - members only); reprographic services (c).
There is a quick reference and referral service for non-members.
Information services: Conventional information services. See entry under 'Metals Information' for on-line searching services, etc.
Publications: Publications list.

Steel Castings Research and 19.61
Trade Association
– SCRATA
Address: 5 East Bank Road, Sheffield, S2 3PT, UK
Telephone: (0742) 28647
Telex: 54281 SCRTA
Enquiries to: Manager of information and library services, John R. Whitehead
Founded: 1952
Subject coverage: Research into all aspects of steel foundry technology; development of steel foundry processes and equipment; commercial centre for United Kingdom steel foundry industry; certified test house.
Library facilities: Open to all users; loan services (c); reprographic services (c).
Largest dedicated steel foundry library in western world. Consultation by non-members by previous arrangement only. Includes comprehensive collection

of steel castings standard specifications (worldwide).
Information services: Bibliographic services (c); no translation services; literature searches (c); no access to on-line information retrieval systems.
Steel grade identification service - supply of chemical composition/mechanical property data from steel grade designation. Equivalent cast steels - information on comparable/equivalent grades in worldwide steel castings specifications.
Consultancy: No consultancy.
Publications: Bibliographic list; publications list.

Welding Institute 19.62

Address: Abington Hall, Abington, Cambridge CB1 6AL, UK
Telephone: (0223) 891162
Telex: 81183
Enquiries to: Head of Technical Enquiry Service, K.G. Richards; Librarian, J.M. Loader
Founded: 1968 (amalgamation of Institute of Welding and British Welding Research Association)
Subject coverage: Welding; brazing; soldering; thermal cutting; surfacing; metal spraying; inspection and non-destructive testing; welding design; fatigue; brittle fracture; corrosion; welding metallurgy; fabrication techniques; welded construction; quality control.
Library facilities: Open to all users (open to research (organizational) and professional (individual) members, other researchers considered on their merits); loan services (to members only and inter-library loans); reprographic services (to members only).
Library holdings: 5000 books; 500 current periodicals; 5000 pamphlets; 10 000 microforms.
Information services: Available to research and professional members only.
The institute operates Welding Institute data base, entitled 'Weldasearch', which is publicly available from Lockheed Dialog (over 70 000 abstracts covering 1967 onwards, annual input of approximately 5000 abstracts). Also, current awareness, bibliographies, and retrospective search services.

Zinc Development Association 19.63

– ZDA
Address: 34 Berkeley Square, London W1X 6AJ, UK
Telephone: (01) 499 6636
Telex: 261286
Enquiries to: Director of marketing and information services, Dr A.H. Wall
Founded: 1938
Subject coverage: All aspects of technology of zinc and its applications - metal, alloys, and compounds.
Library facilities: Open to all users; loan services (nc); reprographic services (c).
Information services: Bibliographic services (c); translation services (c); literature searches (c); access

to on-line information retrieval systems.
Charges may be made for bibliographic services and literature searches, depending on the subject.
The information services are operated jointly with those of the Lead Development Association and Cadmium Association.
All abstracted material from 1970 onwards (1955 onwards for material on zinc coatings) is available on-line as the ZLC data base through Pergamon InfoLine. ZDA library and abstracts service has on-line access to most other appropriate data bases.
Consultancy: Yes (technical advice is generally free).

YUGOSLAVIA

Institut za Varilstvo 19.64

– ZAVAR
[Welding Institute]
Address: Ptujska 19, 61000 Ljubljana, Yugoslavia
Telephone: (061) 346 061
Telex: 31444 yu zavar
Enquiries to: Director, Professor Pavel Štular Documentation Department
Founded: 1956
Subject coverage: Welding and allied processes.
Library facilities: Not open to all users; loan services (nc); reprographic services (c).
Library holdings: 5800 bound volumes; 83 current periodicals; 160 reports.
Information services: Bibliographic services (c); translation services (c); literature searches (c); no access to on-line information retrieval systems.
Consultancy: Telephone answers to technical queries (nc); writing of technical reports (c); no compilation of statistical trade data; no market surveys in technical areas.

Metalurški Institut Hasan Brkić, Zenica * 19.65

– MIZEN
[Hasan Brkić Institute of Metallurgy, Zenica]
Address: Matije Gupca 7, pp 221, 72000 Zenica SR BiH, Yugoslavia
Telephone: (072) 32 866
Telex: 43-125
Affiliation: SOUR Rudarsko-Metalurški Kombinat, Zenica
Enquiries to: Manager, Franciska Hikl-Kogoj Documentation Centre
Founded: 1961

Subject coverage: Metallurgy and connected branches.
Library facilities: Not open to all users; loan services (nc); reprographic services (c).
Library holdings: 40 000 bound volumes; 260 current periodicals; 1500 reports.
Information services: Bibliographic services (nc); translation services (c); literature searches (nc); no access to on-line information retrieval systems.
Consultancy: No consultancy.

20 NON-METALLIC MATERIALS

AUSTRIA

Österreichisches Institut für 20.1
Verpackungswesen an der
Wirtschaftuniversität Wien *
– ÖIV
[Austrian Institute of Packaging]
Address: Gumpendorferstrasse 6, A-1060 Wien, Austria
Telephone: (0222) 57 96 86/279
Enquiries to: Director, Dr Erich F. Ketzler
Founded: 1956
Subject coverage: All topics related to the field of packaging, packaging testing, and self-service (statistics, environmental impact due to packaging).
Library facilities: Open to all users; no loan services; reprographic services.
Information services: No bibliographic services; no translation services; literature searches; no access to on-line information retrieval systems.
Consultancy: Yes.

Österreichisches 20.2
Kunststoffinstitut
– ÖKI
[Austrian Plastics Institute]
Address: Arsenal, Objekt 213, Franz Grill Strasse 5, A-1030 Wien, Austria
Telephone: (0222) 78 16 01
Telex: 613222447
Affiliation: Austrian Research Institute for Chemistry and Technology
Enquiries to: Librarian, Reinhard Soboll
Founded: 1953
Subject coverage: Chemistry, technology, and testing of polymeric materials as well as their starting materials and additives.
Library facilities: Not open to all users; no loan services; reprographic services (nc).
Library holdings: About 1000 bound volumes; 50 current periodicals.

Information services: Bibliographic services (c); no translation services; literature searches (c); no access to on-line information retrieval systems.
Consultancy: Telephone answers to technical queries (nc); writing of technical reports (c); no compilation of statistical trade data; no market surveys in technical areas.
Courses: Training courses on processing and application of plastics (on request).

BELGIUM

Institut Belge de l'Emballage/ 20.3
Belgisch Verpakkingsinstituut
– IBE/BVI
[Belgian Packaging Institute]
Address: Rue Picard 15, B-1210 Bruxelles, Belgium
Telephone: (02) 427 25 83
Telex: 62514 ibebvi b
Enquiries to: J. Victor Machiels
Services Responsible
Founded: 1954
Subject coverage: Packaging - materials; legislation; regulations and standards.
Library facilities: Open to all users (c); no loan services; reprographic services.
Library holdings: About 1000 bound volumes; 60 current periodicals.
Information services: No bibliographic services; translation services (c); literature searches (c); no access to on-line information retrieval systems.
Consultancy: Telephone answers to technical queries (c); writing of technical reports (c); compilation of statistical trade data (c).
Courses: Seminars (in Dutch and French) on packaging.

Laboratoire de la Profession de l'Industrie des Vernis, Peintures, Mastics, Encres d'Imprimerie et Couleurs d'Art
20.4
– Labo IVP
[Professional Laboratory for the Varnish, Paints, Mastics, Painting Inks and Pigments Industries]
Address: Avenue Pierre Holoffe, B-1342 Limelette, Belgium
Telephone: (02) 653 09 86
Enquiries to: Secretary
Founded: 1957
Library facilities: Open to all users; no loan services; reprographic services (c).
Consultancy: Yes.

BULGARIA

Shoe, Leather, Fur and Leather Goods Research Institute *
20.5
Address: Kojarska Str 3, 1202 Sofia, Bulgaria
Telex: 22761 opirin bg
Affiliation: State Economic Association 'Pirin'
Enquiries to: Head of information department, Dr Maria Konstantinova
Founded: 1965
Subject coverage: Technology of the shoe, leather, fur and leather goods industry.
Library facilities: Open to all users; no loan services; reprographic services (c).
Information services: Bibliographic services (c); translation services (c); literature searches (c); no access to on-line information retrieval systems.
Consultancy: Consultancy (c).
Publications: Information bulletin, bi-monthly; technical economic indices book, annually.

CZECHOSLOVAKIA

Výzkumný Ústav Gumárenské a Plastikářské Technologie, Oddělení Vědeckých a Teknicko-Economických Informacía Technická Knihovna
20.6
– VÚGPT
[Rubber and Plastics Technology Research Institute, Technical Information Department and Technical Library]
Address: 764 22 Gottwaldov, Louky, Czechoslovakia
Telephone: 27631-5
Telex: 067 311 VÚGPT c
Founded: 1965

Subject coverage: Rubber and thermoplastics processing.
Library facilities: Open to all users; loan services; no reprographic services.
Information services: Bibliographic services; literature searches.

Výzkumný Ústav Kožedélný
20.7
– VUK
[Shoe and Leather Research Institute]
Address: 762 65 Gottwaldov 2, Czechoslovakia
Telephone: 067 231 51
Telex: 067 337 svit and 067 338 svit
Enquiries to: Director, Josef Horák
Founded: 1952
Library facilities: Open to all users; loan services; reprographic services.
Library holdings: The library covers mainly leather and shoemaking publications.
Information services: Bibliographic services; translation services (c); literature searches.
Consultancy: Yes.
Publications: Two monthly journals.

DENMARK

Garverforsøgsstationen
20.8
– Gf
[Leather Research Station]
Address: Gregersensvej, DK-2630 Taastrup, Denmark
Telephone: (02) 996611
Telex: 33 416 ti dk
Facsimile: (02) 99 54 36
Affiliation: Teknologisk Institut
Enquiries to: Librarian, W. Frendrup
Founded: 1885
Subject coverage: Research in leather production and use for the industry in Denmark, Finland, Norway, and Sweden.
Library facilities: Open to all users (nc); no loan services; reprographic services (c).
Library holdings: About 4000 bound volumes; 15 current periodicals.
Information services: No information services.
Consultancy: Yes (occasional charge).

Nordisk Forskningsinstitut for Maling og Trykfarver *
20.9
– NIF
[Scandinavian Paint and Printing Ink Research Institute]
Address: Agern Allé 3, DK-2970 Hørsholm, Denmark
Telephone: (02) 57 03 55
Enquiries to: Secretary, Helle Jensen
Founded: 1946

Subject coverage: Paint and printing ink.
Library facilities: Open to all users; loan services (nc); reprographic services (c).
Information services: Bibliographic services (c); no translation services; literature searches (c); access to on-line information retrieval systems (c).
Literature searches on-line, individual profiles. Hosts available: Dialog, Pergamon InfoLine, and Echo.
Consultancy: Yes (nc).
Publications: Bibliographic list.

FINLAND

Oy Keskuslaboratorio -   20.10
Centrallabortorium AB
[Finnish Pulp and Paper Research Institute, Technical Information Management]
Address: PO Box 136, SF-00101 Helsinki, Finland
Telephone: 90-460411
Telex: 12-1030 kcl sf
Enquiries to: Head of technical information management, Jorma Paakko
Founded: 1937
Subject coverage: Pulp, paper, board, printing, and packaging.
Library facilities: Open to all users; no loan services; reprographic services (photocopies - c).
Information services: Bibliographic services (c - weekly survey of periodicals); translation services (c - mainly from Finnish and Swedish into English and German); literature searches (c).

FRANCE

Centre Technique de l'Industrie   20.11
des Papiers Cartons et Celluloses
[Paper, Paperboard and Cellulose Industries Technical Centre]
Address: BP 7110, 38020 Grenoble Cedex, France
Telephone: 76 44 82 36
Telex: 980642 F
Enquiries to: Head of documentation service
Founded: 1965
Subject coverage: New processes of pulping; fibre modifications for papermaking; use of waste papers; research on liner board corrugating medium and corrugated board; automatization for paper machines; treatment processes for industrial effluents.
Library facilities: Open to all users; no loan services; reprographic services.
Information services: Bibliographic services (c); translation services (c); literature searches (c).
Publications: *Feuillet bibliographiques*, monthly.

Centre Technique du Cuir *   20.12
– CTC
[Leather Technical Centre]
Address: 181, avenue Jean Jaurès, BP 7001, 69342 Lyon Cedex 07, France
Telephone: 78 869 50 12
Telex: 340497 F
Enquiries to: Documentation head, Micheline Catcel
Founded: 1962
Subject coverage: Applied research and technical assistance for leather industries (tannery, shoe industry, leather goods industry).
Library facilities: Open to all users; no loan services; reprographic services (c).
Information services: Bibliographic services (c); no translation services; literature searches (c); access to on-line information retrieval systems (c).
Private data base - Infocuir; no direct access.
Consultancy: Yes (c).

Confédération des Industries   20.13
Céramiques de France
– CICF
[French Ceramic Industries Confederation]
Address: 44 rue Copernic, 75116 Paris, France
Telephone: (1) 45 00 18 56
Telex: 611913 CERAFRA
Enquiries to: Information officer/Librarian, Ms Rousseau
Founded: 1946
Subject coverage: Ceramic industries (ceramic raw materials, whiteware, heavy clay products, refractory materials, etc).
Library facilities: Open to all users (c); no loan services; reprographic services (c).
Library holdings: About 300 books; about 50 current periodicals.
Information services: Bibliographic services (c); no translation services; literature searches (c); no access to on-line information retrieval systems.
The CICF - Service de Documentation Technique takes the place of the Librarian/Information Service of the Institut de Céramique Française.
Consultancy: No consultancy.

Fédération des Fabricants de   20.14
Peintures et Vernis *
– FIPEC
[Paint and Varnish Makers Federation]
Address: ZI 'Petite Montagne', BP 1416, 91019 Evry, France
Telephone: 60 79 38 38
Telex: 600039F
Enquiries to: Mme Valero
Subject coverage: Paints and varnishes.

Library facilities: Not open to all users; loan services (c); reprographic services (c).

Information services: Bibliographic services (c); no translation services; no literature searches; no access to on-line information retrieval systems.

Consultancy: Consultancy (c).

Institut de Recherches sur le Caoutchouc 20.15

– IRCA

[Rubber Research Institute]

Address: 42 rue Scheffer, 75116 Paris, France

Telephone: (1) 47 04 32 15

Telex: 620 871

Affiliation: Centre de Coopération Internationale en Recherche Agronomique pour le Développement

Founded: 1936

Subject coverage: Raw material from solar energy; mineral nutrition; plant protection; exploitation techniques; degradation of rubber; new types of rubber; processing and producing rubber; collecting latex.

Publications: Annual report.

Institut du Verre 20.16

[Glass Institute]

Address: 34 rue Michel Ange, 75015 Paris, France

Telephone: (1) 46 51 45 68

Enquiries to: Head of documentation service, A. Sellin

Founded: 1945

Library facilities: Open to all users; no loan services; reprographic services (c).

Information services: Bibliographic services; translation services; literature searches; access to on-line information retrieval systems.

Consultancy: Yes (c).

GERMAN DEMOCRATIC REPUBLIC

Forschungsinstitut für Leder- und Kunstledertechnologie 20.17

– FILK

[Leather and Artificial Leather Technology Research Institute]

Address: Thälmannring 1, DDR-9200 Freiberg, German Democratic Republic

Telephone: 4241

Telex: 785 152 FILK dd

Affiliation: VEB Kombinat Kunstleder und Pelzverarbeitung

Enquiries to: Information officer

Founded: 1889

Library facilities: Open to all users; no loan services; reprographic services (c).

Information services: Bibliographic services (c); translation services (c - limited); literature searches (c); no access to on-line information retrieval systems.

Publications: *Informationsblatt für die Lederindustrie,* monthly.

GERMAN FEDERAL REPUBLIC

Bibliothek und Archiv Zellcheming 20.18

[Cellulose Chemistry Library and Archives]

Address: Alexanderstrasse 24, D-6100 Darmstadt, German Federal Republic

Telephone: (06151) 162277

Affiliation: Verein der Zellstoff- und Papier-Chemiker und -Ingenieure, Darmstadt

Enquiries to: Professor Thomas Krause; Librarian, Petra Hase

Founded: 1928

Subject coverage: Cellulose chemistry and technology; pulp and paper chemistry and technology.

Library facilities: Open to all users (nc); loan services (only in Germany - nc); reprographic services (c).

Library holdings: 5242 bound volumes; 62 current periodicals.

Information services: Bibliographic services (c); no translation services; literature searches (c); no access to on-line information retrieval systems.

Monthly quick-documentation service to subscribers (c).

Consultancy: Telephone answers to technical queries (nc); no writing of technical reports; no compilation of statistical trade data; no market surveys in technical areas.

Deutsche Forschungsgesellschaft für Druck- und Reproduktionstechnik eV 20.19

– FOGRA

[German Research Institute for Printing and Reproduction Techniques]

Address: Postfach 8004 69, Streitfeldstrasse 19, D-8000 München 80, German Federal Republic

Telephone: (089) 43182-18

Telex: 529122

Facsimile: FOGRA München

Enquiries to: W. Probst

Founded: 1951

Subject coverage: Printing and allied processes.

Library facilities: Open to all users; loan services (c); reprographic services (c).

Information services: Bibliographic services (c); literature searches (c).

Bibliographic services: *FOGRA- Literaturdienst* (monthly abstracting service); *FOGRA- Literaturprofil* (bi-monthly abstracting service in specific fields); *FOGRA-Patentschau* (patent abstracts, monthly).
Publications: List on request.

Deutsche Glastechnische Gesellschaft eV

20.20

– DGG
[German Society of Glass Technology]
Address: Mendelssohnstrasse 75-77, D-6000 Frankfurt am Main 1, German Federal Republic
Telephone: (069) 749088
Enquiries to: Secretary-general, Professor H.A. Schaeffer
Founded: 1922
Subject coverage: Heat economics; furnace construction; refraction materials; melting and forming of glass; technology of machinery; shaping and finishing of glass; pollution control.
Library facilities: Open to all users (nc); loan services (within West Germany only - c); reprographic services (c).
Library holdings: The library contains approximately 40 000 items, including 18 000 patent specifications and standards, 17 000 bound volumes, and 158 current periodicals. The documentation service analyses approximately 260 journals.
Information services: Bibliographic services; literature searches (c).
Consultancy: Telephone answers to technical queries (nc); no writing of technical reports; no market surveys in technical areas.
Publications: *Glastechnische Berichte*, monthly - includes statistical trade data.

Deutsches Kunststoff-Institut

20.21

– DKI
[German Plastics Institute]
Address: Schlossgartenstrasse 6 R, D-6100 Darmstadt, German Federal Republic
Telephone: (06151) 162106
Affiliation: Forschungsgesellschaft Kunststoffe eV
Enquiries to: Jutta Wierer
Founded: 1953
Subject coverage: Science and technology of polymers and plastics.
Library facilities: Open to all users; no loan services; reprographic services (c).
Information services: Bibliographic services (c); no translation services; literature searches (c); access to on-line information retrieval systems (c).
DKI literature data bank *Kunststoffe Kautschuk Fasern*, access by Euronet and DATEX P.
Consultancy: Telephone answers to technical queries (nc); no writing of technical reports; no compilation of statistical trade data; no market surveys in technical

areas.
Publications: *Literatur-Schnelldienst Kunststoffe Kautschuk Fasern* (abstract journal), monthly; thesaurus; list of journals on request.

Forschungsinstitut der Feuerfest-Industrie

20.22

[Refractory Industry Research Institute]
Address: An der Elisabethkirche 27, D-5300 Bonn 1, German Federal Republic
Telephone: (0228) 211051
Telex: 886533
Affiliation: Verband der Deutschen Feuerfest-Industrie
Founded: 1949
Subject coverage: Refractory raw materials and refractories; properties, testing, and behaviour of materials in practice; standardization.
Library facilities: Open to all users (nc); no loan services; reprographic services (c).
Library holdings: 52 current periodicals.
Information services: Bibliographic services; no translation services; literature searches; no access to on-line information retrieval systems.

Institut der Forschungsgemeinschaft für Technisches Glas *

20.23

– FTG
[Industrial Glass Research Institute]
Address: Postfach 1302, Ferdinand-Hotz-Strasse 6, D-6980 Wertheim, Baden-Württemberg, German Federal Republic
Telephone: (09342) 1033
Affiliation: Forschungsgemeinschaft für Technisches Glas
Enquiries to: Dr H.-H. Fahrenkrog
Founded: 1951
Subject coverage: Glass technology and improvement of laboratory glassware manufacture.
Library facilities: Open to all users; no loan services; reprographic services (c).
Information services: No bibliographic services; translation services (c); no literature searches; no access to on-line information retrieval systems.
Consultancy: Yes (c).

Papiertechnische Stiftung

20.24

– PTS
[Paper Technology Foundation]
Address: Hess-Strasse 130a, D-8000 München 40, German Federal Republic
Telephone: (089) 126001-46
Telex: 5 213 088 ptsd
Enquiries to: Head, Brigitte Bauer Information and Documentation
Founded: 1951

Subject coverage: Pulp, paper and board making and converting technology, materials testing; water and wastewater research and treatment.
Library facilities: Open to all users (nc); no loan services; reprographic services (c).
Library holdings: 3600 bound volumes, excluding bound journals; 85 current periodicals.
Information services: Bibliographic services (c); translation services (c - English/German); literature searches (c); access to on-line information retrieval systems (c).
Papertech bibliographic data base with German abstracts.
Consultancy: Telephone answers to technical queries (c); writing of technical reports (c); compilation of statistical trade data (c); market surveys in technical areas (c).
Courses: Details of courses on request.
Publications: Reports.

Westdeutsche Gerberschule 20.25
[West German Tannery School]
Address: Postfach 303, Erwin-Seizstrasse 9, D-7410 Reutlingen, German Federal Republic
Telephone: (07121) 40056
Telex: 729 868 wgrd
Enquiries to: Professor Pauckner
Founded: 1954
Library facilities: Not open to all users.
Consultancy: Consultancy.
Publications: List on request.

HUNGARY

Papiripari Vállalat 20.26
Kutatóintézete
[Paper Industry Research Institute]
Address: Postafioch 86, Budapest H-1751, Hungary
Telephone: 279-620
Telex: 22-4017
Enquiries to: Head, Zsuzsa Tarján
Technical Information Service
Founded: 1949
Subject coverage: Pulp and paper making and paper converting.
Library facilities: Open to all users; loan services; reprographic services.
Library holdings: 25 000 bound volumes; 129 current periodicals; 2250 reports.
Information services: Bibliographic services (nc); no translation services; literature searches (nc); access to on-line information retrieval systems (c).
Consultancy: No consultancy.

ITALY

Associazione Tecnica Italiana per 20.27
la Cellulosa e la Carta
-ATICELCA
[Italian Technical Association for the Pulp and Paper Industry]
Address: Via Sandro Botticelli 19, 20133 Milano, Italy
Telephone: (02) 295041
Enquiries to: Secretary
Founded: 1967
Library facilities: Open to all users; no loan services; reprographic services.
Consultancy: Yes.

Istituto di Ricerche Tecnologiche 20.28
per la Ceramica
-IRTEC
[Ceramic Technology Research Institute]
Address: Via Granarolo 6, 48018 Faenza, (Ravenna), Italy
Telephone: (0546) 46147
Parent: Consiglio Nazionale delle Ricerche
Enquiries to: Director, Professor P. Bisogno
Subject coverage: Silicon nitride preparation; solid electrolytes; alumina and mullite substrates; floor and wall tiles, bricks and refractories; bioceramics; mechanical properties of ceramics; chemical analysis of ceramics.
Publications: *Ceramurgia*, bimonthly; *Ceramics International* and *Ceramics International News*, both quarterly.

Laboratorio Prove Materie 20.29
Plastiche *
[Plastic Materials Laboratory]
Address: Piazza Leonardo da Vinci 32, 20133 Milano, Italy
Telephone: (02) 230879
Enquiries to: Dr Fabrizio Denini
Founded: 1953
Subject coverage: Polymeric materials - tests and research.
Library facilities: Open to all users; no loan services; no reprographic services.
Library holdings: The laboratory library has about 400 books specializing in polymers.
Information services: Not available.

Stazione Sperimentale del Vetro 20.30
[Glass Research Centre]
Address: Via Briati 10, Murano-Venezia, Italy
Telephone: (041) 739422
Telex: 431447 SPEVET
Facsimile: 30121

Affiliation: Associazione Italiana Biblioteche Unione Stampa Periodica Italiana
Enquiries to: Head of documentation service, Dr Antonio Tucci
Founded: 1954
Subject coverage: Physics and chemistry of silicate.
Library facilities: Open to all users (nc); no loan services; reprographic services (c).
Library holdings: 6000 bound volumes; 133 current periodicals.
Information services: Bibliographic services (c); translation services (c); literature searches (c); access to on-line information retrieval systems (c).
The centre produces a bibliographic data base on glass.
Consultancy: Telephone answers to technical queries (c); writing of technical reports (c); no compilation of statistical trade data; no market surveys in technical areas.
Publications: Reports.

Stazione Sperimentale per l'Industria delle Pelli e delle Materie Concianti *   20.31
– SSIP
[Hides and Tanning Materials Industry Experimental Station]
Address: Via Poggioreale 39, 80143 Napoli, Italy
Telephone: (081) 268322; 265576
Telex: 721160
Affiliation: UNIC (Italian Union for Leather Industry), Milan; AICC (Italian Association of Leather Chemists), Turin
Enquiries to: Librarian, M. Ognissanti
Founded: 1885
Subject coverage: Research and analysis in the field of leather, tanning materials; waste water and sludges removal or reutilization; energy recovery; waste utilization.
Library facilities: Not open to all users; no loan services; reprographic services (c).
Information services: Bibliographic services; translation services; literature searches; no access to on-line information retrieval systems.
Consultancy: Yes.

NETHERLANDS

Instituut TNO voor Verpakking   20.32
– IvV-TNO
[Packaging Research Institute TNO]
Address: Postbus 169, 2600 AD Delft, Netherlands
Telephone: (015) 569330
Telex: 38071 zptno nl

Affiliation: Nederlandse Organisatie voor Toegepast-Natuurwetenschappelijk Onderzoek
Enquiries to: Head, C. Sonneveld
Founded: 1946
Subject coverage: Industrial, retail, agricultural and horticultural packaging; packaging for dangerous goods; environmental studies; testing equipment.
Library facilities: Open to all users; loan services; reprographic services.
Information services: Bibliographic services (bi-weekly bibliography to subscribers); literature searches; access to on-line information retrieval systems.
Consultancy: Yes.

Instituut voor Leder en Schoenen TNO   20.33
– ILS-TNO
[Leather and Shoe Research Institute TNO]
Address: Postbus 135, 5140 AC Waalwijk, Netherlands
Telephone: (04160) 33255
Telex: 35083
Affiliation: Nederlandse Organisatie voor Toegepast-Natuurwetenschappelijk Onderzoek
Enquiries to: G.A. Verhoeven
Founded: 1912
Subject coverage: Technical and scientific research and development work for the leather and shoe, and other leather-consuming, industries.
Library facilities: Open to all users; loan services; reprographic services.
Information services: Bibliographic services; no translation services; literature searches; access to on-line information retrieval systems.
On-line information available from the TNO Centre for Technical and Scientific Information and Documentation (CID-TNO), postal address: Postbus 36, 2600 AA Delft; telex 38071 zptno nl; telephone (015) 569330.
Consultancy: Telephone answers to technical queries (nc); writing of technical reports (c); no compilation of statistical trade data; no market surveys in technical areas.

Kunststoffen- en Rubberinstituut TNO   20.34
– KRITNO
[Plastics and Rubber Institute TNO]
Address: Postbus 71, 2600 AB Delft, Netherlands
Telephone: (015) 569330
Telex: 38071 zptno nl
Affiliation: Nederlandse Organisatie voor Toegepast-Natuurwetenschappelijk Onderzoek
Enquiries to: Director, Dr L.C.E. Struik
Subject coverage: Raw materials; processing; manufacturing; biomedical applications of plastics and rubber; properties of plastics and rubber materials;

product testing; polymer chemistry; polymer physics.

Verfinstituut TNO 20.35
[Paint Research Institute TNO]
Address: Postbus 203, 2600 AE Delft, Netherlands
Telephone: (015) 569330
Telex: 31453 zptno nl
Affiliation: Nederlandse Organisatie voor Toegepast-Natuurwetenschappelijk Onderzoek
Enquiries to: Librarian, M van de Wetering
Founded: 1954
Subject coverage: Manufacture, properties, and uses of paints and related products.
Library facilities: Not open to all users; loan services (nc - limited); reprographic services (c).
The library is open for staff members from the paint industry and other scientific research organizations, when the librarian is on duty. Private persons on request. All visits by appointment only.
Information services: Bibliographic services; no translation services; literature searches (c); access to on-line information retrieval systems.
On-line information available from CID-TNO. Bibliographic services: exchanges between Dutch libraries.
Consultancy: Yes: information from existing knowledge with a charge for services taking longer than about half an hour.
Publications: List on request.

NORWAY

Institutt for Grafisk Forskning 20.36
– INGRAF
[Norwegian Research Institute for the Graphic Arts Industry]
Address: Boks 250, Vinderen, N-0319 Oslo 3, Norway
Telephone: (02) 14 00 90
Telex: 78171 forsk n
Affiliation: Norwegian printing, newspaper and paper industries
Enquiries to: Documentalist, Randi Corneliussen
Founded: 1970
Subject coverage: Graphic arts techniques; information services; research and development.
Library facilities: Open to all users; loan services (nc - not periodicals); reprographic services (c).
Information services: Bibliographic services (c); translation services (c); literature searches (c); no access to on-line information retrieval systems.
Data base containing own literature abstracting references (approximately 1000 references per year).
Consultancy: Yes (c).
Publications: INGRAF-LU (abstracting service, published monthly; contains 40-80 abstracts from

about 80 periodicals); *INGRAF-INFORMASJON ?R (information leaflet, concerning research and development activities and progress at home and abroad, new reports offered to members, and conferences and arrangements offered for member participation);* Instituttrapporter (reports on completed contract work); *Teknisk informasjon* (practical results of work carried out, written in non-scientific language).

Papirindustriens 20.37
Forskningsinstitutt
– PFI
[Pulp and Paper Research Institute]
Address: PB 250 Vinderen, Forskningsveien 3, Oslo 3, Norway
Telephone: (02) 14 00 90
Telex: 18171 Forsk N
Enquiries to: Librarian, Bep Ødegaard
Founded: 1923
Subject coverage: Pulp and paper research.
Library facilities: Open to all users; loan services (nc); reprographic services (c).
Library holdings: 145 current periodicals.
Information services: Bibliographic services; no translation services; literature searches; access to on-line information retrieval systems (c).
Consultancy: Telephone answers to technical queries; writing of technical reports.
Publications: Abstracts in Norwegian covering pulp and paper subjects.

POLAND

Centralny Ośrodek Badawczo- 20.38
Rozwojowy Opakowań
– COBRO
[Polish Packaging Research and Development Centre]
Address: Ulica Konstancińska 11, 02-942 Warszawa, Poland
Telephone: 422011
Telex: 812473 cobropl
Affiliation: Ministry of Materials Economy
Enquiries to: Director, Jan Lekszycki
Founded: 1974
Subject coverage: Packaging materials and packages; design of packing techniques; experimental implementation of packaging materials, packages and packing systems; standardization and information services in packaging.
Library facilities: Open to all users; loan services (c); reprographic services (c).
Library holdings: About 6500 books; 50 periodicals; 7500 standards; 4500 catalogues and prospectuses; 1000 reports.

Information services: Bibliographic services; no translation services; literature searches (c); access to on-line information retrieval systems (from 1988).

Literature information system is based on PIRA *Packaging Abstracts* received in the form of computer printouts, translated into Polish and published in Poland according to agreement concluded with PIRA, the Research Association for the Paper and Board, Printing and Packaging Industries, Leatherhead, United Kingdom.

Bibliographic services - *Packaging Information Survey*, 6 numbers per year each containing approximately 300 items; 100 copies printed and distributed to subscribers.

Consultancy: Consultancy services - charge on man/hour basis.

Publications: List on request.

Instytut Celulozowo-Papierniczy * 20.39
– ICP
[Pulp and Paper Research Institute]
Address: Ulica Gdańska 121, POB 300, 90-950 Łódź, Poland
Telephone: 365300
Telex: 884559 ICEPPL
Enquiries to: Head of Information Centre, Dr Józef Robowski
Founded: 1947
Subject coverage: Pulp and paper industry, including paper converting and paper and board packagings.
Library facilities: Open to all users; loan services (nc); reprographic services (c).
Information services: Bibliographic services (c); translation services (c); literature searches (c); no access to on-line information retrieval systems.
Consultancy: No consultancy.
Publications: *Informacja Ekspresowa, Celuloza i Papier* (Express Information, Cellulose and Paper); *Biuletyn Informacyjny, Celuloza i Papier* (Information Bulletin, Cellulose and Paper).

Instytut Przemysłu Skórzanego * 20.40
[Leather Research Institute]
Address: Ulica Zgierska 73, 90 960 Łódź, Poland
Telephone: (042) 576210
Telex: 886423 ipes pl
Enquiries to: Head, Mirosława Wieczorkowska
Founded: 1951
Subject coverage: Technology of leather, fur, and footwear industry; methods of assessment of auxiliary chemicals, raw materials for leather industry and leather products; environmental pollution.
Library facilities: Open to all users; loan services (nc); reprographic services (c).
Library holdings: The library includes: catalogues - alphabetical and UDC for books, periodicals and unpublished research works; manual systems; accessions lists; abstracts and reference journals;

indexes; microform reading apparatus.
Information services: Bibliographic services (c); translation services (c); literature searches (c); no access to on-line information retrieval systems.

Literature searches - by means of UDC and other catalogues, indexes and card files. Full literature data on leather problems and analytic-synthetic papers may be provided and if necessary completed with information given by experts working in the institute.

Consultancy: Yes (nc).
Publications: *Prace Instytutu Przemysłu Skórzanego - Prace IPS* (Works of the Leather Research Institute), annually; *Informacja Ekspresowa* (Current Review) - includes elaborations of foreign technical literature of the leather and shoe industry), 20 numbers per year; *Przeglad Dokumentacyjny* (Current Abstracts Review - includes abstracts of foreign technical and economic literature of the leather and shoe industry), 10 numbers per year.

Instytut Przemysłu Tworzyw i Farb * 20.41
– IPTiF
[Plastics and Paint Industry Institute]
Address: POB 210, Chorzowska 50, 44-100 Gliwice, Poland
Telephone: 319041 45
Telex: 036215
Enquiries to: Manager, J. Średniawa
Founded: 1950
Subject coverage: Plastics and paints.
Library facilities: Open to all users; loan services (nc); reprographic services (nc).
Consultancy: Yes (c).
Publications: *Biezaca Informacja Chemiczna, Seria: Farby i Lakiery* (Current Chemical Information, Series: Paints and Varnishes), monthly; *Biuletyn Informacyjny* (Information Bulletin), quarterly.

Instytut Szkła i Ceramiki 20.42
[Glass and Ceramics Institute]
Address: Ulica Postepu 9, 02-676 Warszawa, Poland
Telephone: 437421
Telex: ISC 81 23 73
Affiliation: Ministry of Chemistry and Light Industry
Enquiries to: Head, Dr Krystyna Szwejda
Scientific, Technical and Economic Information Branch Centre
Founded: 1952
Subject coverage: Basic research on glass and ceramics; thermal engineering of glass and ceramic furnaces; glass and ceramic raw materials - their ore dressing and application; model research on glass furnaces; refractories; glass technology; glassware forming; low-melting glasses-glass joints; ceramic bodies and ceramic technology; utilization of waste raw materials; research on colours and ceramic stains

and their production.
Library facilities: Not open to all users (nc); loan services (nc); reprographic services (c).
The library is open to members of technical libraries only.
Library holdings: 12 570 bound volumes; 4100 current periodicals; 3030 reports.
Information services: Bibliographic services (c); translation services (c); literature searches (c); no access to on-line information retrieval systems.
Consultancy: No telephone answers to technical queries; writing of technical reports (nc); no compilation of statistical trade data; no market surveys in technical areas.
Publications: Dictionary of Glass-Making.

Resortowy Ośrodek Informacji Naukowej, Technicznej i Ekonomicznej Przemysłu Lekkiego * 20.43
– ROINTE PL
[Departmental Centre of Scientific, Technological and Economic Information for Light Industry]
Address: Ulica Mickiewicza 20, 90-950 Łódź, Poland
Telephone: 364951
Telex: 885416
Enquiries to: Managing director, Wiesław Rapacki
Founded: 1962
Subject coverage: Textile, clothing, leather, and shoe industries.
Library facilities: Open to all users; reprographic services.
Information services: Bibliographic services.
The centre offers SDI services, abstracting and other services.
Consultancy: Yes (on information searches).
Publications: Informacja Ekspresowa (Express Information); Przeglad Dokumentacyjny (Documentation Review); Nabytki Biblioteczne (Library Acquisitions); Wykazy tematyczne projektów wynalazczych (Subject Registers of Rationalizing Projects).

PORTUGAL

Associação dos Industriais de Vidro de Embalagem * 20.44
– AIVE
[Glass Containers' Manufacturers' Association]
Address: Largo de Andaluz 16-1, 1000 Lisboa, Portugal
Telephone: (019) 54 98 10
Affiliation: Confederação da Industria Portuguesa
Enquiries to: Artur Schiappa de Carvalho

Founded: 1975
Subject coverage: Glass container industry.
Library facilities: Open to all users; loan services (nc); reprographic services (c).
Information services: Bibliographic services (c); translation services (c); no literature searches; no access to on-line information retrieval systems.
Consultancy: Yes (nc).

SPAIN

Asociación de Investigación de la Industria Gráfica * 20.45
– AIIG
[Graphic Industry Research Association]
Address: Apartado de Correos, 358, Plaza de los Santos Juanes s/n, Bilbao 5, Spain
Telephone: 94-433 68 51
Affiliation: Graphic Arts Industries
Enquiries to: Librarian
Founded: 1968
Subject coverage: Graphic arts and related subjects.
Library facilities: Not open to all users; no loan services; no reprographic services.
Services open generally for associate members only; non-members only by special permission. Charge made.
Information services: Bibliographic services (c); translation services (c); literature searches (c).
Bibliographic sources: books, periodicals and magazines from industry and manufacturers, technical associations, and research organizations.
Consultancy: Yes (c).

Asociación de Investigación Técnica de la Industria Papelera Española 20.46
– IPE
[Technical Research Association of the Spanish Paper Industry]
Address: Carretera La Coruña, km 7, PO Box 33.045, 28040 Madrid, Spain
Telephone: 91-207 09 77
Telex: 49313 AIPE E
Founded: 1963
Subject coverage: Pulp, paper, and allied industries.
Library facilities: Reprographic services (c).
The library is open for the association's members only, but written requests for access by non-members are considered.
Information services: Bibliographic services (c); translation services (c); literature searches (c).
Consultancy: Yes (nc).
Publications: Publications list.

Asociación Española de Fabricantes de Pinturas *

20.47

– ASEFAPI
[Paintmakers Association of Spain]
Address: Juan Ramon Jimenez 2 - 2, Madrid-16, Spain
Telephone: 91-259 37 96
Affiliation: Federación Empresarial de la Industria Química Española, FEIQUE
Enquiries to: Dr Juan Jose Fortea Laguna
Founded: 1977
Subject coverage: Paints.
Library facilities: Open to all users; loan services (nc); reprographic services (c).
Information services: Bibliographic services (nc); translation services (c); literature searches (c); no access to on-line information retrieval systems.
Consultancy: Yes (nc).
Publications: List available on request.

Departamento de Curtidos del CSIC

20.48

[Leather Department of the CSIC]
Address: Jorge Girona Salgado s/n, Edificio Juan de la Cierva, Barcelona 34, Spain
Telephone: 93-204 06 00
Affiliation: Consejo Superior de Investigaciones Científicas
Enquiries to: Director
Founded: 1956
Subject coverage: Leather technology.
Library facilities: Open to all users; no loan services; reprographic services.
Information services: Bibliographic services; literature searches.
Consultancy: Yes.

Instituto de Plásticos y Caucho

20.49

– IPC
[Plastics and Rubber Institute]
Address: Juan de la Cierva 3, 28006 Madrid 6, Spain
Telephone: 91-262 29 00
Affiliation: Consejo Superior de Investigaciones Científicas
Enquiries to: Director, Dr J. Fontan
Subject coverage: Polymer analysis; polymer physics; plastics technology; macromolecular chemistry; rubber technology.
Publications: *Revista de Plásticos Modernos*, monthly.

Sociedad Española de Cerámica y Vidrio

20.50

– SECV
[Spanish Ceramic and Glass Society]
Address: Carretera de Valencia, km 24 300, Arganda del Rey, Madrid, Spain
Telephone: 91-871 18 00

Enquiries to: General secretary, Professor José Mariá Fernández Navarro
Founded: 1960
Subject coverage: Art and design; basic science; enamel on metal; brick/tiles; whitewares; refractories; glass.
Library facilities: Open to all users; loan services (c); reprographic services (c).
Information services: Bibliographic services (c); translation services (c); literature searches (c); access to on-line information retrieval systems (c).
Consultancy: Telephone answers to technical queries (nc); writing of technical reports (nc); compilation of statistical trade data (nc); market surveys in technical areas (nc).

SWEDEN

Svenska Förpackningsforskningsinstitutet

20.51

– PACKFORSK
[Swedish Packaging Research Institute]
Address: Torshamnsgatan 24, PO Box 9, S-163 93 Spånga, Sweden
Telephone: 08-752 02 80
Enquiries to: Information officer, Stig G. Bergstedt
Founded: 1954
Library facilities: Not open to all users; loan services (nc); reprographic services (c).
Library holdings: 2400 bound volumes; 180 current periodicals; 3500 reports.
Information services: Bibliographic services (c); translation services (c); literature searches (c); access to on-line information retrieval systems (c).
PIRA *Packaging Abstracts* ; PAKLEGIS.
Consultancy: Telephone answers to technical queries (nc); writing of technical reports (c); compilation of statistical trade data (c); market surveys in technical areas (c).

Svenska Silikatforskningsinstitutet *

20.52

[Swedish Institute for Silicate Research]
Address: PO Box 5403, S-402 29 Göteborg, Sweden
Telephone: 031-16 23 18
Enquiries to: Ulla-Britt Jigholm
Founded: 1946
Subject coverage: Glass and ceramic science.
Library facilities: Not open to all users; no loan services; reprographic services (c).
Information services: Bibliographic services (c); access to on-line information retrieval systems (through the main library of Chalmers University of Technology).
Consultancy: Yes.

Sveriges Gummitekniska Förening
20.53

[Swedish Institution of Rubber Technology]
Address: PO Box 11107, S-161 11 Bromma, Sweden
Telephone: 08-98 82 35; 98 70 10
Telex: 11481 IGUM S
Enquiries to: Secretary-general, P. Normelli
Founded: 1950
Library facilities: There is no library; all loans are transferred to the Royal Technology University Library, S-10044 Stockholm.
Information services: Translation services (c).
Consultancy: Yes (c).
Publications: List on request.

SWITZERLAND

Eidgenössische Materialprüfungs- und Versuchsanstalt für Industrie, Bauwesen und Gewerbe *
20.54

– EMPA
[Swiss Federal Laboratory for Materials Testing and Research]
Address: Unterstrasse 11, Postfach 977, CH-9001 St Gallen, Switzerland
Telephone: (071) 20 91 41
Telex: 71278
Enquiries to: Librarian, Gertrud Luterbach EMPA-Bibliothek
Founded: 1880
Subject coverage: Textiles, leather; printing; paper, packaging; oils, fats, detergents; wood preservation.
Library facilities: Open to all users; loan services (c); reprographic services (c).
Information services: Literature searches (c); access to on-line information retrieval systems (c).
The library has on-line access to the following: Dialog; Chemical Information System; Pergamon InfoLine; DataStar; Télésystèmes Questel; DIMDI.

UNITED KINGDOM

British Ceramic Research Limited
20.55

– Ceram Research
Address: Queens Road, Penkhull, Stoke-on-Trent ST4 7LQ, UK
Telephone: (0782) 45431
Telex: 36228
Facsimile: (0782) 412331
Enquiries to: Marketing officer, J.A. Farrell
Founded: 1920

Subject coverage: Contract research, membership programmes for ceramic technology in whitewares, refractories, advanced ceramics, building materials, industrial hygiene and pollution.
Library facilities: Not open to all users (access is only to staff of member companies - nc); loan services (members only - nc); reprographic services (copy charge only).
Literature and market surveys (c).
Library holdings: 40 000 bound volumes; 400 current periodicals; 40 000 reports.
Information services: Bibliographic services (c); translation services (c); literature searches (c); access to on-line information retrieval systems (c).
Charges to subscribing member companies are reduced or waived.
Consultancy: Telephone answers to technical queries (nc); no writing of technical reports; compilation of statistical trade data (c); market surveys in technical areas (c).

British Ceramic Society
20.56

Address: Shelton House, Stoke Road, Shelton, Stoke-on-Trent ST4 2DR, UK
Telephone: (0782) 23116/7
Enquiries to: Assistant secretary, S. Buchanan
Founded: 1900
Subject coverage: Science and technology of ceramics.
Library facilities: Not open to all users; loan services (nc); reprographic services (c).
Information services: No information services.
Consultancy: Yes (nc).

British Glass Industry Research Association
20.57

– BGIRA
Address: Northumberland Road, Sheffield S10 2UA, UK
Telephone: (0742) 686201
Telex: 547676 Chamco G for BGIRA
Enquiries to: Information officer, P.J. Doyle
Founded: 1954
Library facilities: Not open to all users; loan services (members only - nc); reprographic services (c).
Information services: Bibliographic services (c); translation services (c); literature searches (c).
Access to on-line information retrieval systems will soon be available - c.
Consultancy: Telephone answers to technical queries (occasional charge); writing of technical reports (c); compilation of statistical trade data (c); no market surveys in technical areas.
Courses: Courses are arranged, but not on a regular basis.

British Leather Confederation 20.58

Address: Leather Trade House, Kings Park Road, Moulton Park, Northampton NN3 1JD, UK
Telephone: (0604) 494131
Telex: 317124 CORIUM G
Enquiries to: Director, Dr R.L. Sykes; Information officer, I.C. Waring
Founded: 1920
Subject coverage: Leather research.
Library facilities: Open to all users (by appointment - nc); loan services (nc); no reprographic services.
Library holdings: 2500 bound volumes; 125 current periodicals; 250 internal reports.
Information services: No information services.
Consultancy: No consultancy.

Paint Research Association 20.59

Address: Waldegrave Road, Teddington, Middlesex TW11 8LD, UK
Telephone: (01) 977 4427
Telex: 928720
Enquiries to: Librarian, S.C. Haworth
Founded: 1926
Subject coverage: Surface coatings science and technology, including paints, pigments, oils, resins, polymers, additives; painting; health hazards; safety and environmental regulations; paint defects; corrosion and fouling; chemical and physical tests and analysis; microbiology; industrial commercial information.
Library facilities: Open to all users (by appointment only - c); loan services (members only); reprographic services (c).
Information services: Bibliographic services (c); translation services (c); literature searches (c); access to on-line information retrieval systems (c).

Plastics and Rubber Institute 20.60
– PRI
Address: 11 Hobart Place, London SW1W 0HL, UK
Telephone: (01) 245 9555
Telex: 912881 CWUKTX G (Marked ATTN PRI
Enquiries to: Secretary general, G.W. Stockdale; Managing editor, Patricia Battams
Founded: 1975
Subject coverage: Plastics, rubber and allied topics.
Library facilities: Open to all users (nc); no loan services; reprographic services (c).
Information services: No information services.
Consultancy: Telephone answers to technical queries (nc); no writing of technical reports; no compilation of statistical trade data; no market surveys in technical areas.

Rapra Technology Limited 20.61
Address: Shawbury, Shrewsbury, Shropshire SY4 4NR, UK
Telephone: (0939) 250383
Telex: 35134

Enquiries to: Manager, Paul Cantrill. Information Centre
Founded: 1919
Subject coverage: All aspects of information relating to rubbers and plastics (including their synthesis, properties and processing), and to the manufacture of and markets for rubber and plastic products.
Library facilities: Open to all users (c); loan services (c); reprographic services (c).
Members of Rapra can use the library free of charge.
Library holdings: About 20 000 bound volumes; 410 current periodicals (from 30 countries); numerous reports.
Rapra's library contains the world's most comprehensive collection of data on the rubber and plastics industries. Holdings include conference papers, specifications and standards, government reports, trade literature and directories.
Information services: Bibliographic services (c); translation services (Hungarian, Polish, Czech, Japanese, French, Spanish, Italian, German, Russian - c); literature searches (c); access to on-line information retrieval systems (c).
Rapra data bases and data bases held by Pergamon InfoLine, Dialog, SDC, Textline, Datastar and IRS.
Consultancy: Telephone answers to technical queries (c); writing of technical reports (c); compilation of statistical trade data (c); market surveys in technical areas (c).
Consultancy services cover all aspects of the development and manufacture of rubber and plastic products, including products design, material selection, physical testing, chemical analysis, prototype manufacture and marketing research.

Research Association for the 20.62
Paper and Board, Printing and
Packaging Industries
– PIRA
Address: Randalls Road, Leatherhead, Surrey KT22 7RU, UK
Telephone: (0372) 376161
Telex: 929810
Enquiries to: Librarian, Marjorie Thompson
Founded: 1930
Subject coverage: Printing, packaging, paper and board, electronic publishing, management and marketing.
Library facilities: Open to all users (free to members, charge to non-members); loan services (c - to members only); reprographic services (c).
Information services: Bibliographic services (c); translation services (c); literature searches (c); access

to on-line information retrieval systems (c).
PIRA produces four data bases available on Pergamon InfoLine: PIRA (paper and board printing, and packaging); MMA (management and marketing); EPA (electronic publishing); PAKLEGIS (packaging legislation).
Publications: Five monthly abstract journals: *Paper and Board Abstracts; International Packaging Abstracts; Printing Abstracts; Electronic Publishing Abstracts; Management and Marketing Abstracts* .

SATRA Footwear Technology Centre 20.63

Address: SATRA House, Rockingham Road, Kettering, Northamptonshire NN16 9JH, UK
Telephone: (0536) 516318
Telex: 34323
Enquiries to: Head, S.R. Swailes.
Information Centre
Founded: 1919
Subject coverage: Footwear; leather; plastics; rubber testing; management information systems; management consulting.
Library facilities: Not open to all users (nc); loan services (c); reprographic services (c).
Products and services are available to member companies only.
Library holdings: 4000 bound volumes; 200 current periodicals; 1000 reports.
Information services: Bibliographic services (c); translation services (c); literature searches (c); access to on-line information retrieval systems (c).
Services are available to member companies only.
Consultancy: Telephone answers to technical queries (c); writing of technical reports (c); compilation of statistical trade data (c); market surveys in technical areas (c).
Services are available to members only.
Courses: Various training courses on all aspects of footwear technology.

Yarsley Technical Centre Limited 20.64

Address: Trowers Way, Redhill, Surrey RH1 2JN, UK
Telephone: (0737) 65070
Telex: 8951511
Affiliation: Fulmer Research Institute Limited
Enquiries to: Marketing services manager, R.C. Watkins
Founded: 1931
Subject coverage: Consultancy on the development, testing, design and evaluation of all non-metallic materials and the products and components manufactured from them; specialization: chemistry technology and applications of polymers and plastics materials.

Library facilities: No library facilities.
Information services: No information services.
Consultancy: Telephone answers to technical queries (occasional charge); writing of technical reports (c); no compilation of statistical trade data; market surveys in technical areas (c).
Polymer design; contract research, design, development and testing of non-metallic materials - c.

YUGOSLAVIA

Inštitut za Celulozo in Papir Ljubljana 20.65
– ICP
[Pulp and Paper Institute, Ljubljana]
Address: PO Box 366/VII, Bogišičeva 8, Ljubljana, Yugoslavia
Telephone: (061) 221-192; 221-140
Enquiries to: Director, Professor Božo Iglič
Founded: 1947
Subject coverage: Aptitude of wood and plant wastes for manufacture of pulp; problems concerning pulp and paper manufacture and qualities; auxiliary materials; paper coating and printing problems; water and wastewater problems.
Library facilities: Open to all users; loan services (nc - generally to staff only); reprographic services (c).
Information services: Bibliographic services (c); translation services (c); literature searches (c); access to on-line information retrieval systems (c).
Bibliographic services: besides author and subject charts (Archiv Zellcheming) also Abstract Bulletin of the Institute of Paper Chemistry and FOGRA Literature service; bibliography of the researchers. The institute does not yet maintain its own data bank, but has the possibility of access elsewhere.
Consultancy: Yes (c - generally).

Jugokeramika 20.66
Address: Srebrnajak 169, 41040 Zagreb, Yugoslavia
Telephone: (041) 210 010
Telex: 21-286
Enquiries to: Librarian
Founded: 1963
Subject coverage: Ceramics; porcelain; refractory materials.
Library facilities: Open to all users; loan services; reprographic services (c).
Micro-reading facilities.
Information services: Literature searches; access to on-line information retrieval systems.
Consultancy: Yes.

21 SPACE SCIENCES

AUSTRIA

Österreichische Gesellschaft für 21.1
Sonnenenergie und
Weltraumfragen Gesellschaft
mbH *
[Austrian Solar and Space Agency]
Address: Garnisongasse 7, A-1090 Wien, Austria
Telephone: (0222) 4381770
Telex: 116560
Enquiries to: Helmut Hummer
Founded: 1972
Subject coverage: Solar energy, space research and technology.
Information services: No bibliographic services; no translation services; literature searches; access to on-line information retrieval systems.
Quest (IRS of ESA); Dialog (Lockheed).
Consultancy: Yes.

BELGIUM

Institut d'Aéronomie Spatiale de 21.2
Belgique *
– IASB-BIRA
[Space Aeronomy Institute]
Address: Avenue Circulaire 3, B-1180 Bruxelles , Belgium
Telephone: (02) 374 27 28
Telex: 21563
Affiliation: Ministère de l'Éducation Nationale
Enquiries to: Director
Founded: 1964
Subject coverage: Research in space aerdonomy (upper atmosphere, space, planetary atmosphere, terrestrial-solar relationships).
Library facilities: Open to all users; no loan services; reprographic services (c).
Information services: Bibliographic services (nc).

Consultancy: Yes (nc).

Institut von Karman de 21.3
Dynamique des Fluides *
– IVK
[Von Karman Institute for Fluid Dynamics]
Address: Chaussée de Waterloo 72, B-1640 Rhode Saint Genèse, Belgium
Telephone: (02) 358 19 01
Enquiries to: Librarian, N.V. Toubeau
Founded: 1956
Subject coverage: Aerodynamics; physics; wind effects on buildings and structures, laser-doppler velocimetry, computational fluid dynamics, pollution; turbomachinery; axial and centrifugal compressors.
Library facilities: Open to all users; loan services; reprographic services (c).
The institute is the Belgian deposit library for NASA publications.
Information services: Not available.

Observatoire Royal de Belgique 21.4
– ORB
[Royal Observatory of Belgium]
Address: Avenue Circulaire 3, B-1180 Bruxelles, Belgium
Telephone: (02) 374 38 01/02
Telex: 21565
Affiliation: Ministère de l'Éducation Nationale
Enquiries to: Librarian, Mr Dale.
Service de la Bibliothèque
Founded: 1834
Subject coverage: Positional astronomy and geodynamics; astrometry and celestial mechanics; astrophysics; radioastronomy and solar physics; stellar statistics.
Library facilities: Not open to all users; loan services (nc); reprographic services (c).
The loan of books and periodicals is authorized to any researcher after a written request mentioning the aim of the research has been accepted by the director of the observatory and after payment of a deposit.

Library holdings: About 125 000 bibliographical units; 4226 periodicals.
Information services: Bibliographic services (nc - by written request to the director); no translation services; no literature searches; no access to on-line information retrieval systems.
Consultancy: Consultancy (nc - by written request to the director).
Publications: Catalogue of the library's periodicals.

BULGARIA

Bulgarian Astronautical Society 21.5
Address: Tolbukhin 18, Sofia, Bulgaria
Library facilities: Library.

CZECHOSLOVAKIA

Astronomický Ústav ČSAV, 21.6
Středisko Vědeckých Informací
[Astronomical Institute, Scientific Information Centre]
Address: 251 65 Ondřejov, Czechoslovakia
Telephone: (0204) 999 321
Telex: 121579
Affiliation: Czechoslovak Academy of Sciences
Enquiries to: Manager, Dr Josef Zavřel
Subject coverage: Astronomy and related sciences.
Library facilities: Open to all users (nc); loan services (nc); reprographic services (nc).
Library holdings: 90 000 bound volumes; 329 current periodicals; international exchange of publications with 375 institutions all over the world.
Information services: Bibliographic services (nc); no translation services; no literature searches; no access to on-line information retrieval systems.
Consultancy: Telephone answers to technical queries (nc); no writing of technical reports; no compilation of statistical trade data; no market surveys in technical areas.
Publications: *Bulletin of the Astronomical Institute of Czechoslovakia*; research and engineering reports.

Výzkumný a Zkušební Letecký 21.7
Ústav
– VZLÚ
[Aeronautical Research and Test Institute]
Address: Beranových 130, 199 05 Praha 9 - Letňany, Czechoslovakia
Telephone: (02) 826511 and 827041
Telex: 12 1893
Affiliation: AERO - Czechoslovak Aircraft Works
Enquiries to: Information Manager, Jiří Kučera
Subject coverage: Aerospace.

Library facilities: Available through libraries only.
Information services: Bibliographic services; translation services; literature searches; access to on-line information retrieval systems.
Computerized SDI system.
Consultancy: Yes.
Publications: *ARTI Reports* (in English with summaries in Russian, French, German, and Czech); *ARTI Journal* (in Czech with summaries in English and Russian); *Aerospace Reference Journal; Business Literature References; Factografic Information Bulletin.*

DENMARK

Dansk Astronautisk Forening 21.8
[Danish Astronautical Association]
Address: Postbox 31, DK-1002 København K, Denmark
Telephone: (01) 60 12 13
Library facilities: Library.

Dansk Rumforskningsinstitut 21.9
[Danish Space Research Institute]
Address: Lundtoftevej 7, DK-2800 Lyngby, Denmark
Telephone: (02) 882277
Telex: 37198 Danru
Affiliation: Ministry of Education
Founded: 1967
Library facilities: Open to all users; loan services; reprographic services.
Consultancy: Yes.
Publications: Annual report; list on request.

Københavns Universitets 21.10
Astronomiske Observatorium
[Copenhagen University, Astronomical Observatory]
Address: Øster Voldgade 3, DK-1350 København K, Denmark
Telephone: (01) 14 17 90
Telex: 44 155
Enquiries to: Librarian
Founded: 1861
Subject coverage: Astronomy and astrophysics.
Library facilities: Open to all users (nc); reprographic services (limited - c).
In certain cases, books from the library are available (without charge) through University Library, 2nd department, Nørre Allé 49, DK-2200 København N, Denmark.
Library holdings: About 10 000 bound volumes; about 100 current periodicals.
Information services: Bibliographic services (limited - nc); no translation services; literature searches (limited - nc); no access to on-line information retrieval systems.

Consultancy: Telephone answers to technical queries (nc); no writing of technical reports; no compilation of statistical trade data; no market surveys in technical areas.

Staff members are usually willing to answer minor queries from non-staff members, but as a rule services are not offered to the public.

Publications: Reprint list available on request.

FINLAND

Suomen Avaruustutkimusseura 21.11
[Finnish Astronautical Society]
Address: PL 507, SF-00101 Helsinki 10, Finland
Library facilities: Library.

Tähtitieteellinen yhdistys Ursa 21.12
– URSA
[Ursa Astronomical Association]
Address: Laivanvarustajankatu 3, SF-00140 Helsinki 14, Finland
Telephone: 90-174 048
Enquiries to: Librarian
Founded: 1921
Subject coverage: Astronomy.
Library facilities: Open to all users (nc); loan services (in Finland only - nc); reprographic services (limited - c).
Library holdings: About 3000 bound volumes, some 2000 of which are in active use; over 20 current perodicals, through subscription, membership and exchange; some reports.
Information services: No information services.
Consultancy: Telephone answers to technical queries (limited - nc); no writing of technical reports; no compilation of statistical trade data; no market surveys in technical areas.
Archives of astronomical photographs for papers, journals etc (c).
Courses: Several courses held annually for the general public.
Publications: List available.

FRANCE

Centre de Données Stellaires 21.13
– CDS
[Stellar Data Centre]
Address: Observataire Astronomique, 11, rue de l' Université, 67000 Strasbourg, France
Telephone: 88 35 43 00
Telex: 890 506 STAROBS F
Enquiries to: Director, C. Jaschek
Founded: 1972

Subject coverage: Astronomy, excluding solar system.
Information services: Bibliographic services (c); access to on-line information retrieval systems (c).
Details of SIMBAD, CDS's astronomical data base, can be obtained from CDS on request. Recently CDS has initiated the implementation of a reduced astronomical data base for the general public, accessible through the French public TELETEL service.
Publications: *Star Catalogs and Files Available at the Stellar Data Center; International Directory of Astronomical Associations and Societies*; leaflets; free bulletin; list on request.

Centre National d'Études Spatiales 21.14
– CNES
[National Centre for Space Studies]
Address: 129 rue de l'Université, 75007 Paris, France
Telephone: (1) 45 55 91 21
Telex: 204 627 Cnespar
Affiliation: Ministère de l'Industrie et de la Recherche
Enquiries to: Director, programmes and planning, Jean Marie Luton
Founded: 1961
Subject coverage: National policy administration and research in space science.
Publications: Annual report; *Lettre du CNES*, bimonthly; *Espace Information*, quarterly.

Institut National d'Astronomie et de Géophysique 21.15
– INAG
[National Institute of Astronomy and Geophysics]
Address: 77 avenue Denfert Rochereau, 75014 Paris, France
Telephone: (1) 43 20 13 30
Telex: 270070
Affiliation: Centre National de la Recherche Scientifique
Subject coverage: Coordinates research programmes in astronomy and geophysics.
Publications: Annual report.

Observatoire de Paris, Bibliothèque * 21.16
[Paris Observatory, Library]
Address: 61 avenue de l'Observatoire, 75014 Paris, France
Telephone: (1) 43 20 12 10
Telex: 270776 OBS Paris
Enquiries to: Conservateur, A.-M. de Narbonne
Founded: 1667
Subject coverage: Astronomy; astrophysics; history of science.
Library facilities: Not open to all users; loan services; reprographic services (c).

Information services: Bibliographic services (c); translation services (Russian only - c); no literature searches; access to on-line information retrieval systems (c).
Consultancy: Yes (c).

Office National d'Études et de Recherches Aérospatiales *   21.17

– ONERA
[National Aerospace Research Office]
Address: BP 72, 92322 Chatillon Cedex, France
Telephone: (1) 46 57 11 60
Telex: 260 907F
Affiliation: Délégation Générale pour l'Armement, Ministère de la Défense
Enquiries to: Head of Documentation Department, Françoise Lhullier
Subject coverage: Aerodynamics; energetics; flight mechanics; aeroelasticity and structure dynamics; metallic and organic materials; optics; acoustics; electronics; flight experimentation.
Library facilities: Not open to all users; loan services (nc); no reprographic services.
Information services: No information services.
Publications: *La Recherche Aérospatiale.*

Union Astronomique Internationale   21.18

– UAI-IAU
[International Astronomical Union]
Address: 61 avenue de l'Observatoire, 75014 Paris, France
Telephone: (1) 43 25 83 58
Telex: 205671 IAU F
Affiliation: International Council of Scientific Unions
Enquiries to: UAI-IAU Secretariat
Founded: 1919
Subject coverage: The union was founded to provide a forum where astronomers from all over the world can develop astronomy in all its aspects through international cooperation. It has adhering member organizations in 51 countries and more than 6000 members.
Library facilities: No library facilities.
Information services: No information services.
Consultancy: No consultancy.
Publications: *Reports on Astronomy; Proceedings of the General Assembly; Highlights of Astronomy; Symposium;* biannual information bulletin; further details on request.

GERMAN DEMOCRATIC REPUBLIC

Gesellschaft für Weltraumforschung und Raumfahrt der DDR   21.19

[Outer Space Research and Space Travel Society of the GDR Democratic Republic]
Address: Simon-Dach-Strasse 13, DDR-1034 Berlin, German Democratic Republic
Telephone: 589 22 16

GERMAN FEDERAL REPUBLIC

Astronomisches Rechen-Institut *   21.20

[Astronomical Computing Institute]
Address: Mönchhofstrasse 12-14, D-6900 Heidelberg 1, German Federal Republic
Telephone: (06221) 49026
Enquiries to: Dr Lutz D. Schmadel
Subject coverage: Astronomy, astrophysics, and related areas.
Library facilities: Open to all users; no loan services; no reprographic services.
Information services: Bibliographic services (c); no translation services; no literature searches; no access to on-line information retrieval systems.
Consultancy: No consultancy.
Publications: *Astronomy and Astrophysics Abstracts; Apparent Places of Fundamental Stars.*

Deutsche Forschungs- und Versuchsanstalt für Luft- und Raumfahrt eV *   21.21

– DFVLR
[German Aerospace Research Establishment, Libraries Division]
Address: Linder Höhe, D-5000 Köln 90, German Federal Republic
Telephone: (02203) 60 11
Telex: 8 874 410 (dfvw d)
Enquiries to: Dipl-Ing P. Sternemann
Founded: 1968
Subject coverage: Transportation and communications systems; aircraft technology; space technology; remote sensing technology; energy and propulsion systems; future programmes and supporting research and development.
Library facilities: Not open to all users; no loan services; no reprographic services.
The Libraries Division is the organizational combination of the Central Libraries located in the five DFVLR Research Centres in Braunschweig, Göttingen, Köln-Porz, Stuttgart, and

Oberpfaffenhofen near Munich.
Information services: No bibliographic services; no translation services; no literature searches.
Establishment of a data processing link between the five Central Libraries via the in-house computer network is presently underway. In addition, there is an information retrieval system at the division's centre in Köln-Porz with on-line access to the following hosts: INKA-Karlsruhe; ESA-IRS (Frascati); Dialog of Lockheed; Datastar.

Hermann-Oberth-Gesellschaft eV   21.22
[Hermann-Oberth Society]
Address: Hermann-Köhl-Strasse, Eing H, 2800 Bremen 1, German Federal Republic
Telephone: (0421) 55 47 22
Subject coverage: Space technology.
Library facilities: Library.

GREECE

National Observatory of Athens   21.23
Address: PO Box 20048, 118 10 Athinai, Greece
Telephone: (01) 3464 161; 3461 362
Telex: 215530 OBS GR
Parent: Ministry of Research and Technology
Enquiries to: Head, Professor C.J. Macris
Founded: 1842
Subject coverage: There is an astronomical, a geodynamics, an ionospheric, and a meteorological institute.
Publications: Bulletins published by each of the institutes; Annals; *Mémoires.*

HUNGARY

Földmérési és Távérzékelési Intézet   21.24
– FÖMI
[Geodesy, Cartography and Remote Sensing Institute]
Address: Postafió k 546, H-1373 Budapest, Hungary
Telephone: (01) 113 431
Telex: 22-4964
Enquiries to: Deputy scientific director, Dr Tibor Lukács
Founded: 1967
Subject coverage: Geodesy; surveying; photogrammetry; engineering surveying; satellite geodesy; remote sensing.
Library facilities: Open to all users (nc); loan services (nc); reprographic services (c).
Library holdings: 20 500 bound volumes; 94 current periodicals; 1550 reports.

Information services: Bibliographic services (nc); no translation services; literature searches (nc).
Literature searches are made by means of the reference journal *Geodinform* which has an annual input of about 1500 abstracts.
Consultancy: No consultancy.

Napfizikai Obszervatórium MTA   21.25
[Heliophysical Observatory]
Address: PO Box 30, H-4010 Debrecen, Hungary
Telephone: (52) 11-015
Telex: 72517 deobs
Affiliation: Hungarian Academy of Sciences
Enquiries to: Director, Dr B. Kálmán
Founded: 1958
Subject coverage: Solar physics; astronomy.
Library facilities: Not available, but the observatory has a large amount of solar observations from the last 100 years, available on request to interested specialists.
Library holdings: 9665 bound volumes; 93 current periodicals.
Information services: No information services.
Consultancy: No consultancy.
Publications: *Publications of the Debrecen Heliophysical Observatory;* preprints.

IRELAND

Dunsink Observatory   21.26
Address: Castleknock, Dublin 15, Ireland
Telephone: (01) 387911; 387959
Telex: 31 687 DIAS EI
Affiliation: Dublin Institute for Advanced Studies
Enquiries to: Librarian
Founded: 1783
Subject coverage: Astronomy and astrophysics.
Library facilities: Loan services (occasional); reprographic services.
The library is open to academics and approved users.

ITALY

Associazione Italiana di Aeronautica e Astronautica   21.27
[Italian Aeronautical and Astronautical Association]
Address: Via Po 50, 00198 Roma, Italy
Telephone: (06) 84 45 8 94
Library facilities: Library.

European Space Agency, Information Retrieval Service 21.28

– ESA-IRS
Address: CP 64, Via Galileo Galilei, Frascati 00044, Italy
Telephone: (06) 94011
Telex: 610637
Enquiries to: Head of On-line Services Division, Dr G.A. Proca
Founded: 1966
Subject coverage: Science and technology.
Library facilities: Open to all users (as it is computer based, users must have a terminal and password to gain access); no loan services; reprographic services (copies of references can be obtained on-line or off-line; copies of original documents can be ordered through the ESA-IRS system - c).
Information services: No bibliographic services; translation services (c); literature searches (c); access to on-line information retrieval systems (c).
On-line files available: ABI INFORM; ACID RAIN; ACOMPLINE; Areospace Daily; AFEE; AGRIS; ALUMINIUM; ARTIFICIAL INTELLIGENCE; Asian Geotechnology; BIIPAM; Biosis; BNF Metals; BRIX; BUSINESS SOFTWARE; CAB; CAD/CAM; CBA; CETIM; CHEMABS; CHEMABS Training; CISDOC; Compendex; Conference Papers Index; DELFT HYDRO; EDF DOC; EDIN-Inis Training; ELECTRONIC MAGAZINE; EMIS; ENEL; ENERGYLINE; ENERGYNET; ENVIROLINE; EUDISED R&D; Fluidex; Food Science - FSTA; HSELINE; IBSEDEX; Inis; Inspec; Inspec Information Training File; Inspec Training; IPA; IRRD; Ismec; LAB. HAZ. BULLETIN; Labourdoc; LABOUR INFOR (LID); LEDA; MASS SPECTR BULL; MATHSCI; MERLIN-TECH; METADEX; MOLARS; Nasa; NTIS; OCEANIC; Pack ABS; Pascal; Pascal Training; POLLUTION; ROBOMATICS; SATELDATA; SPACE COMPONENTS; SPACE PATENTS; SPACE SOFT; Standards & Specifications; TELECOM-MUNICATIONS; TELEGEN; Textline/Newsline; TRANSDOC; ULRICH's PERIODICALS; WORLD REPORTER; WTI.
Other Services include: electronic mail; creation and maintenance of private files; library automation; creation and maintenance of tree-structured data bases; computer-aided instruction Videotex.
Consultancy: Yes (c).

Istituto di Astrofisica Spaziale * 21.29

[Space Astrophysics Institute]
Address: CP 67, 00044 Frascati, Italy
Telephone: (06) 9425651
Telex: CNR FRA 610261
Affiliation: Consiglio Nazionale delle Ricerche
Enquiries to: Dr T. Tittoni
Founded: 1970

Subject coverage: Theoretical and observational astrophysics; space research; planetology.
Library facilities: Not open to all users; no loan services; reprographic services.
Information services: Bibliographic services.
Consultancy: Yes.

NETHERLANDS

Koninklijke Nederlandse Vereniging voor Luchtvaart 21.30

[Royal Netherlands Aeronautical Association]
Address: Josef Israëlsplein 8, 2596 's Gravenhage, Netherlands
Telephone: (070) 24 54 57
Library facilities: Library.

Nationaal Lucht- en Ruimtevaart Laboratorium 21.31

– NLR
[National Aerospace Laboratory]
Address: Postbus 90502, 1006 BM Amsterdam, Netherlands
Telephone: (020) 5113113
Telex: 11118
Enquiries to: Librarian, C.W. de Jong
Founded: 1919
Subject coverage: Aerospace - fluid dynamics; flight mechanics, flight testing, flight operations; structures and materials; space flight technology; informatics.
Library facilities: Not open to all users; loan services (nc); reprographic services (c).
Library holdings: 10 000 bound volumes; 750 current periodicals; 120 000 reports.
Information services: Bibliographic services (c); no translation services; literature searches (c); access to on-line information retrieval systems (c).
Consultancy: Telephone answers to technical queries (nc); writing of technical reports; no compilation of statistical trade data; no market surveys in technical areas.
Courses: On-the-job training courses on aerospace - related topics (irregular).
Publications: Bibliographic list.

Stichting Radiostraling van zon en Melkweg * 21.32

– SRZM
[Netherlands Foundation for Radioastronomy]
Address: Radiosterrenwacht Dwingeloo, Postbus 2, 7990 AA Dwingeloo, Netherlands
Telephone: (05219) 7244
Telex: 42043 SRZM NL
Enquiries to: J.F. van der Brugge
Founded: 1949

Subject coverage: Radioastronomy: observations; telescopes; data handling; development of electronics and software.
Library facilities: Not open to all users.
Consultancy: Consultancy.
Publications: Research papers; internal technical reports; annual reports.

Technische Hogeschool Delft 21.33
[Delft University of Technology, Department of Aerospace Engineering]
Address: Kluyverweg 1, 2629 HS Delft, Netherlands
Telephone: (015) 781455
Enquiries to: Executive manager, F. Hospers
Founded: 1940
Subject coverage: Aerospace engineering.
Library facilities: Open to all users (nc); loan services (nc); reprographic services (photocopier for self-service only).
Library holdings: 12 500 bound volumes; 184 current periodicals; 70 000 reports.
Information services: No information services.
Consultancy: No consultancy.

NORWAY

Norsk Astronautisk Forening * 21.34
[Norwegian Astronautical Society]
Address: Postboks 43, Blindern, Oslo 3, Norway
Library facilities: Library.

POLAND

Planetarium i Obserwatorium 21.35
Astronomiczne im Mikołaja
Kopernika *
[Mikołaja Kopernika Astronomical Planetarium and Observatory]
Address: Skrytka Pocztwa 10, 41-501 Cherzów, Poland
Telephone: 585149
Enquiries to: Kierownik Biblioteki
Founded: 1955
Subject coverage: Astronomy.
Library facilities: Open to all users; loan services (nc); no reprographic services.
Information services: No information services.
Consultancy: Yes (nc).

Polskie Towarzystwo 21.36
Astronautyczne *
[Polish Astronautical Society]
Address: Skrytka Posztwa 2319, Pałac Kultury i Nauki, 00-901 Warszawa, Poland
Telephone: 200 211
Library facilities: Library.

PORTUGAL

Observatório Astronómico de 21.37
Lisboa
– OAL
[Lisbon Astronomical Observatory]
Address: Tapada da Ajuda, 1300 Lisboa, Portugal
Telephone: (019) 637351
Enquiries to: Director, Dr Ezequiel Cabrita
Founded: 1861
Subject coverage: Time service; meridian observations; latitude variations; occultations.
Library facilities: Open to all users; no loan services; no reprographic services.
Publications: *Dados Astronómicos para os Almanaques*, regular publication; *Bulletin de l'Observatoire Astronomique de Lisbonne (TAPADA)*, irregular. Astronomy; time; coordinates (geography).

SPAIN

Instituto Nacional de Técnica 21.38
Aeroespacial
– INTA
[National Institute of Aerospace Technology]
Address: Carretera de Ajalvir, Km 4.500, Torrejón de Ardoz, Madrid, Spain
Telephone: 91-675 07 00
Telex: 22026 E
Facsimile: 91-675 52 63
Enquiries to: Director, A. García de Castro Informática
Founded: 1942
Subject coverage: Aerospace.
Library facilities: Open to all users (nc); no loan services; reprographic services (c).
Library holdings: 20 000 bound volumes; about 120 current periodicals.
Information services: No bibliographic services; no translation services; no literature searches; access to on-line information retrieval systems (c). National centre of ESA/IRS.
Consultancy: No consultancy.

Observatorio Astronómico Nacional
21.39

– OAN

[National Astronomical Observatory]

Address: Calle de Alfonso XII 3, 28014 Madrid, Spain

Telephone: 91-227 01 07

Affiliation: Ministerio de la Presidencia del Gobierno; Instituto Geográfico Nacional

Enquiries to: Director, Manuel Lopez Arroyo

Founded: 1790

Subject coverage: Astronomy and astrophysics.

Library facilities: Not open to all users; no loan services; reprographic services (nc).

Library holdings: 7000 bound volumes; 30 current periodicals.

Information services: Bibliographic services (nc).

Consultancy: Writing of technical reports.

Publications: Anuario del Observatorio Astronómico de Madrid; Boletín Astronómico del Observatorio de Madrid.

SWEDEN

Flygtekniska Försöksanstalten, Biblioteket
21.40

– FFA

[Aeronautical Research Institute of Sweden, Library]

Address: PO Box 11021, Ranhammarsvägen 12-14, S-161 11 Bromma, Sweden

Telephone: 08-75 91 000

Telex: 10725 FFA S

Enquiries to: Librarian, Gunnel Larsson

Founded: 1939

Subject coverage: Aeronautics; aerodynamics; structures; instrumentation; engineering; noise; wind engineering.

Library facilities: Not open to all users; loan services (nc); no reprographic services.

Information services: No information services.

SWITZERLAND

Eidgenössische Technische Hochschule Zürich, Institut für Astronomie
21.41

– ETH

[Swiss Federal Institute of Technology, Zürich, Astronomy Institute]

Address: ETH-Zentrum, CH-8092 Zürich, Switzerland

Telephone: (01) 256 38 13

Telex: 53 178 ethbi

Enquiries to: S. Weber

Founded: 1855

Subject coverage: Astronomy.

Library facilities: Not open to all users (open to staff, guest scientists, and astronomy students of the ETH and the University of Zürich); no loan services; no reprographic services.

Schweizerische Arbeitsgemeinschaft für Raumfahrt
21.42

[Swiss Astronautics Association]

Address: POB 1011, Lidostrasse 5, CH-6002 Lucerne, Switzerland

Library facilities: Library.

UNITED KINGDOM

Royal Aeronautical Society
21.43

– RAeS

Address: 4 Hamilton Place, London W1V 0BQ, UK

Telephone: (01) 499 3515

Enquiries to: Librarian, A.W.L. Nayler

Founded: 1866

Subject coverage: Aerospace.

Library facilities: Open to all users (charge for non-members); loan services (members only); reprographic services.

Library holdings: 25 000 bound volumes; over 420 current periodicals; 50 000 reports.

Information services: No information services.

Consultancy: Telephone answers to technical queries (nc); no writing of technical reports; no compilation of statistical trade data; no market surveys in technical areas.

Courses: Lectures; symposia; conferences.

Publications: Publications Index 1897-1977; list on request.

Royal Aircraft Establishment, Main Library
21.44

– RAE

Address: Q4 Building, Farnborough, Hampshire GU14 6TD, UK

Telephone: (0252) 24461

Telex: 858134

Affiliation: Ministry of Defence

Enquiries to: Chief librarian

Subject coverage: Aerodynamics; engineering physics; materials; radio and navigation; space; structures; flight systems; instrumentation and trials.

Library facilities: Not open to all users (available only within Ministry of Defence areas).

Royal Greenwich Observatory * 21.45
– RGO

Address: Herstmonceux Castle, Hailsham, East Sussex BN27 1RP, UK
Telephone: (0323) 833171
Telex: 87451 RGOBSY G
Affiliation: Science Research Council
Enquiries to: Librarian, Janet Dudley; J. Hutchins
Founded: 1675
Subject coverage: Astronomy; astrophysics; navigation; chronometry; telescope design and maintenance; optics; instrument design and development; computing.
Library facilities: Open to all users (for reference only on receipt of written application); loan services (to other libraries only, not to individuals - nc); reprographic services (c).
Reproductive services - paper and microfilm copies supplied; archive copying charged at Public Record Office rates; other copying at local rates.
RGO is a place of deposit under the Public Records Acts and maintains the records of the observatory from 1675 to date, together with those of the Board of Longitude. It also has a rare book collection.
Information services: Bibliographic services; no translation services; no literature searches; no access to on-line information retrieval systems.
Consultancy: Consultancy (sunrise, sunset, calendar and almanac information - c).

YUGOSLAVIA

Savez Astronautičkih i Raketnih 21.46
Organizacija Jugoslavije
[Yugoslav Astronautical and Rocket Society]

Address: Bulevar Revolucije 44, 11000 Beograd, Yugoslavia
Telephone: (011) 333 404
Library facilities: Library.

22 TEXTILES, WEAVING AND CLOTHING

AUSTRIA

Österreichisches Textil-   22.1
Forschungs-Institut
– ÖTI

[Austrian Textile Research Institute]
Address: Postfach 117, Spengergasse 20, A-1051 Wien, Austria
Telephone: (0222) 552543
Telex: 136750
Enquiries to: Information officer, Dr Brunhilde Herzog
Founded: 1967
Subject coverage: Textile industry; man-made fibre industry; textile technology; textile physics; textile chemistry; finishing textile floor coverings; international testing standards; technical science of application; physiological properties of clothing.
Library facilities: Not open to all users; no loan services; reprographic services (c).
Library holdings: 520 bound volumes; 20 current periodicals.
Information services: No information services.
Consultancy: No consultancy.

BELGIUM

Centre Scientifique et Technique   22.2
de l'Industrie Textile Belge *
– CENTEXBEL

[Scientific and Technical Centre of the Belgian Textile Industry]
Address: 24 Rue Montoyer, B-1040 Bruxelles, Belgium
Telephone: (02) 233 93 30
Enquiries to: Information officer, G. Bettens.
41 St Pietersnieuwstraat, B-9000 Gent, Belgium (telephone: (091) 23 38 21)
Subject coverage: Textiles and related topics.

Library facilities: Open to all users; no loan services; reprographic services (c).
Information services: Bibliographic services (c); no translation services; literature searches (c); access to on-line information retrieval systems (c).
The library has on-line access to the following data bases: Titus; World Textile Abstracts.
Consultancy: Yes (c).
Publications: List on request.

BULGARIA

Scientific and Technical Union of   22.3
Textiles, Clothing and Leather
Industry
Address: Rakovski 108, 1000 Sofia, Bulgaria
Telex: 22 185 nts bg

CZECHOSLOVAKIA

Státní Výzkumný Ústav Textilní   22.4
[State Textile Research Institute]
Address: Ujezu 2, 460 97 Liberec, Czechoslovakia

DENMARK

Dansk Textil Institut *   22.5
– DTI

[Danish Textile Institute]
Address: Postboks 80, Gregersensvej 5, DK-2630 Taastrup, Denmark
Telephone: (2) 998822
Telex: 33754 dti dk
Affiliation: Department of Industry
Enquiries to: Information officer, Merete Wichfield
Founded: 1959

Subject coverage: Production and end use properties of textiles and clothing.
Library facilities: Open to all users; loan services (nc); reprographic services (c).
Information services: Bibliographic services (c); translation services (c); literature searches (c); no access to on-line information retrieval systems.
Consultancy: Yes (c).

FRANCE

Centre Technique de la Teinture et du Nettoyage, Institut de Recherche Entretien et Nettoyage * 22.6

– CTTN-IREN
[Dyeing and Cleaning Technical Centre, Institute of Research for Dry-cleaning Industries]
Address: Avenue Guy de Collongue, BP 41, 69131 Ecully Cedex, France
Telephone: 78 33 08 61
Enquiries to: Head, information department, Joëlle A. Josserand
Founded: 1954
Subject coverage: Dry-cleaning and laundering subjects.
Library facilities: Open to all users; no loan services; reprographic services (c).
Information services: Bibliographic services (c); translation services (c); literature searches (c); no access to on-line information retrieval systems.
Consultancy: Yes (c).

Institut Textile de France 22.7
– ITF
[Textile Institute of France]
Address: 35 rue des Abondances, BP 79, 92105 Boulogne Billancourt Cedex, France
Telephone: (1) 48 25 18 90
Telex: 250940
Enquiries to: Information and DP department manager, Jean-Marie Ducrot
Founded: 1948
Subject coverage: Textiles: natural and chemical fibres; machinery and technology.
Library facilities: Open to all users; loan services; reprographic services (c).
Information services: Bibliographic services; translation services; literature searches; access to on-line information retrieval systems.
The ITF maintains the textile data base Titus, accessible through Télésystèmes-Questel, containing 160 000 abstracts and references created since 1968. The system includes an automatic translation service, whereby information can be requested on-line in German, English, Spanish or French. The data base is used to provide SDI in 20 standard profiles in textiles technology.
Publications: Bulletin Scientifique de l'Institut Textile de France, quarterly; ITMA 83, a six-volume report on the textiles technology exhibition.

GERMAN FEDERAL REPUBLIC

Zentralstelle für Textildokumentation und - information * 22.8
– ZTDI
[Textile Documentation and Information Centre]
Address: 22 Cromforder Allee, D-4030 Ratingen 1, German Federal Republic
Telephone: (02102) 27051
Telex: 8585 374 vtdi
Affiliation: Verein Textildokumentation und - information eV
Enquiries to: Information officer, Rosethea Schip
Founded: 1956
Subject coverage: Textiles, textile machinery, clothing, fibres, cleaning, testing.
Library facilities: Open to all users; loan services (c); reprographic services (c).
Library holdings: The library's holdings include textile patents (only German from 1956); textile standards (all German, most of Switzerland, some international); access to several data bases, especially textiles.
Information services: Bibliographic services (c); translation services (c); literature searches (c); access to on-line information retrieval systems (c).
The library has access to the following on-line information retrieval systems: SDC, ODAV, Lockheed, Télésystèmes.
As a co-producer of the textile data base Titus, with 150 000 items since 1969-70, and the centre's card index with 175 000 from 1956-70, the library can search all literature in these files. Additionally there are search possibilities in the patent files and several national and international guides.
Consultancy: Yes (c).

HUNGARY

Textilipari Kutató Intézet 22.9
– TKI
[Textile Research Institute]
Address: Pf 6, H-1475 Budapest, Hungary
Telephone: 472 300
Telex: 224698 texki

Enquiries to: Head, József Horváth
Research Organization Department
Founded: 1949
Subject coverage: Textile industry and power engineering research and experiments; production of instruments and experimental equipment for the textile industry; instrument service; technical information.
Library facilities: Open to all users; loan services; reprographic services.
Information services: Bibliographic services; literature searches.
Consultancy: Yes.
Publications: *Textilipari Kutató Intézet Tájékoztatója; Textilipari Kutató Intézet Közleményei; Textilipari Kutató Intézet Tevékenysége.*

Textilipari Müszaki és Tudományos Egyesület   22.10
[Hungarian Society of Textile Technology and Science]
Address: Anker-köz 1, 1016 Budapest VI, Hungary
Library facilities: Library.

ITALY

Istituto di Ricerche e Sperimentazione Laniera, Oreste Rivetti   22.11
[Wool Research Institute, Oreste Rivetti]
Address: Piazza Lamarmora 5, 13051 Biella, Italy
Telephone: (15) 20490; 21655
Affiliation: Consiglio Nazionale delle Ricerche
Enquiries to: Director, Dr Leo Gallico
Founded: 1983
Subject coverage: Wool research; textile fibres; textile industry; water pollution; energy saving; quality control.
Library facilities: Open to all users (nc).
Library holdings: 200 bound volumes; 30 current periodicals.
Information services: Access to on-line information retrieval systems is planned.
Consultancy: Telephone answers to technical queries (nc); writing of technical reports (c). Quality control (c).

Stazione Sperimentale per la Seta   22.12
[Silk Experimental Institute]
Address: Via Giuseppe Colombo 81, 20133 Milano, Italy
Telephone: (02) 299 890; 235 047
Enquiries to: Information officer, Antonella Generali
Founded: 1923

Library facilities: Open to all users; no loan services; no reprographic services.
Information services: Bibliographic services.
Consultancy: Yes.

NETHERLANDS

Nederlands Textielinstituut   22.13
[Dutch Textile Institute]
Address: Postbus 518, de Schutterij 16, 3900 AM Veenendaal, Netherlands

Vezelinstituut TNO   22.14
– VI-TNO
[Fibre Research Institute TNO]
Address: Postbus 110, 2600 AC Delft, Netherlands
Telephone: (015) 569330
Telex: 38071 zptno nl
Affiliation: Nederlandse Organisatic voor Toegepast-Natuuravetenschappelijk Onderzock
Enquiries to: Librarian, W.A. Nienhuis
Subject coverage: Spinning; weaving; knitting; finishing; clothing; pulp and paper.
Library facilities: Open to all users; loan services; reprographic services (c).
Information services: Bibliographic services; literature searches (c).

POLAND

Centralny Ośrodek Badawczo-Rozwojowy Przemysłu Odziezowego   22.15
– COBRPO
[Polish Clothing Industry Research and Development Centre]
Address: Sterling Street 27/29, 90-950 Łódź, Poland
Telephone: 32 82 60
Telex: 88 42 72
Affiliation: Garment Industry Association
Enquiries to: General manager, Dr Bolesław Dzieduszyński
Scientific, Technical and Economic Information Centre
Founded: 1963
Subject coverage: Clothing industry - techniques, technology, economics.
Library facilities: Open to all users (nc); loan services (institutions only - nc); reprographic services (nc).
Library holdings: 4302 bound volumes; 185 current periodicals; 140 reports.
Information services: Bibliographic services (c); translation services (c); literature searches (c); no access to on-line information retrieval systems.

Services are available to institutions only.

Consultancy: Telephone answers to technical queries; no writing of technical reports; no compilation of statistical trade data; no market surveys in technical areas.

Instytut Włókiennictwa 22.16
– IW
[Textile Research Institute]
Address: Skr Poczt 456, Ulica Gdánska 91/93, 90-540 Łódź, Poland
Telephone: 339600
Telex: 886666
Affiliation: Ministry of Chemical and Light Industry
Enquiries to: Director, Z. Wawrzaszek; Director of information, C.-W. Gabryniak, Branzowy Ośrodek Informacji Naukowej Technicznej i Ekonomicznej (BIONTE)
Founded: 1945
Subject coverage: Textile industry.
Library facilities: Open to all users (nc); loan services (nc); reprographic services (c).
Library holdings: 26 556 bound volumes; 339 current periodicals; 5000 reports.
Information services: Bibliographic services (nc); translation services (c); literature searches (c); access to on-line information retrieval systems (c).
Consultancy: Telephone answers to technical queries (nc); writing of technical reports (c); compilation of statistical trade data (c); market surveys in technical areas (c).
Publications: *Prace Instytutu Włókiennictwa*, annual; *Ekspres - Informacja* (summaries of latest world publishing); *Przegląd Dokumentacyjny* (bibliographic information).

Resortowy Ośrodek Informacji 22.17
Naukowej, Technicznej i
Ekonomicznej Przemysłu
Lekkiego *
– ROINTE PL
[Departmental Centre of Scientific, Technological and Economic Information for Light Industry]
Address: Ulica Mickiewicza 20, 90-950 Łódź, Poland
Telephone: 364951
Telex: 885416
Enquiries to: Managing director, Wiesław Rapacki
Founded: 1962
Subject coverage: Textile, clothing, leather, and shoe industries.
Library facilities: Open to all users; reprographic services.
Information services: Bibliographic services.
The centre offers SDI services, abstracting and other services.
Consultancy: Yes (on information searches).

Publications: *Informacja Ekspresowa* (Express Information); *Przeglad Dokumentacyjny* (Documentation Review); *Nabytki Biblioteczne* (Library Acquisitions); *Wykazy tematyczne projektów wynalazczych* (Subject Registers of Rationalizing Projects).

ROMANIA
Institutul de Cercetări Textile 22.18
[Textile Research Institute]
Address: Strada Lucreţiu Pătrăşcanu 3A, Bucureşti Sector 3, Romania
Telephone: (90) 434402
Telex: 10678
Affiliation: Ministry of Light Industry
Enquiries to: Managing director, Dr Valeriu Rusanovschi
Subject coverage: Topics include cotton, wool, silk, and non-woven products.

SPAIN
Instituto de Investigación Textil 22.19
y Cooperación Industrial *
[Textile Research and Industrial Cooperation Institute]
Address: Calle Colón 15, Terrassa, Barcelona, Spain
Telephone: 785 0400
Telex: 56126
Affiliation: Universidad Politécnica de Cataluña
Enquiries to: Head, textile physics research, Dr Arun Naik
Founded: 1954
Subject coverage: Research activities related to all fields of textiles and environmental pollution.
Library facilities: Not open to all users (only to professors and students); no loan services; no reprographic services.
Library holdings: The institute's collection holds about 1000 books, most of them related to textiles and pollution. The library receives about 60 technical journals every month related to textile research, and various leaflets from different research organizations.
Information services: No bibliographic services; no translation services; literature searches (c); no access to on-line information retrieval systems.
Consultancy: Yes (c).

Instituto de Tecnología Química y Textil   22.20
[Chemistry and Textile Technology Research Institute]
Address: Jorge Girona Salgado s/n, Barcelona 34, Spain
Telephone: (93) 203 74 32
Affiliation: Consejo Superior de Investigaciones Científicas
Enquiries to: Information officer
Centro de Investigación y Desarrollo
Founded: 1955
Subject coverage: Chemistry, textiles, engineering, basic science.
Library facilities: Not open to all users; no loan services; reprographic services.
Information services: Bibliographic services; literature searches.
Consultancy: Yes.

SWEDEN

Svenska Textilforskningsinstitutet *   22.21
– TEFO
[Swedish Institute for Textile Research]
Address: Box 5402, S-402 29 Göteborg, Sweden
Telephone: (031) 20 01 75
Enquiries to: Librarian, Håkan Damberg
Founded: 1946
Library facilities: Open to all users; loan services; reprographic services (c - for more than 20 copies).
Information services: Bibliographic services (c); translation services (c); literature searches (c); access to on-line information retrieval systems (c).
The institute has access to Titus-e in Questel, and to WTA in Dialog. SDI services are offered.
Consultancy: Yes (c).

UNITED KINGDOM

HATRA   22.22
Address: 7 Gregory Boulevard, Nottingham NG7 6LD, UK
Telephone: (0602) 623311
Telex: 37605
Enquiries to: Information officer, J.A. Smirfitt
Founded: 1949
Subject coverage: Textiles and clothing; knitting; dyeing and finishing.
Library facilities: Open to all users (free to members only); loan services (free to members only); reprographic services (c).
Library holdings: 5000 bound volumes; 250 current periodicals; 3000 reports.

The library's collection provides: worldwide coverage on knitting and clothing manufacture; special files on industrial knitting machines; standards; US and UK knitting patents; also a collection on machine-made lace.
Information services: Bibliographic services (free to members only); no translation services; literature searches (c); access to on-line information retrieval systems (c).
The library has access to the following on-line information retrieval systems: InfoLine; Dialog. Charges are made to non-members for all services.
Data base of UK knitting companies and their products (c).
Consultancy: Telephone answers to technical queries (free to members only); no writing of technical reports; compilation of statistical trade data (c); market surveys in technical areas (c).
Courses: Programme of courses issued annually.
Publications: Publications list on request.

International Institute for Cotton   22.23
– IIC
Address: Kingston Road, Didsbury, Manchester M20 8RD, UK
Telephone: (061) 434 9821
Telex: 669671 IICTRD G
Enquiries to: Manager, technical press, Lewis P. Miles; Information officer, Isobel Macdonald
Technical Research Division
Founded: 1967
Subject coverage: Cotton.
Library facilities: Not open to all users; no loan services; reprographic services (nc - limited).
Information services: No information services.
Consultancy: Yes (nc).
Project work is carried out in cooperation with industry, and consultancy is offered on research and development.
Publications: *Cotton Technology*, bimonthly technical news-sheet, in English and Spanish; reprints of selected articles on cotton subjects, mainly conference papers by IIC technical staff; occasional pamphlets.

International Wool Secretariat *   22.24
– IWS
Address: Valley Drive, Ilkley LS29 8PB, West Yorkshire, UK
Telephone: (0943) 601555
Telex: 51457
Parents: Australian Wool Corporation; New Zealand Wool Board; South African Wool Board; Secretariado Uruguayo de la Lana.
Enquiries to: Librarian, Jeanette A. Robbie
Technical Centre
Founded: 1937

Subject coverage: Research and development in and marketing of: wool textiles - apparel, carpets and rugs, special finishes and products; wool processing (from raw material to finished product) spinning, weaving, knitting, dyeing and finishing. Design services; standards and testing; marketing and economics (all related to wool textiles and/or processes).

The IWS is funded by the wool industries of Australia, New Zealand, South Africa and Uruguay, and operates in 50 countries worldwide to promote wool products.

Library facilities: Open to all users (by appointment); loan services (through BLLD); reprographic services (c).

The library has microfiche reader/printer facilities.

Library holdings: Over 300 journals; approximately 3000 books.

Information services: Literature searches, translations and access to on-line information retrieval systems (Pergamon InfoLine, SDC Orbit) are available to staff users only; also current awareness service and referral service.

Consultancy: Yes.

Publications: *Wool Science Review*, free irregular publication; *International Wool Survey*, fortnightly current awareness service by subscription; publicity material.

Shirley Institute   22.25

Address: Didsbury, Manchester M20 8RX, UK

Telephone: (061) 445 8141

Telex: 668417 Shirly G

Enquiries to: Head, R.J.E. Cumberbirch
Information Services

Founded: 1918

Subject coverage: Science, technology, technical management and technical economics of textile materials and products, and related materials in civil and mechanical engineering, medicine etc.

Library facilities: Not open to all users (members only); loan services (members only); reprographic services (international - c).

Library holdings: Bound volumes; about 500 current periodicals; reports.

Information services: Bibliographic services (c); translation services (c); literature searches (c); access to on-line information retrieval systems (c).

The library has access to the following on-line information retrieval systems: Shirley Institute's data base World Textiles, other data bases through Dialog and InfoLine. The institute provides the following bibliographic services: indexes, hard-copy abstracts, data bases on magnetic tape, on-line. Translations offered from French, German, Italian, Russian and Spanish into English.

Consultancy: Telephone answers to technical queries (nc if brief); writing of technical reports (c); compilation of statistical trade data (c); market surveys in technical areas (c).

Evaluation; testing; market research (c).

Courses: About 30 courses per annum on textile science and technology for small groups of trainees.

Publications: *World Textile Abstracts; Textile Digest; Textiles; World Textiles* data base on magnetic tape; complete list on request.

Textile Institute *   22.26
– TI

Address: 10 Blackfriars Street, Manchester M3 5DR, UK

Telephone: (061) 834 8457

Telex: 668297

Enquiries to: Information service manager, P. Nichols

Founded: 1910

Subject coverage: Textiles, including clothing, floor coverings, industrial uses, all aspects from raw fibre to the finished article.

Library facilities: Open to all users; loan services (interlibrary loans only); reprographic services (c).

Information services: Bibliographic services (c); no translation services; literature searches (c); access to on-line information retrieval systems.

Consultancy: Yes (c).

Publications: Publications list available on request.

Wira Technology Group   22.27

Address: Wira House, West Park Ring Road, Leeds LS16 6QL, UK

Telephone: (0532) 790188

Telex: 557189 (WIRALS)

Enquiries to: Information officer, Nicola Parry

Founded: 1918

Subject coverage: Wool textiles; carpets; clothing; cleaning and maintenance.

Library facilities: Open to all users (by appointment); loan services (c); reprographic services (c).

Information services: Bibliographic services (c); translation services (c); literature searches (c); access to on-line information retrieval systems (c).

The library has access to Pergamon InfoLine, Dialog, and SDC Orbit.

Consultancy: Yes (c).

Publications: *Wira Scan*, fortnightly; *CAMRASO Scan*, monthly; *Wira News*, six issues a year; *CAMRASO News*, quarterly; *Wira Textile Data Book*; publications list on request.

YUGOSLAVIA

Savez Inženjera; Tehničara   22.28
Tekstilaca Hrvatske
[Textile Engineers and Technicians of Croatia Association]
Address: Novakova 8, 41000 Zagreb, Yugoslavia
Telephone: (041) 34 847
Library facilities: Library.

23 TIMBER AND FURNITURE INDUSTRY

AUSTRIA

Österreichisches Holzforschungsinstitut, Dokumentationsstelle
23.1

– ÖGH-Dok

[Austrian Wood Research Institute, Documentation Centre]

Address: Arsean, Franz-Grill-Strasse 7, A-1030 Wien, Austria
Telephone: (0222) 782623
Affiliation: Austrian Society of Wood Research
Enquiries to: Reinhold Bayer
Founded: 1948
Subject coverage: Chemical and mechanical technology of wood, including pulp and paper.
Library facilities: Open to all users (c); loan services (c); reprographic services (c).
Library holdings: 12 000 books and pamphlets; 750 current journals and series.
Information services: Bibliographic services (c); no translation services; literature searches (c); no access to on-line information retrieval systems.
Consultancy: No consultancy.
Publications: *Card Index Series*, fortnightly.

BELGIUM

Bureau National de Documentation sur le Bois *
23.2

– BNDB

[National Wood Documentation Bureau]

Address: Rue Royale 109-111, 1000 Bruxelles, Belgium
Telephone: (02) 219 28 32
Subject coverage: Wood and its uses.
Library facilities: Open to all users; reprographic services (c).
Publications: *Le Courrier du Bois*, 4 times a year.

CZECHOSLOVAKIA

Štátny Drevársky Výskumný Ústav, Odvetvové Informačné Strediska
23.3

– ŠDVÚ

[State Forest Products Research Institute, Information Centre]

Address: Lamačská 1, 809 59 Bratislava, Czechoslovakia
Telephone: 43551
Telex: 09333
Affiliation: Ministry of Industry of the Slovak Socialist Republic
Enquiries to: Head of Information Centre
Founded: 1947
Subject coverage: Wood science; wood technology; use of wood in the national economy.
The institute provides services to the wood industry in Czechoslovakia and in COMECON countries, also to other industrial sectors, universities, research institutions, and to a wide technical public.
Library facilities: Open to all users; loan services; reprographic services (c).
Library holdings: 20 000 books; about 300 periodicals; 5000 foreign research reports and reprints; other collections including bibliographies, translations, standards, patents, etc.
Information services: Bibliographic services (c); translation services (c); literature searches (c).
Consultancy: Yes.

DENMARK

Afdeling for Traeteknik, Teknologisk Institut *
23.4

[Technological Institute, Department of Wood Technology]
Address: Box 141, DK-2630 Tåstrup, Denmark
Telephone: 2996611
Telex: 33416 tidk
Enquiries to: Senior scientific officer, C. Boye
Founded: 1906
Library facilities: Not open to all users; loan services (c); reprographic services (c).
Information services: Bibliographic services (c); translation services (c); literature searches (c).
Consultancy: Yes (c).

FRANCE

Centre Technique du Bois et de l'Ameublement
23.5

– CTBA
[Wood and Furniture Technical Centre]
Address: 10 avenue de Saint-Mandé, F-75012 Paris, France
Telephone: (1) 43 44 06 20
Telex: 214 280 CTBOIS
Affiliation: Ministère de l'Agriculture; Ministère de l'Industrie
Enquiries to: Librarian, J. Hervo
Founded: 1952
Subject coverage: Forest exploitation and sawmill industries; wood and wood products; wood properties, characteristics, and uses; furniture industries; wood preservation, glueing, and finishing.
Library facilities: Open to all users (nc); no loan services; reprographic services (c).
Library holdings: 5000 bound volumes; 300 current periodicals.
Information services: Bibliographic services (c); no translation services; literature searches (nc); access to on-line information retrieval systems (c). Questel.
Consultancy: Telephone answers to technical queries (nc); writing of technical reports (c); no compilation of statistical trade data; no market surveys in technical areas.
Courses: Details on request.
Publications: *CTBInfo*, bimonthly; *Revue Documentaire*, bimonthly; list on request.

GERMAN FEDERAL REPUBLIC

Bundesforschungsanstalt für Forst- und Holzwirtschaft *
23.6

– BFH
[Federal Research Centre for Forestry and Forest Products]
Address: Leuschnerstrasse 91, D-2050 Hamburg 80, German Federal Republic
Telephone: (040) 739621
Affiliation: Bundesministerium für Ernährung, Landwirtschaft und Forsten
Enquiries to: Librarian
Founded: 1939
Library facilities: Open to all users; loan services (nc); reprographic services (c).
Information services: No bibliographic services; no translation services; literature searches (c); no access to on-line information retrieval systems.
Consultancy: No consultancy.
Publications: *Mitteilungen der Bundesforschungsanstalt für Forst- und Holzwirtschaft*, irregular; *BFH-Nachrichten*, quarterly; *Jahresbericht der Bundesforschungsanstalt für Forst- und Holzwirtschaft*, (annual report).

HUNGARY

Erdészeti és Faipari Egyetem Központi Könyvtára *
23.7

[University of Forestry and Timber Industry, Central Library]
Address: POB 132, Bajcsy-Zsilinszky utca 4, H-9401 Sopron, Hungary
Telephone: 111-00
Telex: 249126
Affiliation: Ministry of Agriculture and Food
Enquiries to: Director general, Dr István Hiller
Founded: 1735
Subject coverage: Forestry, timber, and woodworking industry.
Library facilities: Open to all users; loan services (nc); reprographic services (c).
Exchange of publications; forestry history research.
Information services: Bibliographic services (nc); translation services (c); literature searches; no access to on-line information retrieval systems.
Consultancy: Yes (nc).

Faipari Kutató Intézet
23.8

[Wood Industry Research Institute]
Address: Vörösmarty útca 56, Pesterzsébet 1, Postafiók 64, 1725 Budapest XX, Hungary
Telephone: 572-022
Affiliation: Ministry of Agriculture and Food

Enquiries to: Head, Lajos Szalay, Müszaki Könyvtár
Founded: 1949
Subject coverage: Modernization of the technology for the wood industry; working out experimentally and manufacturing new products in order to substitute and economically utilize wood.
Library facilities: Not open to all users; loan services (nc); reprographic services (c).
Library holdings: 5726 bound volumes; 195 current periodicals; 1572 reports.
Information services: Bibliographic services (c); no translation services; literature searches (c); no access to on-line information retrieval systems.
Consultancy: No consultancy.
Publications: *Tudományos és Müszaki Tájékoztató - Faipar* (Scientific and Technical Information Service - Wood Industry), quarterly.

ITALY

Istituto per la Ricerca sul Legno   23.9
[Wood Research Institute]
Address: Villa Favorita, Piazza Edison 11, 50133 Firenze, Italy
Telephone: (055) 571581
Telex: 572458 ILFI
Affiliation: Istituto per la Tecnologia del Legno; Consiglio Nazionale delle Ricerche
Enquiries to: Librarian, Dr Simonetta Del Monaco
Founded: 1954
Subject coverage: Wood technology (anatomy, chemistry, mechanization, preservation of wood).
Library facilities: Not open to all users; no loan services; reprographic services (c).
Information services: Bibliographic services (c); no translation services; literature searches (c); access to on-line information retrieval systems.
Consultancy: Yes.
Publications: Bibliographic list; *Contributi Scientifico Pratici per una migliore conoscenza ed utilizzazione del legno* .

Istituto per la Tecnologia del   23.10
Legno
[Wood Technology Institute]
Address: Via Biasi 75, 38010 S Michele all'Adige Trento, Italy
Telephone: (0461) 650168; 650485
Telex: ILSMA 400453
Affiliation: Consiglio Nazionale delle Ricerche
Enquiries to: Director, Dr Attilio Arrighetti
Subject coverage: Basic and applied research on wood as a raw material for industry, excluding pulp and paper manufacture; standardization processes of the field; collecting, updating, and providing

documentation in related fields; direct technical assistance.
Publications: *Quaderni ITL* (research reports, irregular); *Bollettino Nuove Acquisizioni* (New Acquisition Bulletin, 5 per year).

NETHERLANDS

Houtinstituut TNO   23.11
[Forest Products Research Institute TNO]
Address: Schoemakerstraat 97, Postbus 151, 2600 AD Delft, Netherlands
Telephone: (015) 569330
Telex: 38071 zptno nl
Affiliation: Nederlandse Organisatie voor Toegepast-Natuurwetenschappelijk Onderzoek
Enquiries to: Head, A. van der Velden
Founded: 1940
Subject coverage: Wood preservation and chemistry; deterioration and conservation; drying and moisture-measurement; carpentry; furniture; timber.
Library facilities: Not open to all users; loan services; reprographic services.
Consultancy: Yes.
Publications: *Houtbulletin*, biannual; *HouTNOtitie* (eight per annum).

NORWAY

Norsk Treteknisk Institutt   23.12
– NTI
[Norwegian Institute of Wood Technology]
Address: Forskningsveien 3, PO Box 337, Blindern, Oslo 3, Norway
Telephone: (02) 46 98 80
Telex: 74864 tsfn
Affiliation: Norwegian Forestry and Forest Industries Research Council
Enquiries to: Research manager, Karl Mørkved
Founded: 1949
Subject coverage: Sawmill and planning technology; technology and anatomy; wood structures and other wood products; timber protection; glueing and surface treatment; chipboard.
Library facilities: Open to all users; loan services (nc); reprographic services (c).
Information services: Bibliographic services (c); no translation services; literature searches (c); no access to on-line information retrieval systems.
Consultancy: Yes (c).
Publications: Bibliographic list.

POLAND

Instytut Technologii Drewna 23.13
– ITD
[Wood Technology Institute]
Address: Winiarska 1, Poznań 60-654, Poland
Telephone: 22 40 81
Telex: 041 35 05 ITEDE
Affiliation: Ministry of Agriculture, Forestry and Food Economy
Enquiries to: Director, Professor Ryszard Babicki
Founded: 1952
Subject coverage: Processing, preservation, drying, surface finishing of wood and wood-based materials; application of wood and wood-based products in building, furniture, and other industries.
Library facilities: Open to all users (nc); loan services (nc); reprographic services (c).
Library holdings: 42 084 bound volumes; 312 current periodicals; 5502 reports.
Information services: Bibliographic services (c); no translation services; literature searches (c); access to on-line information retrieval systems (nc).
Abstracts compilation available on request.
Consultancy: Telephone answers to technical queries (nc); no writing of technical reports; no compilation of statistical trade data; no market surveys in technical areas.
Publications: *Prace Instytutu Technologii Drewna* (Works of the Wood Technology Institute), quarterly; *Przegląd Dokumentacyjny* (Abstracts Bulletin), bimonthly; *Ekspres Informacja Drzewnictwo* (Express Information Bulletin on Wood Science and Technology), monthly.

Ośrdek Badawczo-Rozwojowy 23.14
Meblarstwa *
– OBRoM
[Furniture Industry Research and Development Centre]
Address: Ulica Jeleniogórska 1-5, 60-179 Poznań, Poland
Telephone: 67 56 21
Telex: 0412745
Enquiries to: Director, W. Czarnota; Information chief, M. Kalemba
Scientific, Technical and Economic Information Department
Founded: 1965
Subject coverage: Wood and wood industry sciences in general; furniture design and manufacture in particular; applied research.
Library facilities: Open to all users; loan services; reprographic services.
Library holdings: The library holds over 30 000 specialist publications, and exchanges technical information with foreign centres.
Information services: Bibliographic services.

Consultancy: Yes.
Publications: *Biuletyn Informacyjny* (Information Bulletin), quarterly; *Przeglad Zawartości Czasopism Fachowych* (Review of Contents of Technical Periodicals), bimonthly.

ROMANIA

Central de Documentare 23.15
Technica Pentru Industria
Lemnului
– CDIL
[Woodworking Industry Documentation Centre]
Address: 46 Soseaua Pipera, Sectorul 2, Bucureşti 30, Romania
Telephone: 33 11 31

SPAIN

Asociación de Investigación 23.16
Técnica de las Industrias de la
Madera y Corcho
– AITIM
[Technical Investigation in Wood and Cork Industries Association]
Address: Flora 3, Madrid 13, Spain
Telephone: 91-242 58 64
Enquiries to: Secretary, Venerada Gomez Serrano
Founded: 1962
Library facilities: Open to all users; no loan services; reprographic services.
Information services: Bibliographic services; translation services; literature searches; no access to on-line information retrieval systems.
Consultancy: Yes.

SWEDEN

Möbelinstitutet 23.17
[Swedish Furniture Research Institute]
Address: PO Box 271 98, S-102 52 Stockholm, Sweden
Telephone: 08-67 92 45
Enquiries to: Information and Documentation Department
Founded: 1967
Subject coverage: Furniture - functional and performance tests and test methods for prototypes and finished furniture products.
Library facilities: Open to all users (open primarily to members, manufacturers, distributors, and designers

of furniture); no loan services; reprographic services (c - at cost price).
Consultancy: Yes.

Svenska Träskyddsinstitutet 23.18
[Swedish Wood Preservation Institute]
Address: Box 5607, S-114 86 Stockholm, Sweden
Telephone: 08-22 25 40
Telex: 14375 STURES
Facsimile: (08) 20 00 63
Enquiries to: Director, Jöran Jermer
Founded: 1974
Subject coverage: Wood preservation.
Library facilities: Open to all users; loan services (nc); reprographic services (c).
Information services: Bibliographic services (limited - c); no translation services; no literature searches; no access to on-line information retrieval systems.
Consultancy: Yes (c).
Publications: Bibliographic list.

UNITED KINGDOM

Furniture Industry Research 23.19
Association
– FIRA
Address: Maxwell Road, Stevenage, Hertfordshire SG1 4BT, UK
Telephone: (0438) 313433
Telex: 827653 FIRA G
Enquiries to: Director, business promotion, A.D. Spillard
Founded: 1961
Subject coverage: Furniture, fabrics, particle boards, adhesives, foams, timbers, surface finishes, and lacquers - research, testing, standards; ergonomics, factory, and marketing consultancies; measuring instruments and test equipment supply.
FIRA works for organizations in membership; these include UK furniture manufacturers, their suppliers of materials, components and machinery, colleges, design/architectural consultants, national standards/test authorities, retailers, and large furniture purchasing groups. Test and production consultancy services are available to non-members from allied industries such as shopfitting, toys and leisure, caravans, sports goods.
Library facilities: Not open to all users (free to members only); reprographic services (c - members only).
Library holdings: About 5000 bound volumes; about 300 current periodicals; reports.
Information services: Bibliographic services (c); translation services (c); literature searches (c); no access to on-line information retrieval systems.

Consultancy: Telephone answers to technical queries (nc - members only); writing of technical reports (c); compilation of statistical trade data (c); market surveys in technical areas (c).
Market surveys in consumer areas - furniture. Product design, production and market research for individual manufacturers.
Courses: Sixty courses per annum on topics related to furniture production, design, and materials.
Publications: *FIRA Bulletin*, quarterly; annual report; marketing reports; list of services; publications list on request.

Timber Research and 23.20
Development Association
– TRADA
Address: Stocking Lane, Hughenden Valley, High Wycombe, Buckinghamshire HP14 4ND, UK
Telephone: (0240 24) 3091
Telex: 83292
Enquiries to: Chief information officer, R. Allcorn; Librarian, Anne Peters
Founded: 1934
Subject coverage: Timber and wood-based products and their utilization.
Library facilities: Open to all users (for reference only - nc); loan services (to members only or through another library - BLDSC form required); reprographic services (c).
Library holdings: 10 000 bound volumes; 200 current periodicals; 35 000 reports.
Information services: Bibliographic services (charge for detailed information); no translation services; literature searches (c); access to on-line information retrieval systems.
Mainly TRADA Library's own retrieval system, TINKER (Timber Information Keyword Retrieval). Viewdata facilities (Mistel frame number 3511615).
Consultancy: Telephone answers to technical queries (charge to non-members for lengthy enquiries); writing of technical reports (c); compilation of statistical trade data (nc); no market surveys in technical areas.
Publications: *TRADA Library Bulletin*, monthly; *Timber Pallet and Packaging Digest*, bimonthly; *Timber Frame Press Cuttings Compilation* , monthly; publications list.

YUGOSLAVIA

Institut za Drvo 23.21
[Wood Institute]
Address: 8 maja 82/I, 41000 Zagreb, Yugoslavia
Telephone: 041-448 611
Enquiries to: Librarian
Founded: 1949

Subject coverage: Sawmilling; hydrothermal wood processing; veneer and board technology; final processing of wood; wood chemistry; woodworking production organization.
Library facilities: Open to all users; loan services; reprographic services (c).
Information services: Bibliographic services (c); translation services (c); literature searches (c).
Consultancy: Consultancy (c).
Publications: *Drvna Industrija* (Woodworking Industry) monthly.

24 TRANSPORTATION

AUSTRIA

Forschungsgesellschaft für das   24.1
Verkehrs-und Strassenwesen im
ÖIAV
[Research Society for Traffic and Road Research in ÖIAV]
Address: Eschenbachgasse 9, A-1010 Wien, Austria
Telephone: (0222) 57 05 22/19
Library facilities: Library.

Kuratorium für   24.2
Verkehrssicherheit
[Road Safety Board]
Address: Postfach 190, Ölzeltgasse 3, A-1031 Wien, Austria
Telephone: (0222) 73 15 71
Telex: 32 22 195
Affiliation: Austrian Automobile Association
Enquiries to: Librarian, Wladimir Bereza-Kudrycki
Founded: 1959
Subject coverage: Traffic safety.
Library facilities: Open to all users (nc); no loan services; reprographic services (c).
Library holdings: About 10 000 bound volumes; 130 current periodicals; about 12 000 reports.
Information services: Bibliographic services (c); no translation services; literature searches (c); no access to on-line information retrieval systems.
Consultancy: Telephone answers to technical queries (nc); writing of technical reports (c); no market surveys in technical areas.
Compilation of statistical accident and traffic data (c).
Publications: List on request.

Österreichische Bundesbahnen *   24.3
– ÖBB
[Austrian Federal Railways]
Address: Elisabethstrasse 9, A-1010 Wien, Austria
Telephone: (0222) 5650-5210

Telex: 1377
Enquiries to: Librarian, Dr Erwin Scheftek Bibliothek der Generaldirektion der Österreichischen Bundesbahnen, Praterstem 3, A-1020 Wien, Austria
Founded: 1896
Subject coverage: Transport.
Library facilities: Open to all users; loan services (c); reprographic services (c).
Information services: Bibliographic services (nc); no translation services; literature searches (nc); no access to on-line information retrieval systems.
Consultancy: Yes (nc).
Publications: ÖBB - Kurzinformationen (Zeitschrift-entitel: Kurzinformation) .

Österreichische Gesellschaft für   24.4
Strassenwesen
[Austrian Road Federation]
Address: Marxergasse 10, A-1030 Wien, Austria
Telephone: (0222) 73 62 96
Library facilities: Library.

BELGIUM

Centre de Recherches Routières/   24.5
Opzoekingscentrum voor de
Wegenbouw
– CRR
[Road Research Centre]
Address: Boulevard de la Woluwe, B-1200 Bruxelles, Belgium
Telephone: (02) 771 20 80
Enquiries to: Director, Jean Reichert
Subject coverage: Motorways, roads, airfield runways - safety and efficiency.

Société Nationale des Chemins de Fer Belges, Documentation et Bibliothèque 24.6

– SNCB
[Belgian National Railways, Documentation and Library]
Address: 85 Rue de France, B-1070 Bruxelles, Belgium
Telephone: (02) 525 35 30
Telex: 21526
Enquiries to: Chef de bureau, F. Van Renterghem
Founded: 1926
Library facilities: Open to all users (nc); no loan services; reprographic services (c).
Library holdings: About 50 000 bound volumes; 200 current periodicals; 100 reports.
Information services: Bibliographic services; no translation services; literature searches (nc); no access to on-line information retrieval systems.
Consultancy: No consultancy.
Publications: SNCB Documentation.

Union Internationale des Transports Publics 24.7

[International Union of Public Transport]
Address: Avenue de l'Uruguay 19, B-1050 Bruxelles, Belgium
Telephone: (02) 673 33 25
Telex: 63 916 uitp-b
Enquiries to: Library
Founded: 1885
Subject coverage: Urban public and regional public transport.
Library facilities: Not open to all users (members only); loan services (interlibrary loan - c); reprographic services (c).
Library holdings: 6000 bound volumes; 300 current periodicals.
Information services: Bibliographic services (members only - nc); no translation services; literature searches (members only - nc).
Access to on-line information retrieval systems is planned.
Consultancy: No consultancy.
Publications: Congress reports; statistical handbook; *Biblio-Index* (quarterly bulletin of abstracts).

BULGARIA

Scientific and Technical Transport Union 24.8

Address: Rakovski 108, 1000 Sofia, Bulgaria
Telex: 22 185 nts bg

CZECHOSLOVAKIA

Ceskoslovenske Statni Drahy 24.9

[Czechoslovak State Railways]
Address: Na Prikope 33, 110 05 Praha, Czechoslovakia

DENMARK

Statens Vejlaboratorium 24.10

[National Road Laboratory]
Address: PO Box 235, Elisagaardsvej 5-7, DK-4000 Roskilde, Denmark
Telephone: (02) 35 75 88
Telex: 43209 Vejlab dk
Affiliation: Vejdirektoratet
Enquiries to: Librarian, Lilian Olling
Founded: 1928
Subject coverage: Road and traffic research.
Library facilities: Open to all users (nc); loan services (nc); reprographic services (c).
Library holdings: 30 000 bound volumes; 170 current periodicals; reports.
Information services: Bibliographic services (nc); no translation services; literature searches (charge apart from manual searching and the laboratory's own data base); access to on-line information retrieval systems (c).
IRRD (International Road Research Documentation).
Consultancy: Telephone answers to technical queries (c); no writing of technical reports; no compilation of statistical trade data; no market surveys in technical areas.
Publications: Vejsektorens Fagbibliotek.

FINLAND

Rautatiehallitus * 24.11

[Finnish State Railways, Board of Administration]
Address: Box 488, SF-00101 Helsinki, Finland
Telephone: (90) 7072827
Telex: 12-301151
Enquiries to: Librarian, H. Väyrynen
Subject coverage: Railways.
Library facilities: Open to all users; loan services (nc - with certain restrictions); reprographic services (nc).
Information services: No bibliographic services; no translation services; literature searches (limited - nc); no access to on-line information retrieval systems.
Consultancy: No consultancy.

Suomen Tieyhdistys ry   24.12
[Finnish Road Association]
Address: Vironkatu 6, SF-00170 Helsinki 17, Finland
Telephone: (90) 627 094

Teknillinen Korkeakoulu   24.13
[Helsinki University of Technology, Department of Civil Engineering]
Address: Rakentajanaukio 4, SF-02150 Espoo, Finland
Telephone: 90-4512414
Telex: 125161
Enquiries to: Librarian, Eeva-Liisa Parkkonen
Subject coverage: Highway and railway engineering; bridge engineering; hydraulic engineering; water resources engineering; structural engineering; foundation engineering and soil mechanics; structural mechanics; construction economics and management; traffic and transportation engineering; concrete technology and steel structures.
Library facilities: Loan and reprographic services are available through the main library of Helsinki University of Technology.
Publications: List on request.

FRANCE

Association pour le   24.14
Développement des Techniques de
Transport, d'Environnement et de
Circulation
[Association for the Development of Transport, Environment and Traffic Techniques]
Address: 38 avenue Émile-Zola, 75015 Paris, France
Telephone: (1) 45 75 56 11

Centre de Documentation de la   24.15
SNCF
[French National Railways Documentation Centre]
Address: 163 bis avenue de Clichy, 75017 Paris, France
Affiliation: Société Nationale des Chemins de Fer Français
Founded: 1938
Subject coverage: Railways; transport; history of railways.
Library facilities: Open to all users; no loan services; reprographic services (c).
Information services: Translation services (access to translations already executed for SNCF - c).
Consultancy: Consultancy.

Institut de Recherche sur les   24.16
Transports
[Transport Research Institute]
Address: 1 avenue du Général Malleret Joinville, BP 28, 94114 Arcueil, France
Subject coverage: This government agency is particularly active in the field of fast inter-city links and short distance transport.

OECD International Road   24.17
Research Documentation
– IRRD
Address: 2 rue André-Pascal, 75775 Paris, France
Telephone: (1) 45 24 92 44
Telex: 620160
Parent: Organisation for Economic Cooperation and Development
Enquiries to: Head of Divison, Burkhard E. Horn
Founded: 1965
Subject coverage: Roads and road-related transport aspects.
Information services: Bibliographic services; literature searches; access to on-line information retrieval systems (by on-line searching of File 43 - IRRD, of the Information Retrieval Service of the European Space Agency).

GERMAN FEDERAL REPUBLIC

Bundesanstalt für Strassenwesen   24.18
– BASt
[Federal Highway Research Institute]
Address: Postfach 10 01 50, Brüderstrasse 53, D-5060 Bergisch Gladbach 1, German Federal Republic
Telephone: (02204) 430
Telex: 8878483 bas d
Affiliation: Bundesministerium für Verkehr
Enquiries to: Head of Section Z 2.2, Dr Konrad Murr
Founded: 1951
Subject coverage: Highway construction technology; traffic engineering; road safety and accident research; central services.
Library facilities: Not open to all users; loan services; reprographic services (nc) .
Information services: No bibliographic services; no translation services; literature searches (nc); access to on-line information retrieval systems.
OECD/IRRD: International Road Research Documentation (Coordinating Centre for German-speaking Countries); Geotechnical Abstracts; Conférence Européenne des Ministres des Transports (CEMT), Documentation.
Consultancy: Consultancy (nc).
Publications: Library acquisition lists.

Deutsche Bundesbahn Dokumentationsdienst 24.19
[German Railways Documentation Department]
Address: Kaiserstrasse 48, D-6000 Frankfurt am Main, German Federal Republic

Zentrale Informationsstelle für Verkehr in der DVWG 24.20
– ZIV
[Transport Information Centre]
Address: Brüderstrasse 53, D-5060 Bergisch Gladbach 1, German Federal Republic
Telephone: (02204) 60029
Affiliation: Deutsche Verkehrswissenschaftliche Gesellschaft
Enquiries to: Klaus Thielen
Founded: 1965
Subject coverage: Transport.
Library facilities: Open to all users (nc); no loan services; reprographic services (charge for more than ten copies).
Information services: Bibliographic services (*Schrifttum aus dem Verkehrswesen* - nc); no translation services; no literature searches; access to on-line information retrieval systems (nc).
Pool TRANSDOC of CEMT/CIDET/ICTED, Paris.
Consultancy: No consultancy.

GREECE

Greek Road Federation 24.21
Address: 14-16 Feidiou Street, Athinai 106 78, Greece
Telephone: 36 25 578
Library facilities: Library.

HUNGARY

Közuti Közlekedési Tudományos Kutató Intézet 24.22
– KÖTUKE
[Road Transport Research Institute]
Address: Thán Károly útca 3-5, POB 107, H-1119 Budapest XI, Hungary
Telephone: (01) 666-987
Telex: 226443
Affiliation: Ministry of Transport and Communications
Enquiries to: Head of Information Centre
Founded: 1971
Subject coverage: Roads - planning, operations, maintenance, repair, traffic, safety; vehicles - testing, maintenance, repair, interaction between vehicle and road; transport - general planning, coordination, passenger transport, goods transport; energy; environment.
Library facilities: Open to all users; loan services; reprographic services (c - on non-profitmaking basis).
Information services: Bibliographic services; translation services; literature searches; access to on-line information retrieval systems (c - on non-profitmaking basis).
IRRD.
Consultancy: Yes.
Publications: List on request.

IRELAND

Coras Iompair Eireann * 24.23
– CIE
[Ireland's Transport Company]
Address: Heuston Station, Dublin 8, Ireland
Telephone: (01) 771871
Telex: 25153 CIE EI
Enquiries to: Librarian, Strategic and Corporate Planning Department
14-19 Crow Street, Dublin 2
Founded: 1945
Subject coverage: Public surface transport - passenger and freight.
Library facilities: Each branch of the company has its own collection of technical literature, while the Strategic and Corporate Planning Department holds copies of a range of periodicals, mainly relating to transport. These collections are not generally available for public lending.

ITALY

Azienda Autonoma Ferrovie dello Stato 24.24
– FS
[Italian State Railways]
Address: Direzione Generale FS, Servizio Affari Generali, Piazza della Croce Rossa 1, 00161 Roma, Italy
Telex: 61089
Enquiries to: Head of Documentation Centre
Founded: 1905
Library facilities: Open to all users; loan services; reprographic services.
Information services: Bibliographic services; translation services; literature searches; access to on-line information retrieval systems.
Consultancy: Yes.

Centro per lo Sviluppo dei Trasporti Aerei * 24.25
[Air Transport Development Centre]
Address: Via Sardegna 38, 00187 Roma, Italy
Enquiries to: President, E. Carenini
Library holdings: Over 6000 books and documents.

NETHERLANDS

Instituut voor Wegtransportmiddelen TNO 24.26
– IW-TNO
[Road Vehicles Research Institute TNO]
Address: Postbus 237, 2600 AE Delft, Netherlands
Telephone: (015) 569330
Telex: 38071 zptno nl
Affiliation: Organisatie voor Toegepast Natuurwetenschappelijk Onderzoek
Enquiries to: Librarian, E.G. van Koperen
Founded: 1970
Subject coverage: Research, development and advice in the field of design and application of road vehicles and their parts. Special fields of interest include application of alternative fuels (LPG, LNG and alcohol fuels); safety, crash phenomena and biomechanics; optimization and application of restraint-systems for adults and children; general bicycle technology; approval testing according to national and international regulations.
Library facilities: Open to all users (nc); loan services (nc); reprographic services (nc).
Information services: No bibliographic services; no translation services; literature searches (via CID - TNO - c); access to on-line information retrieval systems (via CID - TNO - c).
Consultancy: No consultancy.
Publications: *By the Way*, newsletter.

Nederlandse Spoorwegen NV 24.27
[Netherlands Railways]
Address: Moreelsepark 1, 3511 EP Utrecht, Netherlands
Telephone: (030) 359111
Telex: 47257
Enquiries to: Head
Documentation and Information Centre
Founded: 1938
Subject coverage: Railways - traffic; engineering.
Library facilities: Loan services (interlibrary only); reprographic services (c).
The library is open to non-staff users after personal introduction only. There is a charge for non-staff users.
Library holdings: 90 000 bound volumes; 1200 current periodicals.
Information services: No information services.

Consultancy: No consultancy.

Stichting Weg * 24.28
[Dutch Road Federation]
Address: Oranjestraat 2A, 2514 JB 's-Gravenhage, Netherlands
Telephone: (070) 64 92 90
Library facilities: Library.

NORWAY

Norges Statsbaner, Hovedadministrasjonens Bibliotek 24.29
– NSB
[Norwegian State Railways, Executive Offices Library]
Address: Storgaten 33, N-0184 Oslo 1, Norway
Telephone: (02) 209550
Telex: 71 168 nsbdc n
Affiliation: Norges statsbaner (Norwegian State Railways)
Enquiries to: Head librarian, Sigrun Tennebø
Founded: 1923, reorganized 1947
Subject coverage: Transport, especially railways.
Library facilities: Open to all users (nc); loan services (nc); reprographic services (nc).
Library holdings: About 43 000 bound volumes; about 800 current periodicals.
Information services: Bibliographic services (nc); no translation services; literature searches (nc); access to on-line information retrieval systems (nc).
Dialog, ESA IRS, national data bases.
Consultancy: Telephone answers to technical queries (no charge for simple questions); no writing of technical reports; no compilation of statistical trade data; no market surveys in technical areas.

Opplysningsrådet for Veitrafikken * 24.30
[Norwegian Road Federation]
Address: Prof Dahlsgate 1, Oslo 3, Norway
Telephone: (02) 60 02 68; 69 57 73

Skipsfartsøkonomisk Institutt 24.31
[Shipping Research Institute]
Address: Helleveien 30, N-5035 Bergen-Sandviken, Norway
Telephone: (5) 25 55 27
Telex: 40642 NHH N
Facsimile: (5) 25 88 74
Affiliation: Norwegian School of Economics and Business Administration
Enquiries to: Director, Professor Arnljot Strømme Svendsen; Tor Wergeland
Founded: 1958

Subject coverage: Applied shipping economics; strategic analysis in shipping; simulation modelling of international economics and shipping markets.
Library facilities: Open to all users (nc); no loan services.
Library holdings: 3500 bound volumes; 1450 current periodicals.
Information services: Bibliographic services (nc); no translation services.
Consultancy: No telephone answers to technical queries; no writing of technical reports; compilation of statistical trade data.
Publications: Bibliographic list.

Transportøkonomisk Institutt   24.32
– TØI
[Transport Economics Institute]
Address: Grensveien 86, Oslo 6, Norway
Telephone: (02) 65 9500
Affiliation: Norwegian Centre for Transport Research
Enquiries to: Librarian
Subject coverage: All aspects of inland transportation: transport engineering; traffic safety; highway engineering; transport economics.
Library facilities: Open to all users; loan services; reprographic services.
Publications: *Samferdsel*, ten per annum; *TØI-Bibliografi*; reports.

POLAND

Ośrodek Informacji Naukowej,   24.33
Technicznej i Ekonomicznej
Komunikacji-Oitek
[Transport Scientific, Technical and Economic Information Centre]
Address: Ulica Chałubinńskiego 6, 00 - 928 Warszawa, Poland
Telephone: 244058
Telex: 813898pkp pl
Affiliation: Ministry of Transport
Enquiries to: Director, T. Bruszewski
Founded: 1961
Subject coverage: Transport.
Library facilities: Open to all users; loan services (nc); reprographic services (c).
Information services: Bibliographic services (nc); translation services (c).
Consultancy: Yes (nc).

ROMANIA

Calle Ferate Romane,   24.34
Departmentul Cailor Ferate
[Romanian Railways Board, Department of Railways]
Address: Bulevardul Dinicu Golescu 38, Bucureşti 7, Romania

Centrul de Documentare si   24.35
Publicatii Technica al
Ministerului Transporturilor
[Technical Documentation and Publications Centre of the Ministry of Transport]
Address: Calea Gritvitei 193B, Bucureşti, Romania

SPAIN

Asociación Española de la   24.36
Carretera
[Spanish Highways Association]
Address: Ciudad Universitaria-Edificio Seminarios Pta 6, Madrid 3, Spain
Telephone: (91) 243 17 04

SWEDEN

Statens Väg- och Trafikinstitut   24.37
– VTI
[Swedish Road and Traffic Research Institute]
Address: S-581 01 Linköping, Sweden
Telephone: 013-11 52 00
Telex: 50125 VTISGIS
Enquiries to: Head of Information and Documentation Section, Tim Sigvard
Subject coverage: Roads, traffic, vehicles, road users and traffic safety.
Library facilities: Open to all users; loan services; reprographic services.
The library is the central Swedish special scientific library in the field of road transport.
Information services: Literature searches (c); access to on-line information retrieval systems (c).
IRRD (International Road Research Documentation) and Roadline (VTI Library) available on-line with TRIP IR-system via Scannet computer network.
The Information and Documentation Section is the Swedish documentation centre for the International Road Research Documentation system (IRRD) within the OECD Road Research Programme.

SWITZERLAND

Dokumentationsdienst SBB 24.38
[Swiss Federal Railways Documentation Service]
Address: Mittelstrasse 43, CH-3030 Bern, Switzerland
Telephone: (031) 60 25 11
Telex: 991212
Enquiries to: Head of documentation service, W. Holzer
Founded: 1923
Subject coverage: Transportation; railroad transport.
Library facilities: Open to all users (nc); loan services (nc); reprographic services (nc, except if required in large quantities).
Library holdings: 45 000 bound volumes; 650 current periodicals; reports.
Information services: Bibliographic services (nc, unless extensive); no translation services; literature searches (nc, unless extensive); no access to on-line information retrieval systems.
Consultancy: Telephone answers to technical queries (nc); no writing of technical reports; no compilation of statistical trade data; no market surveys in technical areas.
Publications: *Jahreskatalog; Bibliographie der Schweizerischen Verkehrswirtschaft.*

International Road Federation 24.39
– IRF
Address: 63 rue de Lausanne, CH-1202 Genève, Switzerland
Telephone: (022) 31 71 50
Telex: 27590 IRF
Enquiries to: Secretary general, Dr J. Schälchli
Founded: 1948
Subject coverage: Construction, improvement, and maintenance of national and international road systems; road construction; road transport.
Library facilities: Open to all users; loan services (only within Switzerland); no reprographic services.
Information services: Bibliographic services; literature searches.
Consultancy: Yes.

Schweizerischer 24.40
Strassenverkehrsverband/
Fédération Routière Suisse *
[Swiss Road Federation]
Address: Schwanengasse 3, Postfach 2299, CH-3001 Bern, Switzerland
Telephone: (031) 22 36 49
Affiliation: International Road Federation
Enquiries to: Secretary general, Dr J. Schälchli
Founded: 1945
Subject coverage: Construction and use of roads.

Library facilities: Open to all users; loan services; no reprographic services.
Information services: Bibliographic services (nc); no translation services; literature searches (nc).
Consultancy: Yes (nc).

TURKEY

Yollar Derneği 24.41
[Turkish Road Association]
Address: Şişli Meydani 364, Istanbul, Turkey
Telephone: 46 70 90

UNITED KINGDOM

Department of the Environment 24.42
and Transport Library
Address: Marsham Street, London SW1P 3EB, UK
Telephone: (01) 212 4847
Telex: 22221
Affiliation: Department of the Environment; Department of Transport
Enquiries to: Librarian
Subject coverage: Housing; town and country planning; local government; pollution; ports; transport (including shipping and civil aviation); water supply; roads; bridges; mineral workings; regional planning.
Library facilities: Not open to all users; loan services; no reprographic services.
Loan services are available only to libraries with BL forms. Library facilities are restricted primarily to officers of both departments; researchers only with prior permission.
Information services: Bibliographic services (c).
Published bibliographic items only.
Publications: *Library Bulletin*, fortnightly abstracts series; *Annual List of Publications; Monthly Supplements*; bibliographies in planning, housing, transport and local government.

Transport and Road Research 24.43
Laboratory *
–TRRL
Address: Old Wokingham Road, Crowthorne, Berkshire RG11 6AU, UK
Telephone: (03446) 77 3131
Telex: 848272
Affiliation: Department of Transport
Enquiries to: Librarian, B.R. Styles
Founded: 1933
Subject coverage: Highway engineering; traffic engineering; road safety; transport planning and operations.

Library facilities: Open to all users (by appointment only, 24 hours notice required); loan services (c - sometimes); reprographic services (by arrangement).
Library holdings: Microfiche collection of pamphlets and some books; 22 000 book titles; 80 000 pamphlet titles; over 1000 journal titles.
Information services: Bibliographic services (c); translation services (c); literature searches (c); access to on-line information retrieval systems (c).
TRACS and PROJEX contributing in part to IRRD and IRS. Other internal data bases: Blaise; Dialog; IRS/ESA; InfoLine. English language coordinating centre for IRRD (OECD) and main contributor to IRF research in progress. SDI computer searches provided to external enquirers for an annual charge: twelve issues, personalized search.
Consultancy: Yes (c).

YUGOSLAVIA

Institut Iker/'Kirilo Savić'/   24.44
Železnički Institut *
[Iker Institute/'Kirilo Savić'/Railway Institute]
Address: 362/ Vojvode Stepe 51, Beograd 11000, Yugoslavia
Telephone: (011) 469322
Telex: 11565
Affiliation: National Library SRS
Enquiries to: Director, Joviša Prokopijević
Founded: 1954
Subject coverage: Research, construction, consulting, engineering, ergonomics, civil engineering, and railway technology.
Library facilities: Open to all users; loan services (c); reprographic services (c).
Information services: Bibliographic services; literature searches.
Consultancy: Yes.

Yugoslav Road Association   24.45
Address: Post Box 470, Kumodraska St 257, 11001 Beograd, Yugoslavia

25 VETERINARY SCIENCES

AUSTRIA

Österreichische Gesellschaft der Tierärzte 25.1
[Austrian Society of Veterinary Surgeons]
Address: Linke Bahngasse 11, A-1030 Wien, Austria
Enquiries to: President, Dr W. Krocza

Veterinärmedizinische Universität Wien, Universitätsbibliothek 25.2
[Veterinary University of Vienna, University Library]
Address: Linke Bahngasse 11, A-1030 Wien, Austria
Telephone: (0222) 735581
Enquiries to: Librarian, Dr Günter Olensky
Founded: 1767
Subject coverage: Veterinary medicine.
Library facilities: Open to all users; loan services; reprographic services.
Library holdings: 110 487 bound volumes; 681 current periodicals.
Information services: Bibliographic services; no translation services; no literature searches; no access to on-line information retrieval systems.
Consultancy: No consultancy.

BELGIUM

Institut National de Recherches Vétérinaires/Nationaal Instituut voor Diergeneeskundig Onderzoek 25.3
– INRV/NIDO
[National Institute for Veterinary Research]
Address: Groeselenberg 99, B-1180 Bruxelles, Belgium
Telephone: (02) 375 44 55
Affiliation: Ministère de l'Agriculture/Ministerie van Landbouw
Enquiries to: Librarian, G. Stiers

Founded: 1930
Library facilities: Open to all users (nc); loan services (nc); reprographic services (c).
Information services: No information services.
Consultancy: Telephone answers to technical queries (nc); no writing of technical reports; no compilation of statistical trade data; no market surveys in technical areas.
Publications: Annual report.

Koninklijke Maatschappij voor Dierkunde van Antwerpen 25.4
[Royal Zoological Society of Antwerp]
Address: Koningin Astridplein 26, B-2018 Antwerpen, Belgium
Telephone: (03) 231 16 40
Enquiries to: Director, F.J. Daman
Founded: 1843
Subject coverage: Zoology, botany, general biology; current research is particularly concerned with pathology of animals in captivity.
Library facilities: Open to all users; no loan services; reprographic services (c).
Loan services - interlibrary only.
Information services: Bibliographic services; no translation services; literature searches; no access to on-line information retrieval systems.
Consultancy: Yes.
Publications: Zoo-Antwerpen; Zoo-Anvers (includes annual report); Acta Zoologica et Pathlogica Antverpiensia.

BULGARIA

Higher Institute of Zootechnics and Veterinary Medicine 25.5
Address: D Blagoev str 62, 6000 Stara Zagora, Bulgaria
Telex: 88 465
Enquiries to: Chief of the library, Ivanka Demireva

Founded: 1974

Subject coverage: Veterinary husbandry, animal husbandry.

Library facilities: Open to all users (nc); loan services (nc); reprographic services (nc).

Library holdings: 139 751 bound volumes; 678 current periodicals.

Information services: Bibliographic services (nc); translation services (nc); literature searches (nc); access to on-line information retrieval systems (nc).

Access to on-line information retrieval systems through the Central Institute for Scientific Information, Sofia.

Consultancy: Telephone answers to technical queries (nc); writing of technical reports (nc).

Publications: Scientific works; reports; bulletin.

CYPRUS

Veterinary Services Department 25.6

Address: PO Box 2006, Nicosia, Cyprus

Telephone: 40 2135

Telex: 4660 MINA GRI CY

Parent: Ministry of Agriculture and National Resources

Enquiries to: Director, K. Polydorou

Subject coverage: Topics covered include virology, parasitology, microbiology, food, biological products, and abortions.

CZECHOSLOVAKIA

Vysoká Škola Veterinární 25.7

[University School of Veterinary Medicine, Central Library]

Address: Palackeho 1-3, 612 42 Brno 12, Czechoslovakia

Telephone: 445

Enquiries to: Librarian

Founded: 1918

Library facilities: Open to all users; loan services; reprographic services.

Information services: Bibliographic services; translation services; literature searches.

Consultancy: Yes.

DENMARK

Danmarks Veterinaer- og 25.8
Jordbrugsbibliotek *
– DVJB

[Danish Veterinary and Agricultural Library]

Address: Bülowsvej 13, DK-1870 København V, Denmark

Telephone: (01) 35 17 88

Telex: 15061 dvj bib dk

Affiliation: Den Kongelige Veterinaer- og Landbohøjskole (Royal Veterinary and Agricultural University)

Enquiries to: Chief librarian, Inge Berg Hansen.

Det veterinaer- og jordbrugsfaglige Dokumentations-center

Founded: 1783

Subject coverage: Agriculture, horticulture, fisheries, food science, veterinary medicine.

Library facilities: Open to all users; loan services (nc); reprographic services (c).

Information services: Bibliographic services; translation services; literature searches (nc for manual searches); access to on-line information retrieval systems (c).

The library has access to Euronet, Lockheed Dialog, AGRIS (Vienna), and IRS (Frascati) for documentation services. For document location there is access to Blaise, Libris (Sweden), and Alis (Denmark). DVJB provides on-line access to its own monograph collection (1979-84) as part of Alis at I/S Datacentralen.

Consultancy: Yes.

Den Danske Dyrlaegeforening 25.9

[Danish Veterinary Association]

Address: Alhambravej 15,DK-1826 Københan V, Denmark

Telephone: (01) 22 17 88

Zoologisk Have 25.10

[Zoological Gardens]

Address: Sdr. Fasanvej 79, DK-2000 København F, Denmark

Telephone: (01) 302555

Enquiries to: Curator, Bengt Holst

Founded: 1859

Subject coverage: Zoo animals, animal behaviour, animal diseases.

Library facilities: Open to all users; no loan services; no reprographic services.

Concentrating on zoo literature (zoo animals, zoo management, zoo periodicals).

Information services: No bibliographic services; no translation services; literature searches (nc); no access to on-line information retrieval systems.

Consultancy: Yes (nc).

FINLAND

Eläinlääketieteellinen Korkeakoulu   25.11
[Veterinary Medicine College, Library]
Address: PO Box 6, SF-00551 Helsinki 55, Finland
Telephone: (90) 711 411
Telex: 123203 ELKK sf
Affiliation: Ministry of Education

Suomen Eläinlääkäriliitto ry   25.12
[Finnish Veterinary Association]
Address: Akavatalo, Rautatieläisenkatu 6,SF-00520 Helsinki 52, Finland
Telephone: (90) 1 50 21

FRANCE

Académie Vétérinaire de France *   25.13
[Veterinary Academy of France]
Address: 60 boulevard Latour-Maubourg, 75007 Paris, France
Enquiries to: Professor Bordet
École Nationale Vétérinaire 94704 Maisons-Alfort Cédex
Founded: 1928
Subject coverage: All aspects of animal science.
Library facilities: Open to all users; no loan services; no reprographic services.
Consultancy: Yes.

Centre National de Recherches Zootechniques   25.14
[National Centre of Zootechnical Research]
Address: Domaine de Vilvert, 78350 Jouy-en-Josas, France
Telephone: 49 56 80 80
Telex: INRACRZ 695431 F
Library holdings: 4000 volumes, 1200 periodicals.

Institut d'Élevage et de Médecine Vétérinaire des Pays Tropicaux *   25.15
– IEMVT
[Animal Production and Veterinary Science of Tropical Countries Institute]
Address: 10 rue Pierre-Curie, 94704 Maisons-Alfort Cedex, France
Telephone: (1) 43 68 88 73
Affiliation: Ministère des Relations Extérieures, Coopération et Développement
Enquiries to: Information officer, Dr J.F. Giovannetti
Founded: 1921
Subject coverage: Research into tropical animal husbandry, development and improvement of livestock and animal production, grazing and nutrition, fodder production and pathology; training; documentation.
Library facilities: Open to all users; no loan services; reprographic services.
Information services: Bibliographic services; literature searches.
Consultancy: Yes.
Publications: *Revue d'Élevage et de Médecine Vétérinaire des Pays Tropicaux* .

Société Vétérinaire Pratique de France   25.16
[Veterinary Practice Society of France]
Address: Maison des Vétérinaires, 10 place Léon Blum, 75011 Paris, France
Library facilities: Library.

GERMAN FEDERAL REPUBLIC

Dokumentationsstelle für Veterinärmedizin   25.17
[Veterinary Medicine Documentation Office]
Address: Koserstrasse 20, D-1000 Berlin 33, German Federal Republic
Telephone: (030) 838 3227
Affiliation: Freie Universität Berlin
Founded: 1969
Subject coverage: Veterinary science.
Library facilities: Not open to all users.
Information services: Literature searches; access to on-line information retrieval systems.
Medlars; SCI-search; Biosis Previews; CAB abstracts/animal; FSTA; Psychological Abstracts; EMbase; Cancerlit.

Tierärztliche Hochschule Hannover *   25.18
[School of Veterinary Medicine School, Hanover]
Address: Bischofsholer Damm 15, D-3000 Hannover, German Federal Republic
Telephone: (0511) 856 464
Telex: 922034 tiho d
Enquiries to: Library director, Dr K. Baresel
Founded: 1778
Subject coverage: Veterinary medicine, laboratory animals, parasitology, zoology.
Library facilities: Open to all users; loan services (c - for international loans); reprographic services.
Information services: Bibliographic services; no translation services; literature searches; access to on-line information retrieval systems (c) .
DIMDI.
Consultancy: Yes.

GREECE

Hellenic Veterinary Medical Society 25.19

Address: PO Box 3546, 102 10 Athinai, Greece
Telephone: (01) 524 46 53
Library facilities: Library.

Veterinary Institute of Thessaloniki 25.20

Address: 66, 26th of October Street, Thessaloniki, Greece
Telephone: (031) 516 608
Affiliation: Ministry of Agriculture
Enquiries to: Director, Dr Dem Giannacoulas
Founded: 1941
Subject coverage: Infectious, parasitic, and other diseases of different animals; hygienic and technological control of foods of animal origin.
Library facilities: Open to all users; loan services (nc - limited period); no reprographic services.
Information services: Bibliographic services (nc); no translation services; literature searches (nc); no access to on-line information retrieval systems.
Consultancy: Yes (nc).

HUNGARY

Állatorvostudományi Egyetem 25.21
[University of Veterinary Science]

Address: Postafiók 2, Landler Jenö u. 2, Budapest, Hungary
Telephone: (01) 222-660
Telex: 224439
Affiliation: Ministry of Agriculture and Food
Enquiries to: Library director, Ilona Bakonyi University central library
Founded: 1787
Library facilities: Open to all users (nc); loan services (nc); reprographic services (c).
Library holdings: 130 000 bound volumes; 610 current periodicals.
Information services: Bibliographic services (nc); no translation services; literature searches (nc); access to on-line information retrieval systems (c).
Consultancy: Telephone answers to technical queries (nc); no writing of technical reports; no compilation of statistical trade data; no market surveys in technical areas.
Publications: Bibliography of Hungarian veterinary literature (annual); list of publications and scientific papers (annual); publications of the central library: *Archives of the University of Veterinary Sciences (1741) 1787-1972; Old Hungarian Books (1574-1850) and Foreign Rarieties (1550-1600) in the Collection of*

Books on Veterinary History; Extractus Protocolli Instituti Veterinarii 1787-1815.

Országos Állategészségügyi Intézet 25.22
[Central Veterinary Institute Library]

Address: Tábornok útca 2, H-1149 Budapest XIV, Hungary
Telephone: 840-100/70
Telex: 224430 aegit h
Affiliation: Ministry of Agriculture and Food
Enquiries to: Librarian, Dr Gy Kovács
Founded: 1928
Subject coverage: Infectious, parasitic, and metabolic diseases of animals; veterinary toxicology.
Library facilities: Open to all users; loan services; reprographic services.
All services to non-staff users from other institutions with official letter only.
Library holdings: 5966 volumes; 88 journals, 38 from abroad.

IRELAND

University College Dublin, Veterinary Medicine Library 25.23

Address: Veterinary College, Ballsbridge, Dublin 4, Ireland
Telephone: (01) 687988
Enquiries to: Librarian, M. McErlean
Founded: 1968
Subject coverage: Veterinary science and related subjects.
Library facilities: Open to all users (under university library regulations); loan services; reprographic services.
Library holdings: 23 000 bound volumes; 470 current periodicals.
Information services: Bibliographic services (nc); no translation services; literature searches (nc); access to on-line information retrieval systems (c).
Consultancy: No consultancy.

ITALY

Società Italiana delle Scienze Veterinarie 25.24
[Italian Society of Veterinary Science]

Address: Via A Bianchi 1, 25100 Brescia, Italy
Telephone: 525 16
Library facilities: Library.

NETHERLANDS

Centraal Diergeneeskundig Instituut 25.25

– CDI

[Central Veterinary Institute]

Address: PO Box 65, 8200 AB Lelystad, Netherlands
Telephone: (03200) 73911
Telex: 40227
Affiliation: Ministry of Agriculture and Fisheries
Enquiries to: Head of Information Department, Dr P.W. van Olm
Founded: 1904
Subject coverage: Veterinary medicine.
Library facilities: Open to all users (by appointment); loan services (c); reprographic services (c).
Information services: Bibliographic services (nc); no translation services; literature searches (c); access to on-line information retrieval systems.
Dialog, ESA, DIMDI.
Consultancy: Yes (nc).

Centraal Proefdierenbedrijf TNO 25.26

– CPB-TNO

[Central Institute for the Breeding of Laboratory Animals TNO]

Address: PO Box 167, Zeist, Utrecht, Netherlands
Telephone: (03439) 1646
Telex: 43413 NEC att. CPB-TNO
Affiliation: Nederlandse Organisatie voor Toegepast - Natuurwetenschapppelijk Onderzoek
Enquiries to: Deputy manager, Dr J.W.M.A. Mullink
Founded: 1954
Subject coverage: Breeding of laboratory animals; genetics; health surveillance.
Library facilities: Open to all users; loan services; reprographic services (limited).
Information services: No information services.
Consultancy: Telephone answers to technical queries; no writing of technical reports; no compilation of statistical trade data; no market surveys in technical areas.

NORWAY

Norges Veterinaerhøgskoles Bibliotek 25.27

[Norwegian College of Veterinary Medicine Library]

Address: PO Box 8146 Dep, Ullevålsveien 72, N-0033 Oslo 1, Norway
Telephone: (02) 693690
Enquiries to: Head librarian, Anne Sakshaug
Founded: 1891

Subject coverage: Veterinary medicine; animal science.
Library facilities: Open to all users (nc); loan services (nc); reprographic services (nc).
Library holdings: 62 400 bound volumes; 496 current periodicals.
Information services: Bibliographic services (nc); no translation services; literature searches (nc); access to on-line information retrieval systems (no charge for persons employed at the college).
Consultancy: Telephone answers to technical queries (nc); no writing of technical reports; no compilation of statistical trade data; no market surveys in technical areas.

POLAND

Instytut Weterynarii 25.28

– IWet

[Veterinary Research Institute]

Address: Al. Partyzantow 57, 24-100 Pulawy, Poland
Telephone: 30-51
Telex: 0642401
Affiliation: Ministry of Agriculture, Forestry and Food Economy
Enquiries to: Secretary for research affairs, Dr Jacek Roszkowski
Founded: 1945
Subject coverage: Development of scientific principles of animal health protection, prophylaxis of zoonoses and hygiene of feedstuffs and animal products.
Library facilities: Open to all users (nc); loan services (nc); reprographic services (nc).
Library holdings: 24 267 bound volumes; 124 current periodicals.
Information services: No information services.
Consultancy: Telephone answers to technical queries (nc); writing of technical reports (nc); no compilation of statistical trade data; no market surveys in technical areas.

Polska Akademia Nauk, Instytut Parazytologi im. W. Stefańskiego, Biblioteka 25.29

[W. Stefański Institute of Parasitology of the Polish Academy of Sciences, Library]

Address: Skr poczt 153, Ulica L. Pasteura 3, 00-973 Warszawa, Poland
Telephone: 22 25 62
Enquiries to: Head librarian, M. Radziejewski
Founded: 1952
Subject coverage: Parasitology, including animal parasitism, its origin, prevalence, manifestations, and effects in natural and experimental parasite-host systems; phylogeny and ontogeny, physiology,

protozoology, immunology, environmental parasitology, parasitic zoonoses.
Library facilities: Open to all users (nc); loan services (nc); reprographic services (c).
Library holdings: 26 720 bound volumes; 646 current periodicals.
Information services: No information services.
Consultancy: No consultancy.
Publications: *Acta Parasitologica Polonica* in English, quarterly; *Polska Bibliografia Parazytologiczna*, annually.

Polskie Towarzystwo Nauk Weterynaryjnych 25.30
[Polish Society of Veterinary Sciences]
Address: Grochowska 272, 03-849 Warszawa, Poland
Telephone: 10 33 97
Library facilities: Library.

PORTUGAL

Escola Superior de Medicina Veterinária Biblioteca * 25.31
[Veterinary School Library]
Address: Rua Gomes Freire, 1199 Lisboa Codex, Portugal
Telephone: (019) 562596/7
Affiliation: Universidade Tecnica de Lisboa
Enquiries to: Librarian, Dr Leopoldo Kamat da Rocha
Founded: 1852
Subject coverage: Veterinary medicine and fields related to medical science.
Library facilities: Open to all users; loan services (nc); reprographic services (c).
Library holdings: The library is the main source of information on Portuguese veterinary medicine; it holds approximately 32 400 monographs and the stock of 'reserved books' is also considerable. It holds author, title, and subject catalogues, a chronological catalogue for old and rare books, and a catalogue of serials.
Information services: Bibliographic services (nc); no translation services; no literature searches; no access to on-line information retrieval systems.
Consultancy: Consultancy (nc).
Publications: Printed bibliographies: *Catálogo de Alguns Reservados da Biblioteca da Escola Superior de Medicina Veterinária; Indice Bibliográfico dos trabalhos escritos pelos Autores Veterinários Portugueses no sector dos animais aquáticos* (Inventory of reports published by Portuguese Veterinaries in Aquatic Animals sector); *Bibliografia Veterinária Portuguesa; Indice Bibliográfico dos escritos produzidos pelos Autores Veterinários Portugueses.*

Sociedade Portuguesa de Ciências Veterinárias 25.32
[Portuguese Society of Veterinary Science]
Address: Rua D Dinis 2-A, 1200 Lisboa, Portugal
Telephone: 68 01 88
Library facilities: Library.

ROMANIA

Institutul de Biologie şi Nutriţie Animală 25.33
[Institute of Biology and Animal Nutrition]
Address: jud. Ilfov, Baloteşti, Romania
Parent: Academia de Stiinte Agricole şi Silvice
Enquiries to: Director, Ion Moldovan

Institutul de Cercetări Veterinare şi Biopreparate 'Pasteur' 25.34
[Pasteur Institute of Veterinary Research and Biological Preparations]
Address: Soseaua Giulesti 333, 77826 Bucureşti, Romania

SWEDEN

Statens Veterinärmedicinska Anstalt * 25.35
– SVA
[National Veterinary Institute]
Address: Box 7073, S-750 07 Uppsala, Sweden
Telephone: 018-16 90 00
Enquiries to: Librarian, Gunnel Martenius Erne
Founded: 1911
Library facilities: Not open to all users; loan services (nc); reprographic services (c).
Information services: Bibliographic services; no translation services; literature searches (manual). Connected to Libris (Swedish Library Information System).
Consultancy: No consultancy.
Publications: Research report; annual bibliography.

SWITZERLAND

Gesellschaft Schweizerischer Tierärzte 25.36
[Swiss Veterinary Society]
Address: Elfenstrasse 18, CH-3000 Bern 16, Switzerland
Telephone: (031) 43 55 43
Library facilities: Library.

TURKEY

Türk Veterinler Hekimleri Birliği 25.37

[Turkish Veterinary Medical Union]
Address: Sağlik Sok 21, Yenisehir, Ankara, Turkey
Telephone: 24 45 58

UNITED KINGDOM

Central Veterinary Laboratory * 25.38
Address: New Haw, Weybridge, Surrey KT15 3NB, UK
Telephone: (092 23) 41111
Telex: 262318
Affiliation: Ministry of Agriculture, Fisheries and Food
Enquiries to: Librarian
Founded: 1917
Subject coverage: Investigation and control of diseases of farm animals; animal husbandry; biochemistry.
Library facilities: Not open to all users; loan services (through British Library); reprographic services (through British Library).
Information services: No information services.

Royal College of Veterinary 25.39
Surgeons, Wellcome Library
– RCVS
Address: 32 Belgrave Square, London SW1X 8QP, UK
Telephone: (01) 235 6568
Enquiries to: Librarian, B. Horder
Founded: 1844
Subject coverage: Veterinary science.
Library facilities: Open to all users (nc - open to non-members upon introduction by a member or by a librarian); loan services (to members and other libraries only); reprographic services (c).
Library holdings: Over 26 000 bound volumes; 310 current periodicals; 310 reports.
Information services: Bibliographic services (nc); no translation services; literature searches (c); access to on-line information retrieval systems (c).
Translation requests may be passed on to a commercial translation agency.
Consultancy: Telephone answers to technical queries (nc); no writing of technical reports; no compilation of statistical trade data; no market surveys in technical areas.

YUGOSLAVIA

Društvo Veterinara SR Srbije 25.40
[Veterinary Society SR, Serbia]
Address: Belevar JNA 18, 11000 Beograd, Yugoslavia
Telephone: (011) 684 597

TITLES OF ESTABLISHMENTS INDEX

SUBJECT INDEX